Chemical Equilibria

Chemical Equilibria

Exact Equations and Spreadsheet Programs to Solve Them

Harry L. Pardue

CRC Press
Taylor & Francis Group
Boca Raton London New York

CRC Press is an imprint of the
Taylor & Francis Group, an **informa** business

CRC Press
Taylor & Francis Group
6000 Broken Sound Parkway NW, Suite 300
Boca Raton, FL 33487-2742

© 2019 by Taylor & Francis Group, LLC

CRC Press is an imprint of Taylor & Francis Group, an Informa business

No claim to original U.S. Government works

Printed on acid-free paper

International Standard Book Number-13: 978-1-138-36722-7 (Paperback)
978-1-138-36725-8 (Hardback)

Library of Congress Cataloging-in-Publication Data

Names: Pardue, Harry L., author.
Title: Chemical equilibria: exact equations and spreadsheet programs to
solve them / Harry L. Pardue.
Description: Boca Raton: CRC Press, Taylor & Francis Group, 2019. | Includes
bibliographical references and index.
Identifiers: LCCN 2018034875| ISBN 9781138367227 (pbk. : alk. paper) |
ISBN 9781138367258 (hardback : alk. paper)
Subjects: LCSH: Chemical equilibrium. | Chemical equilibrium—Mathematics. |
Acid-base equilibrium. | Neutralization (Chemistry)
Classification: LCC QD501 .P29825 2019 | DDC 541/.3920285554—dc23
LC record available at https://lccn.loc.gov/2018034875

Visit the Taylor & Francis Web site at
http://www.taylorandfrancis.com

and the CRC Press Web site at
http://www.crcpress.com

eResource material for this title is available at
http://crcpress.com/9781138367227

Contents

Preface

This book focuses on exact equations for acid-base, solubility and complex-ion equilibria and spreadsheet programs to solve them.

Throughout most of my teaching career I was concerned about approximation procedures used to solve problems involving different types of chemical equilibria. I couldn't help but believe that it should be possible to use computational tools available to us today to solve exact equations for most types of chemical equilibria. This would avoid the need to suffer unacceptably large errors for many situations.

Equilibrium concentrations of reactants and products in most types of equilibria can be expressed in terms of one variable. As examples, concentrations of reactants and products in acid-base, complex-ion, and sparingly salts of weak acids can be expressed in terms of the hydrogen ion concentration. Equilibrium concentrations of reactants and products for sparingly soluble salts of strong acids and hydroxides can be expressed in terms of the solubility.

While working with exact equations for hydrogen ion concentrations in solutions containing strong and/or weak acids and bases, I discovered an important feature of the reactions. I discovered that exact equations can be written in forms that are negative for hydrogen ion concentrations less than the equilibrium concentration and positive for hydrogen ion concentrations larger than the equilibrium concentration. I subsequently discovered that exact equations for hydrogen ion concentrations in complex ion reactions and solubility reactions for salts of weak acids behave the same way. I also observed that ratios of solubility products to solubility product constants for solubilities of salts of strong acids and sparingly soluble metal hydroxides are less than one for solubilities less than the equilibrium solubility and larger than one for solubilities larger than the equilibrium solubility.

These observations have led to a set of iterative spreadsheet programs to solve exact equations for concentrations of reactants and products for acid-base, solubility and complex ion equilibria. For example, given solutions ranging from pure water to mixtures of any or all forms of up to three weak acids with up to six acidic hydrogens each, the weak acids program will start with a hydrogen ion concentration less than the equilibrium concentration and increase it by small amounts until the sign of the exact equation changes from minus to plus. The hydrogen ion concentration that causes the sign change will be the equilibrium concentration resolved to the number of significant figures entered by the user in a cell in each program. Analogous reasoning applies for other types of reactions.

In each case, one enters concentrations, constants, and other parameters corresponding to starting conditions, sets a cell in the program to 1 and waits for the program to calculate equilibrium concentrations. Resolution of concentrations or solubilities to three significant figures is usually achieved within a few seconds to a few minutes. Shorter or longer convergence times are required for resolution to fewer or more significant figures. Hydrogen ion concentrations and solubilities calculated in this way are used to calculate equilibrium concentrations of other reactants and products.

Regarding the programs, one reviewer suggested that the programs should be locked to prevent students from changing them. If a teacher or student thinks that she or he can improve a program, then I believe that she or he should have the opportunity to try. If she or he isn't successful, then the original program will be available on the programs website http://crcpress.com/9781138367227.

This book is a story of how these concepts and equations have been applied to problems involving acid-base, solubility and complex-ion equilibria. I hope the programs will permit teachers and students to spend less time solving problems and more time discussing general concepts and the significance of calculated results. To access Excel files of programs included in this text, (a) type the URL, http://crcpress.com/9781138367227, (b) Click Download/Updates, and (c) click on "Excel files". The Excel files of the programs should appear. Excel is a registered trademark of the Microsoft Corporation.

Author

I was born and grew up in southern West Virginia where I attended public schools and graduated from Chapmanville High School in May 1953. I enrolled as a chemistry major in Marshall College, Marshall University now, in June 1953 and graduated with BS and MS degrees in 1956 and 1957, respectively. I then enrolled as a graduate student in the Department of Chemistry and Chemical Engineering at the University of Illinois in Urbana/Champaign in September 1957. I joined Prof. H. V. Malmstadt's research group during the summer of 1958 and earned a PhD degree with emphasis in analytical chemistry in June 1961. I then joined the faculty in the Department of Chemistry at Purdue University in September of that year.

I served as Head of the Analytical Division in the department from 1969 to 1975 and as Head of the Department of Chemistry from 1983 to 1987. I served as the American Editor for *Analytica Chimica Acta* for more than ten years, as Chairman of the Division of Analytical Chemistry of the American Chemical Society for one year, and on numerous committees for a variety of professional societies.

I continued my teaching and research activities until I retired in 2002. My teaching responsibilities included freshman chemistry, introductory analytical chemistry, instrumental analysis, and graduate-level courses involving advanced analytical chemistry and chemical instrumentation. I was frequently ranked among the top ten teachers in the Purdue University School of Science. My teaching awards included Best Teacher in the School of Science, 2001; the Arthur E. Kelley Award as the Best Teacher in the Purdue University Department of Chemistry, 1992, 2001; and appointment to the Purdue University Teaching Academy in 2001.

My research focused on chemical instrumentation and advances in kinetic-based methods. I directed the thesis research of some sixty masters and doctoral students and was the lead author on more than 175 scientific papers. Research awards include the ACS Award in Analytical Chemistry sponsored by Fisher Scientific, the Anachem Award for distinguished service to analytical chemistry, the Chemical Instrumentation Award, the Association for Clinical Chemistry Award for contributions to clinical chemistry, and the Samuel Natelson Award for Advancements in Clinical Chemistry.

After I retired in 2002 I continued as a visiting professor for two years during which time I worked with two graduate students, Ihab Odeh and Twelde Tesfai, to expand the scope of the introductory course in analytical chemistry and to develop a new set of experiments for the course. Some changes were to decrease emphasis on analyzing unknowns and increase emphasis on several topics, including automated titrations, liquid chromatography, derivative spectroscopy, kinetic-based methods, applications of enzymes as selective reagents, and the use of computers to collect and process laboratory results. I have seldom worked harder, had more fun, or been more pleased with the results of our efforts.

1 Effects of Ionic Strength

This chapter focuses on effects of seemingly inert electrolytes on chemical equilibria. Prior to the early part of the 20th century, most theories regarding ionic equilibria had been based on assumptions of ideal behavior. It was assumed that all solutes behaved as if active concentrations were the same as prepared concentrations. Chemists learned during the early part of the 20th century that secondary interactions that result from attractive forces between oppositely charged ions and between ions and oppositely charged centers on polar molecules such as water can also influence equilibrium reactions and measurement processes.[1,2]

Studies showed that solubility product constants written in terms of equilibrium concentrations were not constant but rather increased with increasing ionic concentrations. Studies also showed that responses of physical properties such as conductance, freezing points, and electrode potentials depended on concentrations of seemingly inert ions.

This chapter is a story about how chemists learned about nonideal effects of low ionic concentrations, about how they developed concepts and terms to describe the behavior, and about what procedures they developed to quantify and correct for nonideal behavior. One of the concepts developed to describe effects of ionic concentrations is ionic strength.[1] We will talk in detail about effects of ionic strength as we progress through this chapter. For now, it is sufficient to note that ionic strength is a measure of ionic concentrations that accounts for both the charges and the concentrations of all ions in a solution.

We will focus on effects of ionic strengths up to about 0.10 M. Although some generalizations in this chapter apply for all ranges of ionic strength, some do not. Therefore, do not assume comments in this chapter apply for ionic strengths larger than about 0.10 M unless you are told explicitly that this is the case. Also, all results and generalizations in this chapter represent a fixed temperature of 25°C.

SOME PRELIMINARY OBSERVATIONS

Many of the earliest observations of nonideal behavior resulted from studies of effects of ionic concentrations on solubilities. Experimental results compiled by G. N. Lewis and M. Randall[1] for effects of electrolyte concentrations on the solubility of thallous chloride and barium iodate, as well as some results for acetic acid and the dissociation of water, are used herein to illustrate the nature of the problem and concepts chemists formulated to explain and quantify it.

Thallous chloride is a sparingly soluble salt that dissolves and dissociates as follows:

$$TlCl \rightleftharpoons Tl^+ + Cl^-$$

The thallous chloride solubility in otherwise pure water was determined to be 0.01607 mol/L. Based on the law of mass action as it was understood near the end of the 19th century, it was assumed that the solubility would be the same in all solutions at the same temperature that did not include common ions. In fact, if your introductory chemistry texts were like mine, that's the way you learned to interpret the law of mass action.

Using the solubility, S = 0.01607 mol/L, and the relationship $[Tl^+] = [Cl^-] = S$, I learned to calculate the solubility product constant as follows:

$$K_{sp} = [\overset{S}{Tl^+}][\overset{S}{Cl^-}] = S^2 = (0.01607\,mol/L)^2 = 2.6 \times 10^{-4}\,M^2$$

TABLE 1.1

Thallous Chloride Solubility in Solutions Containing Potassium Nitrate and Potassium Sulfate[1]

$C_{KNO_3}, C_{K_2SO_4}$	0	0.020	0.050	0.100
S_{TlCl} in KNO_3	0.01607	0.01716	0.01826	0.01961
S_{TlCl} in K_2SO_4	0.01607	0.01779	0.01942	0.02137
K_{sp} (M^2)	2.58×10^{-4}	2.94×10^{-4}	3.33×10^{-4}	3.84×10^{-4}
K_{sp} (M^2)	2.58×10^{-4}	3.16×10^{-4}	3.69×10^{-4}	4.97×10^{-4}

Based on what I learned in my introductory courses, I would have assumed that this solubility and solubility product constant would apply in all solutions at the same temperature that did not include an excess of either Tl^+ or Cl^- as a common ion. Is this a valid assumption?

For example, consider what would happen if we were to add potassium nitrate or potassium sulfate to a saturated solution of thallous chloride. The two salts would dissociate as follows:

$$KNO_3 \rightarrow K^+ + NO_3^- \qquad K_2SO_4 \rightarrow K^+ + SO_4^{2-}$$

Neither K^+ nor NO_3^- nor SO_4^{2-} forms a precipitate with either thallous ion or chloride ion. Based on what I had earned in my introductory courses, I wouldn't have expected either salt to influence the solubility of thallous chloride. I would have been wrong. Me, wrong?????

Let's look at some results.

Table 1.1 summarizes experimental results for thallous chloride solubility in solutions containing increasing potassium nitrate and potassium sulfate concentrations.

The three most important observations regarding these results are that (a) the solubilities are the same in the solutions that don't contain either salt, (b) the solubilities increase with increasing concentrations of both salts, and (c) the solubilities in the solutions containing potassium sulfate increase more than the solubilities in the solutions containing potassium nitrate. Our primary tasks in this chapter are to understand reasons for this behavior and to learn to do calculations needed to quantify the behavior. Let's begin with an analogy to help us visualize reasons for the behavior.

AN ANALOGY

Let's imagine that two very popular singers do a show in your hometown. On the one hand, it would be very easy for the two singers to visit with each other behind stage. On the other hand, if they decided to mingle with the audience after the show, each singer would likely be surrounded by admiring fans wanting to talk, seek autographs, etc., making it more difficult for them to get together.

Ions such as Tl^+ and Cl^- behave in analogous ways. If the ions are alone in solution, they can come together and react without interference from other ions. However, if we add potassium nitrate to the solution, it will dissociate completely, producing K^+ and NO_3^- ions, i.e.,

$$KNO_3 \rightarrow K^+ + NO_3^-$$

Like fans gather around popular performers, electrostatic attractions will cause negative nitrate ions to accumulate around Tl^+ ions and positive K^+ ions to accumulate around chloride ions. The net effect is that the Tl^+ and Cl^- ions would be shielded from each other, reducing the number of ions that could get close enough to react. Consequently, each pair of Tl^+ and Cl^- ions would be less effective in satisfying the solubility product constant, making it necessary for more salt to dissolve to satisfy the constant.

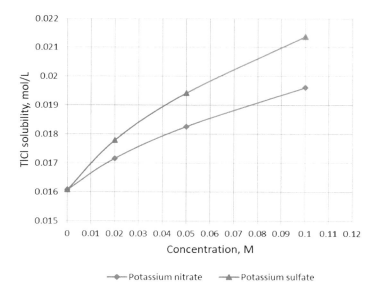

FIGURE 1.1   Effects of potassium nitrate and potassium sulfate concentrations on the solubility of thallous chloride.

TlCl SOLUBILITY IN STRONG ELECTROLYTE SOLUTIONS

As shown in Table 1.1 and Figure 1.1, increasing concentrations of both potassium nitrate and potassium sulfate cause the solubility of thallous chloride to increase. Potassium sulfate concentrations cause larger changes in the solubility than the same concentrations of potassium nitrate.

Based on what we have said to this point, can you think of reasons why concentrations of potassium sulfate cause larger changes in the solubility than equal concentrations of potassium nitrate? Reactions below can help you visualize differences between potassium nitrate and potassium sulfate.

$$KNO_3 \rightarrow K^+ + NO_3^- \qquad K_2SO_4 \rightarrow 2K^+ + SO_4^{2-}$$

Whereas each molecule of potassium nitrate produces one singly charged potassium ion and one singly charged nitrate ion, each molecule of potassium sulfate produces two potassium ions and one doubly charged sulfate ion. Potassium sulfate produces more ions, and one of the ions has a larger charge. As we will see later, both these features influence differences between effects of potassium nitrate and potassium sulfate on the solubility of thallous chloride.

We have focused to this point on effects of inert ions, K^+, NO_3^-, and SO_4^{2-}, on the thallous chloride solubility. Given that behaviors of ions in a solution depend on the total ionic content of the solution, we shouldn't forget that dissociation of thallous chloride also adds ions, Tl^+ and Cl^-, to the solution, i.e.,

$$TlCl \rightleftharpoons Tl^+ + Cl^-$$

These ions must be included in our discussion of effects of ionic concentrations on the behavior of thallous chloride.

CONCEPTS AND EQUATIONS

Most of what we know about effects of ionic concentrations on equilibration reactions was discovered during the first half of the 20th century. Whereas concepts and equations developed during that period involve effects of wide ranges of ionic concentrations, we will focus in this chapter on effects of low ionic concentrations. Let's begin with a quantity called ionic strength.[1]

IONIC STRENGTH

Ionic strength, commonly represented by the symbol μ, is a measure of ionic concentrations that accounts for both the charge, z_i, on each ion and the equilibrium concentration of the ion. More specifically, the contribution of an ion to the ionic strength is one-half the product of the charge squared times the equilibrium concentration of the ion.[1] The ionic strength of a solution is the sum of the contributions of the individual ions. This is expressed in Equation 1.1:

$$\mu = 0.5 \sum_{i=1}^{i=n} z_i^2 [\text{Ion}]_i \tag{1.1}$$

Let's solve an example.

> **Example 1.1 Calculate the ionic strength in a saturated solution of barium iodate containing $C_{KNO_3} = 0.010\,M$ potassium nitrate, assuming that the solubility of barium iodate is $S_{Ba(IO_3)_2} = 1.12 \times 10^{-3}\,M$ and that both salts dissociate completely.**

Step 1: Write the reactions.

$$KCl \rightarrow K^+ + Cl^-; \qquad Ba(IO_3)_2 \rightleftharpoons Ba^{2+} + 2IO_3^-$$

Step 2: Write the concentrations of the ions.

$$\left[K^+\right] = \left[Cl^-\right] = C_{KCl} = 0.010\ M; \qquad \left[Ba^{2+}\right] = S_{Ba(IO_3)_2} = 1.12 \times 10^{-3} M;$$

$$\left[IO_3^-\right] = 2S_{Ba(IO_3)_2} = 2.24 \times 10^{-3} M$$

Step 3: Write an equation for the ionic strength in terms of the equilibrium concentrations of all ions except H^+ and OH^-, concentrations of which are negligibly small.

$$\mu = 0.5\left(1^2\left[K^+\right] + 1^2\left[Cl^-\right] + 2^2\left[Ba^{2+}\right] + 1^2\left[IO_3^-\right]\right)$$

Step 4: Substitute concentrations and solve.

$$\mu = 0.5\left(1^2(0.010\ M) + 1^2(0.010\ M) + 2^2\left(1.12 \times 10^{-3}\ M\right) + 1^2\left(2.24 \times 10^{-3}\ M\right)\right) = 1.34 \times 10^{-2}\ M$$

Steps in this example are judged to be self-explanatory and are not discussed further.

ACTIVITIES AND ACTIVITY COEFFICIENTS

The activity of an ion can be viewed as the effective concentration of the ion. Activities of ions are represented herein as the ion in braces, i.e., {Ion}. The activity coefficient of an ion, represented herein by the symbol f_{Ion}, is the proportionality constant between the activity and the concentration of the ion, i.e.,

$$\{\text{Ion}\} = f_{Ion}[\text{Ion}]$$

in which {Ion} is the activity of the ion, f_{Ion} is the activity coefficient, and [Ion] is the equilibrium concentration of the ion. In other words, an activity coefficient is the proportionality constant between activity and concentration.

Rewriting the foregoing equation as follows:

$$f_{Ion} = \frac{\{Ion\}}{[Ion]}$$

It follows that the activity coefficient of an ion is the fraction of the equilibrium concentration that remains active in the reaction in which the ion is involved. Activity coefficients are often represented by γ. I chose to use the symbol f_{Ion} to represent activity coefficients to emphasize the fact that the activity coefficient of a reactant is the fraction of the reactant that is effective in the reactions(s) in which it is involved.

EQUATIONS FOR ACTIVITY COEFFICIENTS[2,3]

Two common options for calculating activity coefficients are the extended Debye-Hückel equation (EDHE)[2] and the Davies equation.[3] The Debye-Hückel equation is based on theoretical considerations; the Davies equation is based on empirical observations. Each equation is discussed below.

When these concepts were being developed, it was easier to do calculations using logarithms of the quantities rather than the quantities themselves. Computational tools available to us now make it easier to do the calculations using the quantities themselves rather than logarithms of the quantities. Therefore, whereas both the Debye-Hückel and the Davies equations are commonly written in logarithmic forms, I have chosen to write them here in exponential forms that permit more direct calculations of activity coefficients using calculators and computers. The extended Debye-Hückel equation for the activity coefficient of an ion is shown in Equation 1.2a:

$$f_{Ion} = \frac{\{Ion\}}{[Ion]} = 10 \wedge \left[-0.51 z_{Ion}^2 \frac{\sqrt{\mu}}{1 + 3.3 D_{Ion} \sqrt{\mu}} \right] \qquad (EDHE) \qquad (1.2a)$$

in which z_{Ion} is the unsigned charge on the ion, μ is the ionic strength, and D_{Ion} is the hydrated diameter of the ion expressed in nanometers rather than angstroms as reported by J. Kielland.[4] This means that hydrated diameters used herein are 10 times larger than values included in Kielland's paper. If you use Kielland's published values of hydrated diameters, the denominator term should be $1 + 0.33 D_{Ion} \mu^{1/2}$ rather than $1 + 3.3 D_{Ion} \mu^{1/2}$ as in Equation 1.2a.

An equation commonly called the Davies equation[3] is based on empirical observations of large numbers of experimental results. Using averages of activity coefficients for large numbers of situations, Davies arrived at an empirical equation of the form as shown in Equation 1.2b:

$$f_{Ion} = \frac{\{Ion\}}{[Ion]} = 10 \wedge \left[-0.51 z_{Ion}^2 \left(\frac{\sqrt{\mu}}{1 + \sqrt{\mu}} - 0.3\mu \right) \right] \qquad (Davies) \qquad (1.2b)$$

in which f_{Ion} is the activity coefficient, z is the charge on the ion, and μ is the ionic strength.

The following example illustrates applications of the Debye-Hückel and Davies equations.

Example 1.2 **Given $z_{Ba} = 2$ and $D_{Ba} = 0.5$ nm, calculate the activity coefficient of the Ba^{2+} ion at an ionic strength of 0.10 M using the EDHE and Davies equations.**

$$f_{EDHE} = \frac{\{Ba^{2+}\}}{[Ba^{2+}]} = 10 \wedge \left[\frac{-0.51(2^2)\sqrt{0.10}}{1+3.3(0.5)\sqrt{0.10}} \right] = 0.38$$

$$f_{Davies} = \frac{\{Ion\}}{[Ion]} = 10 \wedge \left[-0.51(2^2)\left(\frac{\sqrt{0.1}}{1+\sqrt{0.1}} - 0.3 \times 0.1 \right) \right] = 0.37$$

As you can see, the two equations give similar values of activity coefficients for this situation.

What do the activity coefficients mean? They mean that the effective Ba^{2+} concentration is about 38% of the actual concentration.

Results for other ionic strengths calculated using the Debye-Hückel and Davies equations are summarized in Table 1.2. As shown, differences are close to 1% for ionic strengths up to about 0.08 M and increase rapidly for ionic strengths larger 0.1 M. Differences are smaller for singly and doubly charged ions.

There is a simpler version of the Debye-Hückel equation called the Debye-Hückel limiting law (DHLL). However, given that the simpler equation is more prone to errors than the complete equation and that the complete equation is easily solved using calculators and computers available to us now, the simpler equation is not discussed herein.

CHOOSING BETWEEN THE DEBYE-HÜCKEL AND DAVIES EQUATIONS

It is generally agreed that the Davies equation is applicable to a wider range of ionic strengths than the Debye-Hückel equation. Therefore, my original plan was to use the Davies equation in my programs. However, I soon learned that the iterative programs based on the Davies equation worked well for most situations but failed for others. Despite extensive efforts, I wasn't able to find the causes of the failures or ways to avoid them. I will continue to discuss both equations in the remainder of this chapter and have included options based on the Davies equation in some programs. However, the primary focus in all programs discussed in subsequent chapters is on the Debye-Hückel equation.

MEAN ACTIVITY COEFFICIENTS

On the one hand, it is possible to calculate single-ion activity coefficients using equations described above. On the other hand, given that it isn't possible to differentiate between contributions to activities of individual ions in a solution, it isn't possible to determine single-ion activity coefficients experimentally. The concept of mean activity coefficients was introduced to account for this fact.

TABLE 1.2

Comparison of Single-Ion Activity Coefficients Calculated Using the Debye-Hückel and Davies Equations for Triply Charged Ions

Ion. Str.	0	0.02	0.04	0.06	0.08	0.1	0.2
DH[a]	1.00	0.298	0.204	0.158	0.130	0.111	0.066
Davies	1.00	0.295	0.201	0.157	0.131	0.113	0.076
Dif. (%)	0.00	1.0	1.5	0.6	0.7	1.7	13

[a] Debye-Hückel equation with the hydrated diameter set at 0.5 nm.

Commonly used symbols for mean concentrations and mean activity coefficients are $C_{\pm}$ and $f_{\pm}$, respectively. The equation editor used to prepare equations herein displayed the symbols C_+ and f_+ for $C_{\pm}$ and $f_{\pm}$. Therefore, I used the symbols $C_{+/-}$ and $f_{+/-}$ in place of the usual symbols.

Mean activity coefficients for pairs of ions on either side of a reaction can always be expressed in terms of single-ion activity coefficients. As an example, consider the solubility of lanthanum carbonate:

$$La_2(CO_3)_3 \rightleftharpoons 2La^{3+} + 3CO_3^{2-}$$

Let's begin by representing the *sum of the balancing coefficients* as follows:

$$\nu = 2 + 3 = 5$$

Using this relationship with the rules of a geometric mean, coefficients in the balanced reaction can be used to write relationships between mean and single-ion activity coefficients as shown in Equation 1.3:

$$f_{+/-} = \left(f_{La}^2 f_{CO_3}^3\right)^{1/(2+3)} = \left(f_{La}^2 f_{CO_3}^3\right)^{1/5} \tag{1.3}$$

in which $f_{+/-}$ is the mean activity coefficient, f_{La} and f_{CO_3} are activity coefficients of La^{3+} and CO_3^{2-}, and 2 and 3 are coefficients of the La^{2+} and CO_3^{2-} ions in the balanced reaction.

The mean activity coefficient could be calculated by first calculating single-ion activity coefficients and then using these to calculate the geometric mean as in Equation 1.3. Alternatively, for any pair of ions for which products of unsigned charges and balancing coefficients are equal, the mean activity coefficient can be calculated using modified forms of the Debye-Hückel or Davies equation.

The form of the Debye-Hückel equation used to calculate mean activity coefficients is shown in Equation 1.4a:

$$f_{+/-} = 10 \wedge \left[\frac{-0.51 z_+ z_- \sqrt{\mu}}{1 + 3.3 D_{Av} \sqrt{\mu}}\right] \qquad (EDHE_{\pm}) \tag{1.4a}$$

in which z_+ and z_- are the *unsigned charges* on the cation and anion, respectively; μ is the ionic strength; and D_{Av} is the average hydrated diameter of the two ions. The form of the Davies equation used to calculate mean activity coefficients is shown in Equation 1.4b:

$$f_{+/-} = 10 \wedge \left[-0.51 z_1 z_2 \left(\frac{\sqrt{\mu}}{1 + \sqrt{\mu}} + 0.3\mu\right)\right] \qquad (Davies) \tag{1.4b}$$

Example 1.3 illustrates an application of Equation 1.4a.

Example 1.3 Calculate the mean activity coefficient of lanthanum carbonate at an ionic strength of 0.10 M. For the La^{3+} and CO_3^{2-} ions: $z_{La} = 3$, $z_{CO_3} = 2$, $D_{La} = 0.9$ nm, and $D_{CO_3} = 0.45$ nM.

Step 1: Calculate the average hydrated diameter.

$$D_{Av} = \frac{D_{La} + D_{CO_3}}{2} = (0.9 \text{ nm} + 0.45 \text{ nM})/2 = 0.68 \text{ nm}$$

Step 2: Calculate the mean activity coefficient using the Debye-Hückel equation.

$$f_{+/-} = 10 \wedge \left[\frac{-0.51 z_+ z_- \sqrt{\mu}}{1 + 3.3 D_{Av} \sqrt{\mu}} \right] = 10 \wedge \left[\frac{-0.51(3)(2)\sqrt{0.1}}{1 + (3.3)(0.68)\sqrt{0.1}} \right] = 0.27$$

I will trust you to show that the mean activity coefficient calculated using the Davies equation for the same situation is $f_{+/-} = 0.272$.

It will be informative to compare the mean activity coefficient calculated from single-ion activity coefficients with that in Example 1.3.

EXERCISE 1.1

Show that single-ion activity coefficients for La^{3+} and CO_3^{2-} for $\mu = 0.10$ M are $f_{La} = 0.17_8$ and $f_{CO_3} = 0.36_4$, respectively, and that the mean activity coefficient calculated from these numbers is $f_{+/-} = 0.27_4$. *Hint:* Use the EDHE to calculate activity coefficients of La^{3+} and CO^{2-} with $D_{La} = 0.9$ nm and $D_{CO_3} = 0.45$ nm.

Agreement between this result and that in Example 1.3 is quite good.

The mean activity coefficient of 0.27 means that the combined effects of ionic strength on the La^{3+} and CO_3^{2-} ions are such that about 27% of the mean concentration of the ions remains active in the solubility reaction. This means that calculation procedures that ignore effects of ionic strength will introduce significant errors.

ADDITIONAL EXPERIMENTAL RESULTS

Having described the concept of ionic strength, Lewis and Randall[1] stated, "In dilute solutions the activity coefficient of a given strong electrolyte is the same in all solutions of the same ionic strength" (p. 1141). Let's test this assertion using the thallous chloride results.

ACTIVITY COEFFICIENTS VS. IONIC CONCENTRATIONS

Figure 1.2A includes plots of activity coefficients vs. potassium nitrate and potassium sulfate concentrations. As you can see, the plots for the two salts are very different, with potassium sulfate

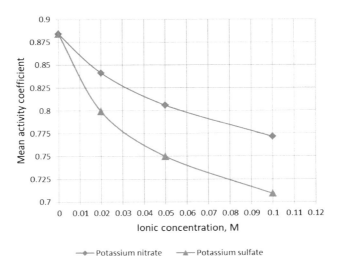

FIGURE 1.2A Effects of potassium nitrate and potassium sulfate concentrations on the mean activity coefficients of thallous chloride.

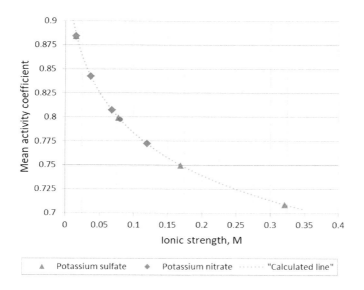

FIGURE 1.2B   Effects of ionic strength on mean activity coefficients of thallous chloride.

concentrations producing larger changes than potassium nitrate. As discussed earlier, whereas potassium nitrate produces two singly charged ions per molecule, potassium sulfate produces two singly charged ions and one doubly charged ion. Both these factors influence differences between the effects of the two salts on solubilities of thallous chloride.

Figure 1.2B includes plots of the same activity coefficients vs. ionic strength. Unlike plots of activity coefficients vs. potassium nitrate and potassium sulfate concentrations, the plots of the activity coefficients vs. ionic strength follow the same pattern.

Unlike plots of activity coefficients vs. potassium nitrate and potassium sulfate concentrations, the plots of the activity coefficients vs. ionic strength are superimposed on each other. This supports Lewis and Randall's suggestion that activity coefficients for a strong electrolyte should be the same in solutions with the same ionic strength.

Activity Coefficients from Solubilities

Lewis and Randall[1] described a procedure to determine activity coefficients from solubility reactions using the thallous chloride solubility in the presence of potassium chloride as an example (p. 1129). I will paraphrase their treatment using slightly different symbols.

Let's use the reactions below to identify symbols used herein:

$$TlCl \rightleftharpoons \overset{S}{Tl^+} + \overset{S}{Cl^-} \qquad KCl \rightarrow \overset{C_{KCl}}{K^+} + \overset{C_{KCl}+S}{Cl^-}$$

in which S is the solubility and C_{KCl} is the potassium chloride concentration. We can write Tl^+ and Cl^- concentrations as

$$\left[Tl^+ \right] = S \qquad \left[Cl^- \right] = C_{KCl} + S$$

Next, replacing the authors' $m_\pm$ with $C_{+/-}$, let's write the mean concentration of the ions, $C_{+/-}$, as the square root of the product of the concentrations, i.e.,

$$C_{+/-} = \left(\left[Tl^+ \right] \left[Cl^- \right] \right)^{1/2} = \left[S \left(C_{KCl} + S \right) \right]^{1/2}$$

The authors included results showing that the TlCl solubility in a solution containing 0.050 M KCl was S = 0.00590 M. Substituting these numbers into the foregoing equation gives

$$C_{+/-} = \left(\left[Tl^+ \right] \left[Cl^- \right] \right)^{1/2} = \left[S(C_{KCl} + S) \right]^{1/2} = \left[0.0059 (0.05 + 0.0059) \right]^{1/2} = 0.0181 \, M$$

The analogous equation for the mean concentration for a solution containing a salt of thallous ion such as TlNO$_3$ is

$$C_{+/-} = \left(\left[Tl^+ \right] \left[Cl^- \right] \right)^{1/2} = \left[(C_{KCl} + S)(S) \right]^{1/2}$$

The authors then plotted reciprocals of mean concentrations calculated in this way, i.e., $1/C_{+/-}$, for a large number of analogous situations vs. the square root of the ionic strength. By extrapolating linear portions of the plots to zero ionic strength as in their Figure 4, they obtained $1/C_{+/-} = 70.3$, corresponding to $C_{+/-} = 1/70.3 = 0.0142$ M at zero ionic strength. At zero ionic strength, $C_{+/-} = (S \times S)^{1/2} = S$, corresponding to a solubility at zero ionic strength of $S_0 = 0.0142$ M.

The authors then calculated mean activity coefficients as ratios of the solubility at zero ionic strength divided by mean concentrations for ionic strengths larger than zero, as shown in Equation 1.5:

$$f_{+/-} = \frac{S_0}{C_{+/-}} \tag{1.5}$$

Using the numbers $S_0 = 0.0142$ M and $C_{+/-} = 0.0181$ M given above, the mean activity coefficient for thallous chloride in the solution containing 0.050 M KCl is $f_{+/-} = S_0/C_{\pm} = 0.0142 \, M/0.0181 \, M = 0.785$. The authors reported $f_{+/-} = 0.784$.

RESULTS FOR A BIVALENT-UNIVALENT SALT, BARIUM IODATE

The authors also included results for barium iodate and lanthanum(III) iodate in solutions containing barium nitrate concentrations from 0 M to 0.10 M. Selected results for barium iodate are summarized in Table 1.3.

Using the 1:2 stoichiometry of the salt, the mean concentration of barium iodate is calculated as $C_{+/-} = ([Ba^{2+}][IO_3^-]^2)^{1/3} = (S(2S)^2)^{1/3}$. Using the solubility for $C_{Ba(IO_3)_2} = 0$, the mean concentration is

---

TABLE 1.3

Solubility Results for Barium Iodate in Solutions Containing Barium Nitrate

$C_{Ba(NO_3)_2}$	$S_{Ba(IO_3)_2}$	$1/C_{+/-}$	Ion. Str., μ	$f_{+/-}$[a]	$f_{+/-}$[b]
0	0.000790	797	0.00237	0.89	—
0.0005	0.000681	770	0.00354	0.86	—
0.001	0.000606	751	0.00482	0.83	0.83
0.0025	0.000488	706	0.00896	0.78	—
0.01	0.000337	597	0.0310	0.66	0.64
0.025	0.000307	472	0.0759	0.52	—
0.05	0.000283	396	0.1508	0.44	—
0.1	0.000279	317	0.3009	0.35	—

[a] Calculated as $C_{+/-}/C_{+/-,\text{Intercept}}$.
[b] From the authors' Table XXI, p. 1142.

calculated as $C_{+/-} = \{0.00079[2(0.00079)]^2\}^{1/3} = 1.254 \times 10^{-3}$, corresponding to $1/C_{+/-} = 797$. Other reciprocals of mean concentrations were calculated in the same way. A plot of $1/C_{+/-}$ vs. $\mu^{1/3}$ was linear with a slope of -924 and an intercept on the ordinate of $1/C_{+/-} = 900$. Activity coefficients calculated as $C_{+/-}/900$ are included in the fifth column of Table 1.3. Mean activity coefficients from 0.89 to 0.35 indicate that the mean concentrations behave as if they are 89% to 35% as effective as the actual concentrations.

Whereas the authors included interpolated values of activity coefficients for several rounded barium nitrate concentrations in their Table XXI, they included numerical values for only two of the actual barium nitrate concentrations included in the study. Those values are included in the last column of Table 1.3 for comparison purposes. Activity coefficients calculated as described above differ from these reported values by less than 5%.

CONDITIONAL CONSTANTS AND THERMODYNAMIC CONSTANTS

At some time during the early part of the 20th century, someone decided to identify two types of equilibrium constants called *conditional constants* and *thermodynamic constants* to account for differences in behaviors at finite ionic strengths larger than and equal to zero. Situations corresponding to zero ionic strength are commonly referred to as infinite dilution.

Conditional constants are constants based on results obtained for ionic strengths larger than zero. Conditional constants are represented with subscripted μ's, e.g., $K_{sp\mu}$ or $K_{a\mu}$, to emphasize the fact that they depend on ionic strength.

As examples, solubility product constants calculated from solubilities in Table 1.1 are called conditional constants, implying that they depend on solution conditions. For example, using the solubility $S = 0.01607$ M for the solution corresponding to a potassium nitrate concentration of zero, the conditional solubility product constant is $K_{sp\mu} = [Tl^+][Cl^-] = S^2 = 0.01607^2 = 2.58 \times 10^{-4}$. I will trust you to use solubilities in Table 1.1 to confirm that conditional constants for potassium nitrate concentrations of 0.02 M, 0.050 M, and 0.1 M are $K_{sp\mu} = 2.94 \times 10^{-4}$, 3.33×10^{-4}, and 3.84×10^{-4}, respectively.

Thermodynamic constants are constants at zero ionic strength, commonly referred to as *infinite dilution*. Given that concentrations at zero ionic strengths are activities, I find it useful to think of thermodynamic constants as activity-based constants. With a few exceptions, thermodynamic constants are the constants in tables in texts and handbooks and, more recently, on the internet. Thermodynamic constants are usually determined by extrapolating equilibrium constants or quantities related to equilibrium constants for ionic strengths larger than zero to zero ionic strength. Let's talk about some examples.

THERMODYNAMIC CONSTANTS FROM CONDITIONAL CONSTANTS

As noted earlier, most equilibrium constants given in tables in texts and handbooks and on the internet correspond to ionic strengths of zero. Have you ever wondered how those constants were determined? Let's talk about a few examples.

Chemists have devised a variety of extrapolation procedures to extract information from conditional constants that can be used to calculate thermodynamic constants. For example, for solubility reactions with balancing coefficients m and b for the metal ion and basic parts of solubility reactions, Lewis and Randall[1] used plots of reciprocals of mean concentrations, i.e., $1/C_\pm$, vs. ionic strength to the $1/(m + b)$ power to linearize solubility data. Intercepts of the lines with the ordinate were used to calculate activity coefficients that could be used to calculate thermodynamic constants.

A simpler option is to plot conditional constants vs. a power of the ionic strength that gives plots that can be extrapolated to zero ionic strength. I will use this option below to quantify the thermodynamic solubility product constant for thallous chloride, the dissociation constant for acetic acid, and the dissociation constant for water.

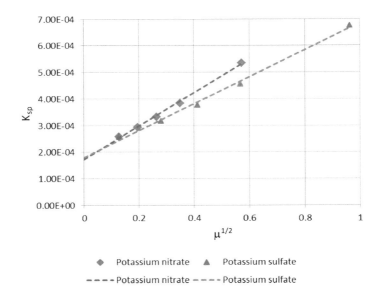

FIGURE 1.3   Effects of ionic strength on the solubility product constant for thallous chloride.

Solid points are experimentally determined constants; dashed lines are least squares fits of K_{sp} vs. $\mu^{1/2}$.

Solubility Product Constant for Thallous Chloride

Figure 1.3 is a plot of conditional solubility product constants for thallous chloride vs. the square root of ionic strength. Solid points are experimental values of conditional constants from Table 1.1, and dashed lines are least squares fits of K_{sp} vs. $\mu^{1/2}$. Intercepts of the plots with the ordinate correspond to the thermodynamic solubility product constant for thallous chloride. Least squares values of the intercepts for thallous chloride in potassium nitrate and potassium sulfate solutions correspond to solubility product constants of $K_{sp} = 1.7_3 \times 10^{-4}$ for the potassium nitrate solution and $1.7_9 \times 10^{-4}$ for the potassium sulfate solutions. These constants are consistent with tabulated values of the constants, which vary from 1.7×10^{-4} to 1.8×10^{-4}.

For the record, it isn't unusual for constants determined using different options to differ by several percent. This is a reason tabulated values of many constants are limited to two significant figures.

Dissociation Constant for Acetic Acid

Harned and Ehlers[5] used results of potentiometric measurements to determine conditional dissociation constants for acetic acid at a variety of ionic strengths and temperatures. Conditional constants they reported at 25°C and for ionic strengths from 0.00591 M to 0.1799 M are summarized in Table 1.4.

Figure 1.4 includes plots of the conditional constants vs. the ionic strength. Solid points are experimentally determined constants, and the dashed line represents calculated values using the slope,

---

TABLE 1.4
Conditional Dissociation Constants[4] for Acetic Acid at 25°C

Ion. Str.	0.00951	0.02403	0.04175	0.09871	0.16095	0.17994
$K_{a,\mu}$ (10^{-5})	1.752	1.747	1.743	1.734	1.724	1.726

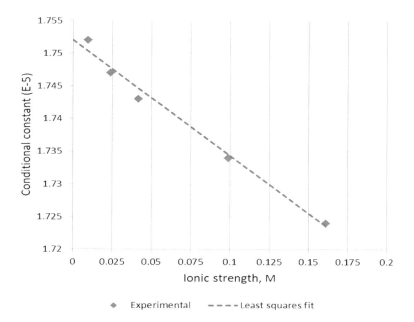

FIGURE 1.4 Effects of ionic strength on conditional constants for acetic acid.

Solid points are experimental results; the dashed plot is the least squares fit.

−0.177, and intercept, 1.75×10^{-5}, obtained from a linear least squares fit of conditional constants vs. the ionic strength. The intercept corresponds to a value of $K_a = 1.8 \times 10^{-5} \; M^2$ for the thermodynamic dissociation constant for acetic acid. This is consistent with tabulated values of the constant.

DISSOCIATION CONSTANT FOR WATER

Harned and Donelson[6] used results of potentiometric measurements to determine conditional dissociation constants for water at a variety of ionic strengths and temperatures. Potentials they reported for ionic strengths up to 0.11 M are summarized in Table 1.5.

Based on the authors' Equation 6, cell potentials are related to the ionization constant for water as follows:

$$E_{Cell} + 0.05916\log\frac{C_{LiBr}}{C_{LiOH}} - E_0 = -0.05916\log K_w$$

in which E_{Cell} is the cell voltage, C_{LiBr} and C_{LiOH} are molar concentrations of lithium bromide, E_0 is the standard potential of the Ag/AgBr electrode, and K_w is the dissociation constant. Using the

TABLE 1.5
Experimental Quantities[a] Used to Calculate Conditional Dissociation Constants for Water

Ion. Str.	0.02	0.03	0.04	0.06	0.08	0.11
C_{LiBr}, M	0.01	0.02	0.03	0.05	0.07	0.10
Potential, V	0.89884	0.88080	0.87017	0.85660	0.84751	0.83770
$K_{w,\mu}$ ($10^{-14}\; M^2$)	1.028	1.039	1.049	1.069	1.089	1.117

[a] $C_{LiOH} = 0.010$ M for all ionic strengths.

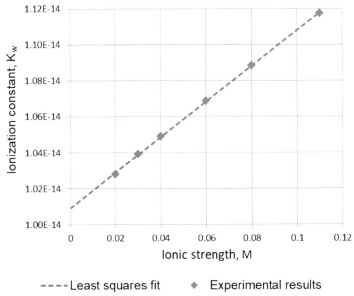

FIGURE 1.5   Effects of ionic strength on conditional ionization constants for water.

rules of logarithms with $E_0 = 0.07131$ V, conditional dissociation constants, $K_{w,\mu}$, are related to cell potentials as follows:

$$K_{w,\mu} = 10 \wedge \left(-\frac{E_{Cell} + 0.05916 \log \dfrac{C_{LiBr}}{C_{LiOH}} - 0.07131}{0.05916} \right)$$

in which $K_{w,\mu}$ is the conditional dissociation constant for water, E_{Cell} is the potential of the cell, and C_{LiBr} and C_{LiOH} are concentrations of lithium bromide and lithium hydroxide, respectively.

Conditional constants calculated using this equation with cell potentials in Table 1.5 are summarized in the last row of the table.

As a reminder, the thermodynamic constant included in tables is the value of the constant at zero ionic strength. Figure 1.5 includes plots of conditional dissociation constants for water vs. the ionic strength. Solid points are constants calculated as described above, and the dashed line is a least squares fit of conditional constants vs. the ionic strength, i.e.,

$$K_w = 9.87 \times 10^{-15} \left(K_{w,\mu} \right) + 1.009 \times 10^{-14}$$

The intercept corresponds to $K_w = 1.009 \times 10^{-14}$ at 25°C. This value rounded to three significant figures is close to the generally accepted value of the constant.

A NONLINEAR CURVE-FITTING OPTION FOR DETERMINING THERMODYNAMIC CONSTANTS

As illustrated in the preceding discussion, our predecessors obtained information at infinite dilution by devising ways to obtain linear plots of dependent variables such as solubilities vs. independent variables such as ionic strength. In so doing, they were able to use linear extrapolations to obtain activity coefficients and equilibrium constants at infinite dilution. We now have data-processing

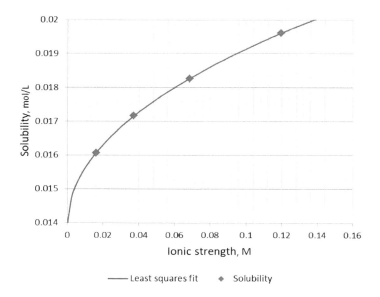

FIGURE 1.6 Nonlinear fit of the Debye-Hückel equation to solubility data for thallous chloride in potassium nitrate solutions.

capabilities that don't require linear relationships to extrapolate results to infinite dilution. One such option is illustrated below.

Points in Figure 1.6 represent thallous chloride solubilities from Table 1.1 for solutions containing potassium nitrate. The solid line is a least squares fit of a rearranged form of Equation 1.6a to the solubilities. The equation (1.6a) is as follows:

$$S = \frac{S_0}{f_\pm} = \frac{P_1}{10^{\wedge}\left[-1^2 P_2\left(\dfrac{\sqrt{\mu}}{1 + 3.3 P_3 \sqrt{\mu}}\right)\right]} \tag{1.6a}$$

in which P_1, P_2, and P_3 are fitting parameters representing the solubility, S_0, at zero ionic strength; the constant, 0.51; and the hydrated diameter, D, from the Debye-Hückel equation.

The solid line was obtained by using a curve-fitting program (Origin 9.1, OriginLab Corp., One Roundhouse Plaza, Northampton, MA 01060) to obtain a least squares fit of the equation to the experimental solubilities. Best-fit values of the fitting parameters are $P_1 = 0.0140$ M, $P_2 = 0.523$, and $P_3 = 0.197$. These parameters correspond to $P_1 = S_0 = 0.0140$ M, $P_2 = 0.523$, and $P_3 = D = 0.197$. Substituting these parameters into Equation 1.6a gives Equation 1.6b:

$$S = \frac{S_0}{f_\pm} = \frac{0.0140}{10^{\wedge}\left[-1^2 (0.523)\left(\dfrac{\sqrt{\mu}}{1 + 3.3(0.197)\sqrt{\mu}}\right)\right]} \tag{1.6b}$$

As you can see from the plot, this equation fits the data very well.

The value of $S_0 = 0.0140$ M is very close to that calculated from the zero intercept of Figure 4 in Lewis and Randall's paper,[1] i.e., $1/70.3 = S_0 = 0.0142$ M. The thermodynamic, solubility product constants calculated using S_0 obtained by these two sets of results are $K_{sp} = (0.014)^2 = 1.9_6 \times 10^{-4}$ and $K_{sp} = (0.0142)^2 = 2.0_2 \times 10^{-4}$.

Although I used the Debye-Hückel equation as a first trial, the procedure should be adaptable to the Davies equation as well as other solubility reactions and other types of equilibria.

Whereas calculations using the Debye-Hückel or Davies equation are more useful when experimental data aren't available, the curve-fitting option could be useful for anyone working extensively with one or more sets of reactions for which they have experimental data for different ionic strengths. It not only gives the solubility at zero ionic strength but also gives more reliable values of the fitting parameters for individual situations than can be calculated using either the Debye-Hückel or Davies equation.

SUMMARY

The following are some of the most important conclusions from this chapter.

- The ionic strength of a solution is a measure of concentration that accounts for the charges and concentrations of all ions in a solution.
- The activity of an ion is the concentration that represents the way the ion behaves in solution.
- The activity coefficient of an ion is the fraction of the total concentration of an ion that remains active. As such, it is the decimal equivalent of the percentage of the total concentration that remains active.
- The mean activity coefficient of an electrolyte is the geometric mean of the single-ion activity coefficients.
- Zero ionic strength, also called infinite dilution, is the only condition for which activity coefficients are equal to unity and activities are equal to concentrations.
- Mean activity coefficients can be determined experimentally; single-ion activity coefficients cannot.
- For ionic strengths up to about 0.050 M, single-ion activity coefficients and, in some special cases, mean activity coefficients can be calculated using the Debye-Hückel equation.
- Davies equation is more reliable for ionic strengths over wider ranges.
- Equilibrium constants determined at zero ionic strength are called thermodynamic equilibrium constants. Thermodynamic equilibrium constants
  - are included in tables such as appendices in handbooks and other texts,
  - are written in terms of activities, and
  - represent intrinsic behaviors of reactions in the absence of external interferences.
- Equilibrium constants for ionic strengths larger than zero are called conditional equilibrium constants. Conditional equilibrium constants
  - are expressed in terms of equilibrium concentrations,
  - vary with ionic strength,
  - can be determined experimentally or calculated from thermodynamic constants and activity coefficients, and
  - are the constants that should be used to calculate equilibrium concentrations.

These observations have stood the test of time and merit your effort to remember them.

REFERENCES

1. G. N. Lewis and M. Randall, *J. Am. Chem. Soc.* **1921,** *43,* 112–1154.
2. P. Debye and E. Hückel, *Phys. Z.* **1923,** *24,* 185–206.
3. C. W. Davies, *Ion Association*; Butterworths: London, 1962; p 41.
4. J. Kielland, *J. Am. Chem. Soc.* **1937,** *59,* 1675–1678.
5. H. S. Harned and R. W. Ehlers, *J. Am. Chem. Soc.* **1932,** *54,* 1350–1357.
6. H. S. Harned and J. G. Donelson, *J. Am. Chem. Soc.* **1937,** *59,* 1280–1284.

2 Monoprotic Acid-Base Equilibria: Exact and Approximate Options

Author's note: My primary purpose in this chapter is to show you in the simplest possible way how modern technology can be used to solve problems that were extremely difficult to solve using technologies available to our predecessors. An alternative approach that is more generally applicable to polyprotic acids and bases is discussed in later chapters. Therefore, much as it pains me to say it, I believe that it will serve you better to focus more attention on calculations using the equations and associated programs described herein than on the concepts used to develop them.

Acid-base equilibria are important in many areas that influence our everyday lives. Those areas include agriculture, health care, pharmaceuticals, food sciences, environmental studies, and industrial processes. Also, acid-base equilibria are important in virtually all the major areas of chemistry, including analytical chemistry, biological chemistry, inorganic chemistry, organic chemistry, and physical chemistry as well as other disciplines, including biology, engineering, and material sciences. This chapter focuses on the basis for and implementation of a program to solve an exact equation for monoprotic acids and bases.

The exact equation for the hydrogen ion concentration for any weak acid, H_nB, with n acidic hydrogens involves the hydrogen ion concentration to the n + 2 power. As examples, exact equations for hydrogen ion concentrations for acids containing 1, 2, and 6 acidic hydrogens involve hydrogen ion concentrations to the third, fourth, and eighth powers—i.e., $[H^+]^3$, $[H^+]^4$, and $[H^+]^8$.

Before electronic calculators and computers were available, it was very difficult to solve third- and higher-order equations. To compensate, chemists used simplifying assumptions to develop approximate equations that were more easily solved using tools available to us. The resulting procedures, called *conventional approximations* herein, are the procedures included in most introductory texts. Whereas conventional approximations give concentration errors less than the usual target of 5% for many situations, they can give errors much larger than 5% for many other situations. I will introduce an alternative approach to approximate calculations, called *unified approximations*, that reduces significantly the errors associated with conventional approximations.

Some goals for this chapter are to help you understand the basis for an exact equation for the hydrogen ion concentration in solutions containing one monoprotic acid-base pair and implement a program to solve the equation. I hope that discussions of these simpler situations will help you prepare for discussions in later chapters of concepts, equations, and programs used to solve problems involving solutions containing concentrations equal to or larger than zero of HCl, NaOH, and mixtures of up to three weak acids and bases with up to six acidic hydrogens per acid without and with correction for ionic strength.

BACKGROUND INFORMATION

We will talk in a later chapter about a systematic approach based on concepts developed by J. E. Ricci.[1] However, to conform as closely as possible to what you have likely learned about acid-base equilibria in introductory courses, I will use a more conventional approach in this chapter. This section describes some background information that is useful for different parts of subsequent discussions.

A Theory of Acids and Bases and Some Related Terminology and Symbols

In 1923, two chemists, J. N. Brønsted in Denmark and J. M. Lowry in England, proposed a common theory of acids and bases. According to their theory, an acid is a *proton donor*, and a base is a *proton acceptor*. When an acid such as acetic acid donates a proton, it is *deprotonated*. When a base such as acetate ion accepts a proton, it is *protonated*. As shown below for acetic acid and acetate ion, the loss of a proton by a weak acid—e.g., acetic acid—is called a *deprotonation reaction* herein, and the gain of a proton by a weak base—e.g., acetate ion—is called a *protonation reaction* herein.

$$\overbrace{CH_3COOH \underset{K_b}{\overset{K_a}{\rightleftharpoons}} H^+ + CH_3COO^-}^{\text{Deprotonation, Left to Right}}_{\text{Protonation, Right to Left}}$$

The corresponding constants, K_a and K_b, are called *deprotonation constants* and *protonation constants* herein.

You may be more familiar with the terms *acid dissociation constants* and *base association constants* for these constants. Feel free to use those terms in place of deprotonation constants and protonation constants if it helps your understanding. However, I have chosen the simpler and less ambiguous deprotonation and protonation terminology to reduce the possibility of confusion when we talk in later chapters about mixtures of acids and bases with other types of reactions, such as complex-formation reactions that also involve dissociation and association reactions.

Generalized Deprotonation Reaction

To simplify and generalize subsequent discussions, uncharged weak acids such as acetic acid are represented by the symbol HB. Charged conjugate bases such as acetate ion are represented by the symbol B^-. The simplified and generalized deprotonation reaction for an uncharged acid is

$$HB \overset{K_a}{\rightleftharpoons} H^+ + B^-$$

in which HB is the weak acid, H^+ is the hydrogen ion, and B^- is the basic part of the acid—e.g., acetate ion. The equation for the deprotonation constant (Equation 2.1) is

$$K_a = \frac{\left[H^+\right]\left[B^-\right]}{\left[HB\right]} \tag{2.1}$$

in which K_a is the deprotonation constant, or dissociation constant if you prefer; $[H^+]$ is the equilibrium hydrogen ion concentration; and $[B^-]$ and $[HB]$ are equilibrium weak base and weak acid concentrations, respectively.

A Way to Simplify Our Lives

Regardless of how we develop an exact equation for the hydrogen ion concentration for a monoprotic weak acid-base pair, the equation will be valid for all combinations of each acid-base pair. Therefore, to simplify our lives, I will assume that the reaction of interest approaches equilibrium by deprotonation of the weak acid and that the resulting solution is acidic: i.e.,

$$HB \rightharpoonup H^+ + B^-; \quad \left[H^+\right] > 1.0 \times 10^{-7} \text{ M}$$

Whereas these assumptions will simplify our lives, the resulting equation will be valid for all combinations of any monoprotic acid-base pair, including zero concentrations of one or both forms of the acid.

IDENTIFYING SOURCES OF THE HYDROGEN ION CONCENTRATION

There will be two sources of the hydrogen ion concentration when a weak acid is deprotonated in water—namely, water and the weak acid. It will be important in subsequent discussions to identify each source of the hydrogen ion concentration. As shown in the reactions below, hydrogen ion concentrations from water are represented by the symbol $[H^+]_w$, and concentrations from the weak acid are represented by the symbol $[H^+]_{HB}$.

$$H_2O \rightleftharpoons \overbrace{H^+}^{[H^+]_w} + OH^- \qquad HB \rightleftharpoons \overbrace{H^+}^{[H^+]_{HB}} + B^-$$

Whereas these representations are useful in deriving the exact equation, the equation is expressed in terms of the equilibrium hydrogen ion concentration, $[H^+]$, which is the sum of the two sources; i.e., $[H^+] = [H^+]_w + [H^+]_{HB}$.

ACID-BASE PROPERTIES OF WATER

Presumably you will have seen the reaction and autoprotolysis constant for water, K_w, written in one of the following forms:

$$H_2O \rightleftharpoons H^+ + OH^-; \quad K_w = [H^+][OH^-]; \quad 2H_2O \rightleftharpoons H_3O^+ + OH^-; \quad K_w = [H_3O^+][OH^-]$$

Either way the reaction and autoprotolysis constants are written, the activity-based autoprotolysis constant for water at 25 °C is $K_w = 1.0 \times 10^{-14}$. The simpler forms of the reaction and equation are used herein.

It follows from the autoprotolysis constant equation that the hydrogen ion and hydroxide ion concentrations are related to each other as shown in Equations 2.2a and 2.2b:

$$[H^+] = \frac{K_w}{[OH^-]} \qquad (2.2a) \qquad\qquad [OH^-] = \frac{K_w}{[H^+]} \qquad (2.2b)$$

As mentioned earlier, we will assume that the solution of interest is acidic. Any excess hydroxide ion in an acidic solution will react with excess hydrogen ion to produce water until the autoprotolysis constant for water is satisfied. This means that the only hydroxide ion concentration in an acidic solution will be that from the autoprotolysis of water.

Therefore, given that all the hydroxide ion concentration in any acidic solution results from the autoprotolysis of water, it follows that the OH^- concentration from water in acidic solutions is the same as the total OH^- concentration: i.e., $[OH^-]_w = [OH^-]$. Also, given that the hydrogen ion and hydroxide ion concentrations from water are the same—i.e., $[H^+]_w = [OH^-]_w$—the H^+ concentration from water in acidic solutions is as follows:

$$[H^+]_w = [OH^-]_w = [OH^-] = \frac{K_w}{[H^+]} \qquad (2.2c)$$

Equation 2.2c gives us a way to express the hydrogen ion concentration from water, $[H^+]_w$, in terms of the equilibrium hydrogen ion concentration, $[H^+]$.

EXACT EQUATION FOR THE HYDROGEN ION CONCENTRATION

Information in the preceding section is used below to develop an exact equation for hydrogen ion concentrations in solutions containing a monoprotic weak acid and/or its conjugate base.

DERIVATION OF AN EXACT EQUATION FOR THE HYDROGEN ION CONCENTRATION

Example 2.1 illustrates a stepwise procedure used to develop an exact equation for the hydrogen ion concentration in a solution containing diluted concentrations, C_{HB} and C_B, equal to or larger than zero of an uncharged or charged weak acid with deprotonation constant K_a.

> **Example 2.1 Derive an exact equation for the hydrogen ion concentration in a solution containing concentrations, C_{HB} and C_B, of the conjugate acid-base pair of a weak acid such as acetic acid with a deprotonation constant, K_a, by assuming that the reaction approaches equilibrium by deprotonation of the weak acid.**

Step 1: Write reactions that produce hydrogen ions in the solution—namely, deprotonation of the weak acid and autoprotolysis of water—and identify the two sources of hydrogen ions.

$$HB \rightleftharpoons \overbrace{H^+}^{[H^+]_{HB}} + B^- \qquad H_2O \rightleftharpoons \overbrace{H^+}^{[H^+]_w} + OH^-$$

Step 2: Write an equation for the total hydrogen ion concentration, $[H^+]$, as the sum of that from the weak acid, $[H^+]_{HB}$, plus that from water, $[H^+]_w$, and rearrange the equation to give an equation for the hydrogen ion concentration from the weak acid.

$$\left[H^+\right] = \left[H^+\right]_{HB} + \left[H^+\right]_w \Rightarrow \left[H^+\right]_{HB} = \left[H^+\right] - \left[H^+\right]_w = \left[H^+\right] - \frac{K_w}{\left[H^+\right]}$$

Step 3: Write equations for equilibrium weak acid and conjugate base concentrations by assuming that the reaction approaches equilibrium by deprotonation of the acid: i.e., $HB \rightarrow H^+ + B^-$.

$$[HB] = C_{HB} - \left[H^+\right]_{HB} = C_{HB} - \left[H^+\right] + \frac{K_w}{\left[H^+\right]} \qquad \left[B^-\right] = C_B + \left[H^+\right]_{HB} = C_B + \left[H^+\right] - \frac{K_w}{\left[H^+\right]}$$

Step 4: Rearrange the deprotonation constant equation for the acid-base pair, $K_a = [H^+][B^-]/[HB]$, into the form $K_a[HB] = [H^+][B^-]$, and substitute equations for the weak acid and weak base concentrations from Step 3 into the rearranged equation for the deprotonation constant.

$$K_a[HB] = \left[H^+\right]\left[B^-\right] = K_a\left(C_{HB} - \left[H^+\right] + \frac{K_w}{\left[H^+\right]}\right) = \left[H^+\right]\left(C_B + \left[H^+\right] - \frac{K_w}{\left[H^+\right]}\right)$$

Step 5: Convert this equation to the form below by (a) multiplying through on the right side by $[H^+]$, (b) multiplying both sides by $[H^+]$, (c) collecting terms for hydrogen ion concentrations with powers from 0 to 3, and (d) transposing terms on the right side of the equation to the left side, leaving zero on the right side.

$$\left[H^+\right]^3 + \left(K_a + C_B\right)\left[H^+\right]^2 - \left(K_a C_{HB} + K_w\right)\left[H^+\right] - K_a K_w = 0 \qquad (2.3)$$

Step 1 summarizes reactions that produce hydrogen ions. Step 2 uses the sum of hydrogen ion concentrations from the weak acid and water to develop an equation for the hydrogen ion concentration produced by deprotonation of the weak acid, $[H^+]_{HB}$, in terms of the total hydrogen ion concentration, $[H^+]$, and the hydrogen ion concentration from water, $[H^+]_w$.

Step 3 reflects the facts that the weak acid concentration is decreased by an amount equal to the hydrogen ion concentration produced by deprotonation of the weak acid and the weak base concentration is increased by the same amount. In other words, *equilibrium weak acid and weak base concentrations are diluted concentrations minus and plus the hydrogen ion concentration produced by deprotonation of the weak acid.*

Step 4 involves substitutions of equations for equilibrium weak acid and weak base concentrations into a rearranged form of the deprotonation constant equation. I will trust you to work through the process in Step 5 to confirm the third-order equation for the hydrogen ion concentration.

SCOPE OF EQUATION 2.3

On the one hand, Equation 2.3 was developed by assuming that an uncharged acid approached equilibrium by deprotonation of the acid in an acidic solution. On the other hand, that assumption was merely a matter of convenience. Equation 2.3 applies for diluted concentrations from zero to any practical upper limit of a weak acid and/or the conjugate base of an uncharged or charged acid approaching equilibrium by deprotonation of the weak acid or protonation of the conjugate base. In other words, Equation 2.3 is applicable to virtually any situation involving a monoprotic weak acid and/or a monoprotic weak base.

A SPREADSHEET PROGRAM TO SOLVE AN EXACT EQUATION FOR THE HYDROGEN ION CONCENTRATION

We will work together to write a spreadsheet program to (a) solve the exact equation for the hydrogen ion concentration, (b) use the hydrogen ion concentrations to calculate fractions of monoprotic acids in protonated and deprotonated forms, and (c) compare hydrogen ion concentrations calculated using exact and approximate equations.

LAYOUT OF THE PROGRAM

To begin, open an unused Excel spreadsheet, and organize Sheet 1 as in Table 2.1A, paying attention to significant figures associated with each number.

The three parts of the program are now discussed individually. Numbers in this table are what you should see when you have completed the three parts of the program with Cells C15–C18 and E15–E17 as in the table.

Parts of the program in Rows 14–20 are used to solve the exact equation for hydrogen ion concentrations. We will talk about that part of the program shortly. Parts of the program in Rows 21–23 use the hydrogen ion concentration calculated in Rows 14–20 to calculate fractions, α_{HB} and α_B, of the weak acid in the protonated and deprotonated forms, respectively. Parts of the program in Rows 24–26 are used to calculate and compare hydrogen ion concentrations calculated using two approximation procedures and to compare the approximate concentrations to those calculated using the exact equation.

PROGRAM TO CALCULATE THE HYDROGEN ION CONCENTRATION

The program used to calculate the hydrogen ion concentration is based on a discovery made during my work with Equation 2.3 and other analogous equations. I discovered that the equation is *negative* for hydrogen ion concentrations smaller than the equilibrium concentration and *positive* for hydrogen ion concentrations larger than the equilibrium concentration. Therefore, our strategy for

TABLE 2.1A

Layout of an Excel Spreadsheet Program to Solve an Exact Equation for the Hydrogen Ion Concentration in Any Solution Containing a Monoprotic Weak Acid and/or Its Conjugate Base

A-13	B	C	D	E	F	G
	Set E15 to 0 to change input information and to 1 to solve problems.					
14	Input quantities		Control settings		Low resolution	
15	C_{HB}	1.00E-01	**Rset/Strt**	**0**	Cubic eq.	-1.8E-19
16	C_B	0.00E+00	Sig. Figs.	3	$C_{H(LoRes)}$	1.00E-15
17	K_a	1.80E-05	Initial C_H	1.00E-15	High resolution	
18	K_W	1.00E-14			Cubic eq.	-1.8E-19
19					[H$^+$]	**1.00E-16**
20	Are calculations complete?			No	Iterations	**0**
21	Fractions		C_T		Equilibrium C's	
22	α_{HB}	α_B	$C_{HB} + C_B$	[HB]	[B]	
23	5.56E-12	1.00	1.00E-1	5.56E-13	1.00E-1	
24	Unified approx.			Both	Conventional approx.	
25	Q^0/K_a	[H$^+$]	Error (%)	K_b	[H$^+$]	Error (%)
26	0.0E+00	**1.33E-03**	1.3E+15	5.56E-10	**1.34E-03**	1.3E+15

solving Equation 2.3 is to start with a hydrogen ion concentration less than the equilibrium concentration and to increase the hydrogen ion concentration by small increments, ΔC_H, until the sign of Equation 2.3 changes from minus to plus. The hydrogen ion concentration that produces the change from minus to plus is accepted as the equilibrium concentration.

Having organized an Excel sheet as in Table 2.1A, left click on File/Options/Formulas, and set Workbook Calculations to Automatic; also check the box for Enable iterative calculation, and set Maximum Iterations to 1,000 and Maximum Change to 1E-4. This prepares the program to do iterative calculations.

COMPLETING THE PART OF THE PROGRAM TO SOLVE EQUATION 2.3

Having set conditions for iterative calculations as described above, type the formulae from Table 2.1B into the indicated cells in the order—left to right, top to bottom—shown in the table, and press Ctrl/Shift/Enter simultaneously after each entry is complete.

If you enter the formulae in the order left to right and top to bottom as in this table and press Ctrl/Shift/Enter simultaneously after each entry, the numbers should appear as in Table 2.1A. I suggest that you save your program using an easily recognizable name, such as *Monoprotic Exact_Approximate*, after each formula is entered successfully.

TABLE 2.1B

Formulae to Be Entered into Selected Cells to Complete a Program to Solve Equation 2.3 for the Hydrogen Ion Concentration

G16 =IF(E15=0,E17,IF(G15<0,10*G16,G16))   G15 =G16^3+(C17+C16)*G16^2-(C17*C15+C18)*G16-C17*C18

G19 =IF(G15<0,0.1*G16,IF(G18<0,G19+10^-   G18 =G19^3+(C17+C16)*G19^2-(C17*C15+C18)*G19-C17*C18
E16*G16,G19))

E20 =IF(G18<0,"No","Yes")   G20 =IF(E15=0,0,if(G18<0,G20+1,G20))

Numbers in Cells B15–B18 correspond to a 0.10 M solution of acetic acid, $K_a = 1.8 \times 10^{-5}$.

To calculate the hydrogen ion concentration for conditions in Table 2.1A, set E15 to 1, wait for E19 to change from "No" to "Yes," and read the hydrogen ion concentration from Cell G19. Press F9 one or more times if the iteration counter in G20 stops changing before E20 changes to "Yes." Cell C19 should read 1.34×10^{-3}, corresponding to $[H^+] = 1.34 \times 10^{-3}$ M, when iterations in G20 stop changing and E20 changes to "Yes."

Numbers you should see in relevant cells when the process is complete are summarized and discussed below.

G15	G16	G18	G19	E20	G20
9.8E-7	1.00E-2	2.6E-11	1.34E-3	Yes	47

If numbers in any cell(s) differ significantly from those in the above table, you should check for consistency between equations in Table 2.1B and equations in the corresponding cells in your program.

LOW-RESOLUTION AND HIGH-RESOLUTION STEPS

As you probably know, hydrogen ion concentrations in strongly basic to strongly acidic solutions can vary over ranges from near 10^{-14} M to 1.0 M or higher. To avoid very long convergence times, the program is implemented in two general steps, called *low-resolution* and *high-resolution* steps herein.

The low-resolution step is used to find the upper limit of a 10-fold range of hydrogen ion concentrations that includes the equilibrium concentration. This step starts with a very low hydrogen ion concentration—e.g., 1.0×10^{-15} M—in G16 and increases it in 10-fold steps until the sign of Equation 2.3 in G15 changes from − to +. The concentration that causes the − to + change is the upper limit of a 10-fold range of H^+ concentrations that includes the equilibrium concentration. Let's call this the *upper-limit concentration*.

The high-resolution step starts with an H^+ concentration in G19 equal to one-tenth of the upper-limit concentration in G19 and increases it by small amounts, ΔC_H, until the sign of Equation 2.3 in G18 changes from − to +. The H^+ concentration that produces the − to + change is the equilibrium concentration resolved to within the size of the increment, ΔC_H.

The size of the concentration change, ΔC_H, is controlled by the number of significant figures in E16. With E16 set to 3, the size of the change is 10^{-3} times the upper-limit concentration calculated in the low-resolution step. For conditions in Table 2.1A, the upper-limit concentration is 0.010 M, and with E16 set to 3, the concentration change is $\Delta C_H = 10^{-3} \times 0.010 = 1.00 \times 10^{-5}$. This means that the equilibrium H^+ concentration is resolved to the third significant figure—i.e., 1.34×10^{-3} M.

With E16 set to 2 or 4, the hydrogen ion concentration will be resolved to two or four significant figures. Resolution to three significant figures is a good compromise between calculation time and uncertainties associated with equilibrium constants.

SOME ILLUSTRATIVE NUMBERS

Let's use the hydrogen ion concentration in a 0.10 M acetic acid solution to associate some numbers with the foregoing description.

The equilibrium hydrogen ion concentration is 1.34×10^{-3} M; i.e., $[H^+] = 1.34 \times 10^{-3}$ M. The upper limit of the 10-fold range that includes 1.34×10^{-3} M is 0.010 M. Therefore, when the sign of the exact equation in G15 changes from − to +, the low-resolution step will stop with 0.010 in G16. The − to + change at the end of the low-resolution step triggers the start of the high-resolution step.

With E16 set to 3, the high-resolution step starts with a trial H^+ concentration equal to one-tenth of the upper-limit concentration—i.e., $0.1 \times 0.010 = 0.0010$—in G19 and increases it by increments,

$\Delta C_H = 10^{-3}(0.010 \text{ M}) = 10^{-5}$ M, until Equation 2.3 changes from $-$ to $+$. The H^+ concentration that causes the $-$ to $+$ change is accepted as the equilibrium concentration resolved to within the size, ΔC_H, of the increments—here, 10^{-5} M.

Analogous considerations apply for other hydrogen ion concentrations. Whatever the equilibrium hydrogen ion concentration, it will be resolved to the number of significant figures in E16.

Numbers of iterations required to solve any problem depend on the number of significant figures in E16 and how far the hydrogen ion concentration is from the start of the 10-fold range resolved in the high-resolution step. The larger the hydrogen ion concentration relative to the start of each 10-fold range, the larger the number of iterations required to resolve the H^+ concentration.

I suggest that you compare numbers of iterations with E16 set to 2, 3, and 4, respectively.

OTHER COMBINATIONS OF CONJUGATE ACID-BASE PAIRS

The program can be used for any combination of diluted weak acid and weak base concentrations and deprotonation constants in Cells C15–C17. The same general process applies for any hydrogen ion concentration larger than the starting value in E17—e.g., 10^{-15} M. Whatever the equilibrium concentration, the algorithms in Cells G15 and G16 will find the upper limit of a 10-fold range that includes the equilibrium concentration. Then the algorithms in G18 and G19 will resolve concentrations within the 10-fold range to the number of significant figures in E16.

Contents of Cells C15–C17 needed to do calculations for solutions containing 0.10 M acetate ion and 0.10 M each of acetic acid and acetate ion are summarized below along with expected hydrogen ion concentrations in G19.

0.10 M Acetate Alone				0.10 M Acetic Acid + 0.10 M Acetate			
C15	C16	C17	G19, [H^+]	C15	C16	C17	G19, [H^+]
0	0.10	1.8E-5	(1.35E-9)	0.10	0.10	1.8E-5	(1.80E-5)

To implement these calculations, set E15 to 0, enter the indicted numbers into C15–C17, reset E15 to 1, wait for E20 to change to "Yes," and read the H^+ concentration from G19.

SOME ATTRACTIVE FEATURES OF THE PROGRAM

Unlike common Solver programs, the program described above doesn't need a first estimate of the hydrogen ion concentration that is reasonably close to the equilibrium concentration. All it needs is a starting concentration less than the equilibrium concentration. Also, the exact number of significant figures to which the hydrogen ion concentration is resolved is controlled precisely by a small number entered by the user into a Cell E16. Although it may be difficult to interpret the results, the number entered into E16 need not be a whole number. It can be a fractional number, such as 2.5 or 3.5.

These same features are included in programs for polyprotic acids and bases discussed in later chapters.

FRACTIONS AND CONCENTRATIONS OF CONJUGATE ACID-BASE PAIRS

We will talk in the next chapter about equations for fractions of weak acids in different forms. Having calculated the hydrogen ion concentration, it is very easy to use it to calculate fractions of the acid-base pair in the fully protonated and deprotonated forms, HB and B. Having calculated the fractions, they can be used with the total diluted weak acid–weak base concentration, $C_T = C_{HB} + C_B$, to calculate equilibrium concentrations in the individual forms.

Example 2.2 is a typical problem that will be used later to illustrate some types of calculations associated with concentration fractions.

Example 2.2 **Given a solution containing 0.10 M dichloroacetic acid and 0.050 M dichloroacetate ion, calculate, (A) fractions of the total concentration in the protonated and deprotonated forms, (B) equilibrium concentrations in the two forms, and (C) the percent deprotonated. For dichloroacetic acid: $K_a = 0.050$.**

CONCENTRATION FRACTIONS

Fractions of a monoprotic weak acid in the fully protonated and fully deprotonated forms are represented in this chapter as α_{HB} and α_B. We will show in Chapter 3 that the fractions depend on the deprotonation constant and hydrogen ion concentration as shown in Equations 2.4a and 2.4b:

$$\alpha_{HB} = \frac{[H^+]}{[H^+] + K_a} \qquad (2.4a) \qquad\qquad \alpha_B = \frac{K_a}{[H^+] + K_a} \qquad (2.4b)$$

in which α_{HB} is the fraction in the protonated form, α_B is the fraction in the deprotonated form, and $[H^+]$ and K_a are the hydrogen ion concentration and the deprotonation constant, respectively.

If you prefer to think in terms of percentages rather than fractions, percentages in the two forms are 100 times the fractions.

EQUILIBRIUM CONCENTRATIONS

Equilibrium concentrations, $[HB]$ and $[B]$, of the protonated and deprotonated forms of a monoprotic weak acid are products of the fractions, α_{HB} and α_B, times the total concentration, C_T, of both forms of the acid. Equations 2.5a to 2.5c are as shown below

$$[HB] = \alpha_{HB} C_T \qquad (2.5a) \qquad\qquad [B] = \alpha_B C_T \qquad (2.5b) \qquad\qquad C_T = C_{HB} + C_B \qquad (2.5c)$$

in which $[HB]$ and $[B]$ are equilibrium concentrations of the two forms of the acid, C_T is the sum of concentrations of the diluted concentrations, C_{HB} and C_B, of the two forms of the acid, and other symbols are as defined above.

Example 2.2 is a typical problem involving concentration fractions and equilibrium concentrations.

Example 2.2 **Assuming that the hydrogen ion concentration in a solution containing 0.10 M dichloroacetic acid and 0.050 M dichloroacetate ion is $[H^+] = 0.0367$ M, calculate (A) fractions of the total concentration in the protonated and deprotonated forms, (B) equilibrium concentrations in the two forms, and (C) the percentage deprotonated. For dichloroacetic acid: $K_a = 0.050$.**

Step 1: Calculate the concentration fractions.

$$\alpha_{HB} = \frac{[H^+]}{[H^+] + K_a} = \frac{0.0367}{0.0367 + 0.050} = 0.423; \quad \alpha_B = \frac{[H^+]}{[H^+] + K_a} = \frac{0.050}{0.0367 + 0.050} = 0.577$$

Step 2: Calculate the total concentration of the two forms of the acid.

$$C_T = C_{HB} + C_B = 0.10 \text{ M} + 0.050 \text{ M} = 0.150 \text{ M}$$

Step 3: Calculate the equilibrium concentrations.

$$[HB] = \alpha_{HB} C_T = 0.423 \times 0.15 = 0.063 \text{ M}; \quad [B^-] = \alpha_B C_T = 0.577 \times 0.15 = 0.087 \text{ M}$$

Step 4: Calculate the percentage deprotonated as 100 times the fraction deprotonated.

$$\text{Depr}(\%) = 100\alpha_B = 100(0.58) = 58\ \%$$

The starting solution contained 66% dichloroacetic acid and 34% dichloroacetate as calculated using the initial concentrations, 0.10 M and 0.050 M, respectively. The equilibrated solution contains 42% dichloroacetic acid and 58% dichloroacetate. These results correspond to concentration changes from 0.10 M to 0.063 M for dichloroacetic acid and from 0.050 M to 0.087 M for dichloroacetic acid and 58% deprotonation of dichloroacetate. The relatively large changes result from the large value of the deprotonation constant; i.e., $K_a = 0.050$.

EXTENSION OF THE ITERATIVE PROGRAM TO CALCULATE CONCENTRATION FRACTIONS AND EQUILIBRIUM CONCENTRATIONS

The part of the program in Rows 21–23 of Table 2.1A is designed to calculate concentration fractions, α_{HB} and α_B, and equilibrium concentrations, [HB] and [B$^-$], of weak acids in the protonated and deprotonated forms. Table 2.2A summarizes the contents of Cells B23–F23 used to calculate concentration fractions and equilibrium concentrations of monoprotic weak acids in the protonated and deprotonated forms.

To implement this part of the program, set E15 to 0, and enter the formulae into the appropriate cells. The fractions, α_{HB} and α_B, are calculated in Cells B23 and C23, respectively. The total concentration of the two forms of the acid is calculated in D23. Equilibrium concentrations are calculated in E23 and F23. As usual, set E15 to 1 to run the program.

The contents of Cells B23–F23 before and after the program has been run with starting conditions as in Table 2.1A are summarized in Table 2.2B.

TABLE 2.2A

Cell Contents for Calculations of Concentration Fractions and Equilibrium Concentrations of Monoprotic Acid-Base Pairs

B23	=G19/(G19+C17)	C23	=C17/(G19+C17)	D23	=C15+C16
E23	=B23*D23	F23	=C23*D23		

TABLE 2.2B

Cell Contents before and after the Program Is Run for Conditions in Table 2.1A—i.e., 0.10 M Acetic Acid, $K_a = 1.8 \times 10^{-5}$

A	B	C	D	E	F	G
			Before—i.e., E15 = 0			
23	5.56E-12	1.00E+00	1.00E-01	5.56E-13	1.00E-01	
			After—i.e., E15 = 1			
23	9.87E-1	1.27E-2	1.00E-1	9.87E-2	1.27E-3	

If the contents of any cells are different than in Table 2.2B, you should look for and correct errors in the formulae entered from Table 2.2A.

I suggest that you use your program to test the results of the calculations in Example 2.2. To do this, set E15 = 0, C15 = 0.10, C16 = 0.050, and C17 = 0.050. Then set E15 to 1, wait for E20 to change to "Yes," and read the fractions and concentrations from Cells B23–F23.

PROGRAM TO COMPARE HYDROGEN ION CONCENTRATIONS CALCULATED USING EXACT AND APPROXIMATE EQUATIONS

Given the ease with which exact equations for the hydrogen ion concentration, concentration fractions, and concentrations of conjugate acid-base pairs can be solved using programs such as those described above, I believe that your time and efforts will be spent more effectively by proceeding to subsequent chapters. However, for the sake of completeness, I will discuss two options for approximate calculations involving monoprotic acids and bases, called *conventional approximations* and *unified* approximations[2] herein.

Conventional approximations are the types of approximations introduced in most introductory texts. Unified approximations involve an approach developed by my colleagues and me[2] to avoid some problems associated with conventional approximations.

Calculations using conventional approximations are simpler than those using unified approximations but give very large errors for some situations. Calculations using unified approximations are somewhat more complicated than conventional approximations but give very small errors for ranges of situations for which conventional approximations give very large errors.

I will begin with some additions to the program described earlier to extend it to calculations involving conventional and unified approximations.

The formulae needed to extend the program described above to include conventional and unified approximations are summarized in Table 2.3.

The layout of the program you should see when you have completed the foregoing changes is shown below for a solution containing 0.10 M each of acetic acid, $K_a = 1.8 \times 10^{-5}$, and sodium acetate.

A	B	C	D	E	F	G
14	Input quantities		Control settings		Low resolution	
15	C_{HB}	1.00E-01	Rset/Strt	1	Cubic eq.	8.2E-10
16	C_B	1.00E-01	Sig. Figs.	3	$C_{H(LoRes)}$	1.00E-04
17	K_a	1.80E-05	Initial C_H	1.00E-15	High resolution	
18	K_W	1.00E-14			Cubic eq.	1.2E-14
19					[H$^+$]	1.80E-05
20	Are calculations complete?			Yes	Iterations	91
21	Fractions		Total C	Eq. concentrations		
22	α_{HB}	α_B	$C_{HB} + C_B$	[HB]	[B]	
23	5.00E-01	5.00E-01	2.00E-01	1.00E-01	1.00E-01	
24	Unified approx.			Both	Conventional approx.	
25	Q^0/K_a	[H$^+$]	Error (%)	K_b	[H$^+$]	Error (%)
26	5.6E-03	1.80E-05	-3.6E-02	5.56E-10	1.80E-05	-8.9E-14

Clearly, all three options give the same hydrogen ion concentration for the solution containing 0.10 M each of acetic acid and sodium acetate. You can show that the same is true for a solution containing 0.10 M acetic acid alone by running the program with C15 = 0.10 and C16 = 0.00 and for one containing 0.10 M acetate ion alone by running the program with C15 = 0.00 and C16 = 0.100. To run the program for these or other situations, set E15 to 0, set C16–C17 to desired values, set E15

TABLE 2.3

Formulae Used to Extend the Program for Monoprotic Acid-Base Equilibria to Conventional and Unified Approximations

B26	=SQRT(C18)/C17*((C16)/(C15+C18))
C26	=IF(B26<1,(-(C17+C16)+SQRT((C17+C16)^2+4*(C17*C15+C18)))/2,2*C18/(-(E26+C15)+SQRT((E26+C15)^ 2+4*(E26*C16+C18))))
F26	=IF(AND(C15>0,C16=0),SQRT(C17*C15),IF(AND(C15>0,C16>0),C17*C15/C16,IF(AND(C15=0,C16>0), SQRT(C17*C18/C16))))
D26	=100*(C26/G19-1) E26 =C18/C17 G26 =100*(F26/G19-1)

TABLE 2.4

Hydrogen Ion Concentrations Calculated Using Exact and Approximate Options for Solutions Containing (A) $C_{HB} = 0.10$ M Trichloroacetic Acid and $C_{NaB} = 0.00$ M Sodium Trichloroacetate, (B) $C_{HB} = 0.00$ M Trichloroacetic Acid and $C_{NaB} = 0.10$ M Sodium Trichloroacetate, and (C) $C_{HB} = 0.10$ M Trichloroacetic Acid and $C_{NaB} = 0.10$ M Sodium Trichloroacetate. For Trichloroacetic Acid: $K_a = 0.22$; for Trichloroacetate Ion: $K_b = K_w/K_a = 4.55 \times 10^{-14}$

Part	A			B			C		
Procedure	Set E15 to 0, enter the concentrations and the deprotonation constant into C15–C17, set E15 to 1, wait for E20 to change to "Yes," and read the H⁺ concentrations from Cells G19, C26, and F26 and the percentage errors from Cells D26 and F26.								
Contents of Input Cells									
Cell numbers	C15	C16	C17	C15	C16	C17	C15	C16	C17
Cell contents	0.10	0.00	0.22	0.00	0.10	0.22	0.10	0.10	0.22
Hydrogen Ion Concentrations									
Exact equation: G19	$[H^+] = 0.0747$ M			$[H^+] = 8.30 \times 10^{-8}$ M			$[H^+] = 0.0582$ M		
Conventional approx.: F26	$[H^+] \cong 0.148$ M			$[H^+] \cong 1.48 \times 10^{-7}$ M			$[H^+] \cong 0.22$ M		
Unified approx.: C26	$[H^+] \cong 0.0747$ M			$[H^+] \cong 8.29 \times 10^{-8}$ M			$[H^+] \cong 0.0582$ M		
Approximation Errors									
Conventional approx.: G26	99 %			79 %			280 %		
Unified approx.: D26	< 0.1 %			0.1 %			< 0.1 %		

to 1, wait for E20 to change to "Yes," and read the hydrogen ion concentrations calculated using the different options from Cells G19, C26, and F26.

We will talk below about the basis for conventional and unified approximations. First, however, I will use results in Table 2.4 to show you why I prefer the unified approximations.

I suggest that you use your program to confirm these results. However, whether you do or don't do the calculations, it is clear that unified approximations give much smaller errors than conventional approximations for these situations. We will show later that unified approximations give very small errors for other situations for which conventional approximations give unacceptably large errors.

CONVENTIONAL APPROXIMATIONS

To simplify your lives, authors of introductory texts, myself included, used very simple situations to introduce quantitative aspects of acid-base equilibria in particular and chemical equilibria in general. The resulting methods are called *conventional approximations* herein.

Conventional approximations for monoprotic acids and bases are based on assumptions that (a) the hydrogen ion concentration from water is negligibly small and (b) changes in reactant concentrations are negligibly small.

Using the symbols HB and B^- for an uncharged monoprotic weak acid and its conjugate base, the deprotonation reaction and constant are as follows:

$$HB \overset{K_a}{\rightleftharpoons} H^+ + B^- \qquad K_a = \frac{[H^+][B^-]}{[HB]}$$

in which K_a is the deprotonation constant and $[H^+]$, $[HB]$, and $[B^-]$ are equilibrium concentrations of the reactants.

The protonation reaction and protonation constant for a weak base are as follows:

$$B^- + H_2O \overset{K_a}{\rightleftharpoons} OH^- + HB \qquad K_b = \frac{[OH^-][HB]}{[B^-]} = \frac{K_w[HB]}{[H^+][B^-]} = \frac{K_w}{K_a}$$

In the conventional approach to a reaction that approaches equilibrium by deprotonation of the weak acid, one writes equations for the weak acid and weak base concentrations in terms of diluted concentrations of the reactants, substitutes these equations into the equation for the deprotonation constant, and rearranges the resulting equation into a form that can be solved for the hydrogen ion or hydroxide ion concentration. I will walk you through a process used to do this.

We will assume in the example below that 0.050 M sodium acetate dissociates completely—i.e., $NaB \rightarrow Na^+ + B^-$—to produce $C_B = 0.050$ M acetate ion.

Example 2.3 Calculate the hydrogen ion concentrations in solutions containing (A) $C_{HB} = 0.10$ M acetic acid, (B) $C_B = 0.10$ M sodium acetate, and (C) $C_{HB} = 0.10$ M acetic acid and $C_{NaB} = 0.050$ M sodium acetate. For acetic acid: $K_a = 1.8 \times 10^{-5}$; for acetate ion: $K_b = K_w/K_a = 5.7 \times 10^{-10}$.

Weak acid alone $(C_{HB} > 0, C_B = 0)$	Weak base alone $(C_{HB} = 0, C_B > 0)$	Conjugate acid-base pair $(C_{HB} > 0, C_B > 0)$
$[HB] \cong C_{HB} - [H^+] \cong C_{HB}$	$[B^-] \cong C_B - [OH^-] \cong C_B$	$[HB] \cong C_{HB} - [H^+] \cong C_{HB}$
$[B^-] \cong C_B + [H^+] \cong [H^+]$	$[HB] \cong C_{HB} + [OH^-] \cong [OH^-]$	$[B^-] \cong C_B + [H^+] \cong C_B$
$K_a = \dfrac{[H^+][B^-]}{[HB]} \cong \dfrac{[H^+][H^+]}{C_{HB}}$	$K_b = \dfrac{[OH^-][B^-]}{[HB]} \cong \dfrac{[OH^-][OH^-]}{C_B}$	$K_a = \dfrac{[H^+][B^-]}{[HB]} \cong \dfrac{[H^+]C_B}{C_{HB}}$
$[H^+] \cong \sqrt{K_a C_{HB}}$ (2.6a)	$[OH^-] \cong \sqrt{K_b C_B}$ (2.6b)	$[H^+] \cong K_a \dfrac{C_{HB}}{C_B}$ (2.6c)

I will trust you to confirm the calculated concentrations below for equations 2.6a, 2.6b, and 2.6c.

$$[H^+] \cong \sqrt{K_a C_{HB}} \cong 1.34 \times 10^{-3} \qquad [OH^-] \cong \sqrt{K_b C_B} \cong 5.3 \times 10^{-6} \qquad [H^+] \cong K_a \frac{C_{HB}}{C_B} \cong 3.6 \times 10^{-5}$$

$$[H^+] \cong \frac{K_w}{[OH^-]} \cong 1.90 \times 10^{-9}$$

Concentrations calculated using the program based on an exact equation.

$$[H^+] = 1.34 \times 10^{-3} \qquad\qquad [H^+] = 1.90 \times 10^{-9} \qquad\qquad [H^+] = 3.60 \times 10^{-5}$$

As shown by hydrogen ion concentrations in the last row, which were calculated using the program based on an exact equation, hydrogen ion concentrations calculated using conventional approximations differ very little from those calculated using the exact equation. However, as shown earlier in Table 2.4, there are situations for which conventional approximations give much larger errors than obtained using unified approximations. Let's talk next about unified approximations.

UNIFIED APPROXIMATIONS

Unified approximations make use of one decision step and two quadratic equations to solve problems involving solutions containing one monoprotic acid and/or the conjugate base of the acid. As noted above, unified approximations give concentration errors within the usual target of ±5% for concentrations in which at least one member of the acid-base pair is equal to or larger than 2×10^{-6} M and much smaller errors for larger concentrations.

BASIS FOR UNIFIED APPROXIMATIONS

The exact equation for the hydrogen ion concentration discussed earlier is the starting point for equations used for unified approximations. That equation is included here for your convenience:

$$\left[H^+\right]^3 + \left(K_a + C_B\right)\left[H^+\right]^2 - \left[K_a C_{HB} + K_w\right]\left[H^+\right] - K_a K_w = 0$$

in which C_B is the conjugate base concentration, C_{HB} is the weak acid concentration, K_a is the deprotonation constant for the weak acid, and K_w is the autoprotolysis constant for water.

Acidic Solutions

Our studies[2] showed that the last term in this equation, $K_a K_w$, is small relative to the remainder of the equation in acidic solutions. After dropping the last term and dividing both sides of the resulting equation by [H^+], the equation can be written as in Equation 2.7a:

$$\left[H^+\right]^2 + \left(K_a + C_B\right)\left[H^+\right] - \left[K_a C_{HB} + K_w\right] \cong 0 \tag{2.7a}$$

in which all symbols are as defined earlier. This equation can be solved using the quadratic formula.

Our studies also showed that the usual assumption that K_w in Equation 2.7a is negligibly small fails for some situations that produce basic solutions. For such situations, it is necessary to use an equation in terms of the hydroxide ion concentration.

Basic Solutions

Although procedures similar to that used to develop the exact equation for the hydrogen ion concentration can be used to develop an analogous equation in terms of the hydroxide concentration, there is a simpler way. Equation 2.7a can be converted to an equation in terms of the hydroxide ion concentration by substituting

$$\left[H^+\right] \Rightarrow \left[OH^-\right], \ K_a \Rightarrow K_b, \ C_{HB} \Rightarrow C_B, \text{ and } C_B \Rightarrow C_{HB}$$

The resulting equation is

$$\left[OH^-\right]^2 + \left(K_b + C_B\right)\left[H^+\right] - \left[K_b C_{HB} + K_w\right] \cong 0 \tag{2.7b}$$

Equation 2.7b can also be solved using the quadratic formula. For basic solutions, the H^+ concentration can be calculated as [H^+] = K_w/[OH^-]—i.e., K_w/Equation 2.7b. Computational forms of Equations 2.7a and 2.7b for hydrogen ion concentrations are shown in Equations 2.8a and 2.8b:

$$\left[H^+\right] \cong \frac{-\left(K_a + C_B\right) + \sqrt{\left(K_a + C_B\right)^2 + 4\left[\left(K_a C_{HB}\right) + K_w\right]}}{2} \tag{2.8a}$$

$$\left[H^+\right] \cong \frac{2K_w}{-\left(K_b + C_{HB}\right) + \sqrt{\left(K_b + C_{HB}\right)^2 + 4\left[\left(K_b C_B\right) + K_w\right]}} \tag{2.8b}$$

All symbols are as defined earlier.

SCENARIO FOR CHOOSING BETWEEN EQUATIONS 2.8A AND 2.8B

A relatively simple scenario for choosing between the equations is summarized as follows:

$$\text{Use Equation 2.8a if } \frac{\sqrt{K_w}}{K_a} \frac{C_B}{\left(C_{HB} + K_w\right)} \le 1; \text{ otherwise, use Equation 2.8b.}$$

The K_w term in the denominator is used to avoid a divide-by-zero error when $C_{HB} = 0$ when this scenario is programmed into a calculator or computer.

This *selection process* and the approximate equations will give errors of 5% or less for any situation involving either a weak-acid or a conjugate-base concentration equal to or larger than 2×10^{-6} M. The basis for the selection scenario is discussed below.

As a reminder, the reaction quotient for any reaction has the same general form as the equilibrium constant with equilibrium concentrations replaced by nonequilibrium concentrations. Any equilibration reaction will approach equilibrium from left to right if the reaction quotient is less than the equilibrium constant—i.e., if Q/K < 1—and from right to left if the reaction quotient is larger than the equilibrium constant—i.e., if Q/K > 1. For an acid-base reaction, it follows that the ratio of the initial reaction quotient to the deprotonation constant, Q^0/K_a, can be used to determine if a reaction will approach equilibrium by deprotonation of a weak acid or protonation of its conjugate base.

Assuming that reactants are added to otherwise pure water, the initial H^+ concentration can be calculated as the square root of the autoprotolysis constant; i.e., $[H^+]_0 = K_w^{1/2} \cong 10^{-7}$ M. It follows that the ratio of the initial reaction quotient to the deprotonation constant will be as shown in Equation 2.9:

$$\frac{Q_0}{K_a} = \frac{1}{K_a} \frac{\left[H^+\right]_0 C_B}{C_{HB}} \cong \frac{10^{-7}}{K_a} \frac{C_B}{C_{HB}} \cong \frac{\sqrt{K_w}}{K_a} \frac{C_B}{C_{HB} + K_w} \tag{2.9}$$

in which Q_0 is the reaction quotient for a hydrogen ion concentration of $[H^+]_0 = 1.0 \times 10^{-7}$ M; K_a is the deprotonation constant for the weak acid; C_{HB} and C_B are diluted concentrations of the weak acid and its conjugate base, respectively; and K_w is the autoprotolysis constant for water.

This quantity is calculated in B23 of the program described above and is used in C23 to choose between the two equations.

If the reaction quotient is less than the deprotonation constant—i.e., if $Q_0/K \le 1$—the reaction will approach equilibrium by deprotonation of the weak acid corresponding to Equation 2.8a. If the reaction quotient is larger than the deprotonation constant—i.e., if $Q_0/K > 1$—the reaction will approach equilibrium by protonation of the weak base corresponding to Equation 2.8b.

EXAMPLE CALCULATIONS

Believing that you learn by doing, I will guide you through solutions to selected examples and trust you to correlate numbers with symbols in the equations and do the calculations. The H^+ concentration calculated using the exact equation for each example is included in parentheses for comparison purposes.

Example 2.4 illustrates calculations for solutions containing a *not-so-weak acid* alone, a mixture of the acid and its conjugate base, and the conjugate base alone.

Example 2.4 Calculate the hydrogen ion concentration in solutions containing (A) 0.10 M trichloroacetic acid (TCA), (B) 0.10 M trichloroacetic acid and 0.050 M sodium trichloroacetate, and (C) 0.10 M sodium trichloroacetate. For trichloroacetic acid: $K_a = 0.22$, $K_b = K_w/K_a = 4.6 \times 10^{-14}$. (Concentrations in parentheses were calculated using the exact equation discussed earlier.)

Part A: $C_{HB} = 0.1$ M, $C_B = 0.0$ M

Step 1A: Select the equation to be used.

The solution contains a weak acid alone, meaning that $C_b = 0$; use Equation 2.8a. To confirm this, calculate $\dfrac{Q_0}{K_a} = \dfrac{10^{-7}}{0.22}\dfrac{C_B}{C_{HB}+K_w} = 0 < 1$.

Step 2A: Calculate the H^+ concentration using Equation 2.8a.

$$\left[H^+\right] \cong \frac{1}{2}\left[-(0.22) + \sqrt{(0.22)^2 + 4\left(0.22 \times 0.10 + 1 \times 10^{-14}\right)}\right] \cong 0.0747\ \text{M}\ \left(0.0747\ \text{M}\right)$$

Part B: $C_{HB} = 0.1$ M, $C_B = 0.05$ M

Step 1B: Select the equation to be used.

$\dfrac{Q_0}{K_a} = \dfrac{10^{-7}}{0.22}\dfrac{C_B}{C_{HB}+K_w} = 2.3 \times 10^{-7} < 1$, so you should use Equation 2.8a with $C_{HB} = 0.10$ M and $C_B = 0.050$ M.

Step 2B: Calculate the H^+ concentration using Equation 2.8a.

$$\left[H^+\right] \cong \frac{1}{2}\left[-(0.22 + 0.050) + \sqrt{(0.22 + 0.050)^2 + 4\left(0.22 \times 0.10 + 1 \times 10^{-14}\right)}\right]$$

$$\cong 0.0656\ \text{M}\ \left(0.0656\ \text{M}\right)$$

Part C: $C_{HB} = 0.0$ M, $C_B = 0.10$ M

Step 1C: Select the equation to be used.

$$\frac{Q_0}{K_a} = \frac{10^{-7}}{0.22}\frac{C_B}{C_{HB}+K_w} = 4.5 \times 10^6 > 1, \text{ so you should use Equation 2.8b.}$$

Step 2C: Calculate the H^+ concentration using Equation 2.8b.

$$\left[H^+\right] \cong \frac{2 \times 1.0 \times 10^{-14}\ \text{M}}{\left[-\left(4.6 \times 10^{-14} + 0.0\right) + \sqrt{\left(4.6 \times 10^{-14} + 0\right)^2\ 4\left(4.6 \times 10^{-14} \times 0.1 + 1.0 \times 10^{-14}\right)}\right]}$$

$$\cong 8.29 \times 10^{-8}\ \text{M}\ \left(8.30 \times 10^{-8}\ \text{M}\right)$$

I will trust you to correlate numerical values with quantities in Equations 2.8a and 2.8b and to confirm H^+ concentrations using the monoprotic acids program. As you can see, results calculated

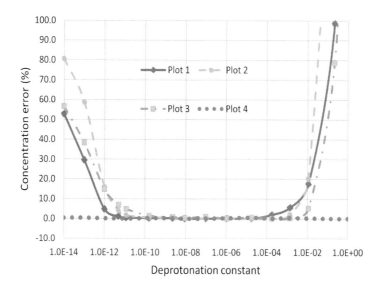

FIGURE 2.1 Percentage concentration errors vs. deprotonation constants for applications of conventional and unified approximations to different combinations of acid-base pairs.

Plots 1–3: Conventional approximations for 0.10 M concentrations of weak acids alone (plot 1), weak acid-base pairs (plot 2), and weak bases alone (plot 3).
Plot 4: Unified approximations applied to each of the three situations.

using unified approximations differ very little from those calculated using the exact equation. I encourage you to use the program described earlier to confirm the concentrations calculated in this example. For any who may want to use a programmable calculator, it is quite easy to program one with iterative capabilities to do calculations using the exact equation and to program one with or without iterative capabilities to do calculations for unified approximations.

GRAPHICAL COMPARISONS OF APPROXIMATION ERRORS

Whereas conventional approximations give very large errors for situations in Example 2.4, unified approximations give hydrogen ion concentrations virtually the same as those calculated using the exact equation. The same applies for many other situations for which conventional approximations give errors much larger than the usual target of 5% or less. For example, I used procedures similar to those described above to calculate approximation errors for deprotonation constants from 1.0×10^{-14} to 1.0 for situations involving weak acids alone, weak bases alone, and mixtures of weak acid-base pairs. Results are plotted in Figure 2.1. Whereas conventional approximations give very large errors for both small and large constants, unified approximations give very small errors for both small and large constants.

I also did calculations to evaluate effects of weak acid concentrations on approximation errors. Results are plotted in Figure 2.2.

On the one hand, unified approximations give consistently smaller errors for wider ranges of conditions than do conventional approximations. On the other hand, given the ease with which the exact equation can be solved using current technologies, I strongly prefer the exact option.

SUMMARY

This chapter begins with a brief discussion of reasons why approximation procedures for hydrogen ion concentrations have been popular in the past. It continues with a suggestion that technologies are now available to solve equations involving hydrogen ion concentrations to the power of 3 to 8—or higher if needed.

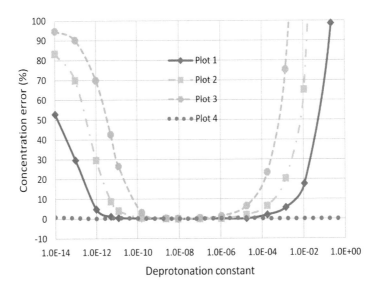

FIGURE 2.2   Percentage concentration errors vs. deprotonation constants for applications of conventional and unified approximations to different weak acid concentrations.

Plots 1–3: Conventional approximations for 0.10 M weak acid (plot 1), 0.010 M weak acid (plot 2), and 0.0010 M weak acid (plot 3).

Plot 4: Unified approximations applied to each of the three situations.

The discussion continues with some background information used to develop an exact equation for the hydrogen ion concentration in solutions containing monoprotic acid-base pairs. Topics discussed include the behaviors of weak acids and bases and the autoprotolysis of water.

Concepts developed in the background section are used to write exact equations for weak acid and conjugate base concentrations. Those equations are used with a rearranged form of the deprotonation constant to develop an exact equation for the hydrogen ion concentration in terms of the deprotonation constant for the weak acid, the autoprotolysis constant for water, and diluted concentrations of the weak acid and weak base.

The equation for the hydrogen ion concentration is used as the basis for an iterative spreadsheet program to calculate hydrogen ion concentrations free of the errors associated with conventional approximation procedures. Two features of the program are that (a) it can start with an initial hydrogen ion concentration that is orders of magnitude less than the equilibrium concentration and (b) it includes a way to control the number of significant figures to which the H^+ concentration is resolved.

It is also shown how hydrogen ion concentrations calculated with the program can be used to calculate fractions and concentrations of acid-base pairs in the protonated and deprotonated forms.

Finally, it is shown how the program can be adapted to do approximate calculations using two options called conventional approximations and unified approximations. It is shown that unified approximations give results with errors of less than 1% for situations for which conventional approximations give errors of 100% or more.

REFERENCES

1. J. E. Ricci, *Hydrogen Ion Concentration, New Concepts in a Systematic Treatment*, Princeton U.P.: Princeton, N.J. 1952.
2. H. L. Pardue, I. N. Odeh and T. M. Tesfai, Unified Approximations: A New Approach for Monoprotic Weak Acid-Base Equilibria, *J. Chem. Ed.*, **2004**, 81, 1367–1375.

3 Concepts and Equations Relevant to a Systematic Approach to Acid-Base Equilibria

Author's note: Concepts and equations in this chapter are critical to discussions of a systematic approach to acid-base equilibria in subsequent chapters.

Calculations involving acid-base equilibria usually focus on calculations of hydrogen ion concentrations and pH. Hydrogen ion concentrations are used to calculate a variety of quantities, including pH, fractions, and concentrations of different forms of weak acids and rate constants for H^+-dependent reactions. pH is used in a variety of ways, including plotting of titration curves.

Calculations of hydrogen ion concentrations and pH in solutions containing different forms of weak acids are complicated by the fact that exact equations involve hydrogen ion concentrations to powers of three and higher. Specifically, for a weak acid with n acidic hydrogens, H_nB, the exact equation for the hydrogen ion concentration involves the hydrogen ion concentration to a power of $n + 2$, i.e., $[H^+]^{n+2}$. As examples, exact equations for hydrogen ion concentrations in solutions containing different forms of monoprotic, HB, or hexaprotic, H_6B. acids involve hydrogen ion concentrations to powers of three and eight, respectively, i.e., $[H^+]^3$ and $[H^+]^8$.

Of the many options for understanding and solving the high-order equations associated with acid-base equilibria, the most promising one in my view and that of others involves an approach called a *systematic approach*. The systematic approach to acid-base equilibria was described initially by Ricci[1] and elaborated upon by several others.[2–4] In the systematic approach, several types of equations, including the autoprotolysis constant for water, mass-balance equations, concentration-fraction equations, and charge-balance or proton-balance equations are used to develop exact equations for the hydrogen ion concentration.

The most common approach to using the systematic approach has been to use it with plots of log concentration vs. pH to resolve the pH and hydrogen ion concentrations.[2,3] On the one hand, these procedures can yield pH's and H^+ concentrations with errors less than those imposed by uncertainties in experimental parameters such as deprotonation constants. On the other hand, it is relatively easy to combine equations based on the systematic approach with currently available spreadsheet programs to calculate hydrogen ion concentrations resolved to three significant figures with calculation times seldom exceeding several seconds. This is the option used in this text.

Briefly, I discovered during the preparation of this text that exact equations for hydrogen ion concentration can be written in forms that are negative for hydrogen ion concentrations less than the equilibrium concentration and positive for hydrogen ion concentrations larger than the equilibrium concentration. If we start with a trial hydrogen ion concentration less than the equilibrium concentration, e.g., 10^{-15} M, and increase it by known amounts, ΔC_H, the trial concentration that causes the sign of the exact equation change from − to + will be the equilibrium H^+ concentration resolved to within the size of the increments, ΔC_H, in the concentration.

Spreadsheet programs described in subsequent chapters permit the user to control the number of significant figures to which hydrogen ion concentrations are resolved by changing a number in one cell of each program. For example, with a significant figures cell set to 3, the programs resolve the hydrogen ion concentration to three significant figures. Calculation times for resolution to three significant figures with and without correction for ionic strength seldom exceed 100 seconds.

This chapter describes background information associated with the systematic approach. Topics discussed include the acid-base properties of water, behaviors of strong acids and bases, equilibration reactions and equilibrium constants for monoprotic, diprotic, and polyprotic weak acids, a systematic set of symbols based on numbers, i, of protons by which different forms of acids differ from the fully protonated acid, cumulative deprotonation constants, mass-balance equations, and concentration fractions and distribution plots for weak acids.

As my friend, Prof. Stanley Deming, used to say, "Listen up." It will be assumed in subsequent chapters that you are familiar with and/or will review concepts and equations discussed in this chapter.

A THEORY OF ACIDS AND BASES AND SOME RELATED TERMINOLOGY AND SYMBOLS

In 1923, two chemists, J. N. Brønsted in Denmark and J. M. Lowry in England, proposed a common theory of acids and bases. According to their theory, an acid is a proton donor and a base is a proton acceptor. When an acid such as acetic acid donates a proton, it is *deprotonated*. When a base such as acetic acid accepts a proton it is *protonated*. As shown below for acetic acid and acetate ion, the loss of a proton by a weak acid is called a *deprotonation reaction* herein and the gain of a proton by a weak base called a *protonation reaction* herein.

$$\overbrace{CH_3COOH \underset{K_b}{\overset{K_a}{\rightleftharpoons}} H^+ + CH_3COO^-}^{\text{Deprotonation, Left to Right}}_{\text{Protonation, Right to Left}}$$

The corresponding constants, K_a and K_b, are called *deprotonation constants* and *protonation constants* herein.

Although described in the context of a monoprotic weak acid here, the same applies for polyprotic acids. For example, the stepwise constants, K_{a1}, K_{a2}, and K_{a3}, for a triprotic weak acid such as phosphoric acid or citric acid are called *stepwise deprotonation constants* herein.

You may be more familiar with the terms, acid dissociation constants and base association constants, for these constants. Feel free to use those terms if it helps your understanding. However, I have chosen the simpler and less ambiguous deprotonation and protonation terminology to reduce the possibility of confusion when we talk in later chapters about mixtures of acids and bases with other types of reactions such as complex-formation reactions that also involve dissociation and association reactions.

ROLES OF HYDROGEN ION CONCENTRATION AND pH

This chapter includes background information for procedures used to calculate hydrogen ion concentrations and pH to be discussed in subsequent chapters. As I'm sure you know, pH is a logarithmic transformation of the hydrogen ion concentration, i.e., $pH = -\log [H^+]$.

Briefly, given that most of the properties of chemical reactions in which we are interested depend more directly on the hydrogen ion concentration than pH, the hydrogen ion concentration is more useful than pH in calculations of these quantities. Given that the pH transformation compacts very wide ranges of hydrogen ion concentrations to smaller ranges, pH is more useful when we want to display or plot properties of chemical reactions over wide ranges of hydrogen ion concentrations.

It will be shown later that fractions of acids in different forms are functions of the hydrogen ion concentration and deprotonation constants. For example, the stepwise deprotonation reactions for an uncharged diprotic acid such as oxalic acid are as follows:

$$H_2B \underset{}{\overset{K_{a1}}{\rightleftharpoons}} H^+ + HB^- \underset{}{\overset{K_{a2}}{\rightleftharpoons}} H^+ + B^{2-}$$

It will be shown later that the fractions of the acid in the three forms are as follows:

$$\alpha_{H_2B} = \frac{\left[H^+\right]^2}{D_2}; \qquad \alpha_{HB^-} = \frac{K_{a1}\left[H^+\right]}{D_2}; \qquad \alpha_{B^{2-}} = \frac{K_{a1}K_{a2}}{D_2}$$

in which $[H^+]$ is the hydrogen ion concentration, K_{a1} and K_{a2} are stepwise deprotonation constants, and D_2 is as follows:

$$D_2 = \left[H^+\right]^2 + K_{a1}\left[H^+\right] + K_{a1}K_{a2}$$

As an example, deprotonation constants for oxalic acid are $K_{a1} = 5.6 \times 10^{-2}$ and $K_{a2} = 5.4 \times 10^{-5}$. For a hydrogen ion concentration of 1.00×10^{-3} M, it is easily shown that the denominator term is 6.00×10^{-5} and that the fractions of oxalic acid in the three forms are $\alpha_{H_2B} = 0.017$, $\alpha_{HB} = 0.93$, and $\alpha_B = 0.050$. These fractions mean that, for $[H^+] = 1.0 \times 10^{-3}$ M, 1.7%, 93%, and 5.0%, respectively, of the oxalic acid are present as oxalic acid, H_2B, bioxalate ion, HB^-, and oxalate ion, B^{2-}. We will talk later more about the basis for and applications of these and other fractions.

For now, suppose that we want to calculate and plot the fractions for hydrogen ion concentrations from 10^{-14} M to 1.0 M. To view the distributions of fractions most clearly, should we plot them vs. the H^+ concentration or pH? A plot of fractions vs. the H^+ concentration would illustrate differences among fractions for hydrogen ion concentrations larger than about 0.010 M clearly with fractions for H^+ concentrations less than about 0.010 M compressed together near the Y axis. In contrast, as shown in Figure 3.1, plots of fractions vs. pH are well separated for pH's from 0 to 14, i.e., $[H^+] = 1.0$ M to 10^{-14} M.

Plots of concentration fractions vs. pH are commonly called *distribution plots*. We will talk in more detail about distribution plots later in this chapter.

In summary, the hydrogen ion concentration is more useful than pH for calculations of concentration fractions and other quantities such as rate constants that depend more directly on the H^+ concentration than the pH. Conversely, pH is more useful than the H^+ concentration for plotting the results. The same conclusions apply for other quantities associated with chemical equilibria and kinetics. Programs described in subsequent chapters are designed to calculate both hydrogen ion concentrations and pH.

Let's talk next about the acid-base behavior of water, strong acids, strong bases, weak acids, and weak bases.

BEHAVIORS OF WATER, STRONG ACIDS, STRONG BASES, WEAK ACIDS, AND WEAK BASES

This section describes an overview of the different types of acids and bases to be discussed in this and subsequent chapters. Given that all situations to be considered involve aqueous solutions, let's start with acid-base properties of water.

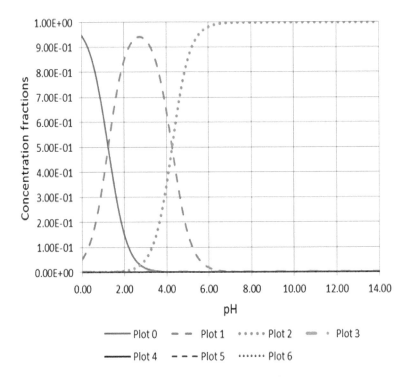

FIGURE 3.1   Fractions of oxalic acid in the forms, H_2B, HB^-, and B^{2-} vs. pH.

Plots 0, 1, and 2: Fractions $\alpha_0 - \alpha_2$, respectively.

ACID-BASE PROPERTIES OF WATER

Presumably you will have seen the reaction and autoprotolysis constant for water, K_w, written in one of the following forms.

$$H_2O \rightleftharpoons H^+ + OH^-; \quad K_w = \left[H^+\right]\left[OH^-\right]; \quad 2H_2O \rightleftharpoons H_3O^+ + OH^-; \quad K_w = \left[H_3O^+\right]\left[OH^-\right]$$

Either way the reaction and autoprotolysis constant are written, the activity-based autoprotolysis constant for water at 25 °C is $K_w = 1.0 \times 10^{-14}$ M^2. The simpler forms of the reaction and equation are used herein.

It follows from the autoprotolysis constant equation that the hydrogen-ion and hydroxide-ion concentrations are related to each other as shown in Equation 3.1a:

$$\left[H^+\right] = \frac{K_w}{\left[OH^-\right]} \qquad (3.1a) \qquad\qquad \left[OH^-\right] = \frac{K_w}{\left[H^+\right]} \qquad (3.1b)$$

Equation 3.1b is used in later chapters to express the hydroxide ion concentration in terms of the hydrogen ion concentration.

It is assumed that all salts, strong acids, and strong bases are strong electrolytes meaning that they dissociate completely in water. Examples of each are discussed below.

STRONG ACIDS AND STRONG BASES

Examples of strong acids are hydrochloric acid, HCl, and perchloric acid, $HClO_4$. Examples of strong bases are sodium hydroxide, NaOH, and potassium hydroxide, KOH. As the names imply, strong acids and bases tend to dissociate completely in water. As examples, dissociation reactions for HCl and NaOH are as follows:

$$HCl \rightarrow H^+ + Cl^- \qquad NaOH \rightarrow Na^+ + OH^-$$

The single left to right arrow is used to represent the fact that the salts dissociate completely.

Unless stated otherwise, hydrochloric acid and sodium hydroxide are used as examples of monoprotic strong acids and strong bases throughout this text. All that we say about HCl and NaOH applies to other monoprotic strong acids and strong bases.

Given that strong acids and strong bases dissociate completely, amounts of the ions produced in solution are the same as amounts of the acid or base added to the solutions. Therefore, equilibrium concentrations of ions such as Na^+ and Cl^- that are not involved in subsequent reactions are equal to the diluted concentrations of the acid and base, respectively. Given a solution prepared to contain concentrations, C_{HCl} and C_{NaOH}, of HCl and NaOH, chloride ion and sodium ion concentrations are expressed in Equations 3.2a and 3.2b:

$$\left[Cl^- \right] = C_{HCl} \qquad (3.2a) \qquad\qquad\qquad \left[Na^+ \right] = C_{NaOH} \qquad (3.2b)$$

For example, given a solution containing 0.050 M HCl and 0.060 M sodium hydroxide and no other reactants, chloride ion and sodium ion concentrations will be

$$\left[Cl^- \right] = C_{HCl} = 0.050 \text{ M} \quad \text{and} \quad \left[Na^+ \right] = C_{NaOH} = 0.060 \text{ M}$$

Relationships such as these are used to correlate ionic concentrations with prepared concentrations.

REACTIONS FOR WEAK ACIDS AND WEAK BASES

We will talk in this and later chapters about situations ranging from as simple as monoprotic weak acids such as formic acid to as complex as hexaprotic acids such as protonated ethylenediaminetetraacetic acid, EDTA for short. To simplify our lives and to generalize the concepts we discuss, most weak acids and weak bases are represented by symbols, H_nB^z, in which n is the number of acidic hydrogens, H^+ is the acidic part of the acid, B is the basic part of the acid and z is the charge on the fully protonated acid. EDTA is a special case. It is common practice to represent the basic part of EDTA by Y^{4-} and the fully protonated form as H_6Y^{2+}. We will conform to that practice herein.

Acetic acid, an uncharged monoprotic acid, protonated ethylenediamine, a doubly charged diprotic acid, phosphoric acid, an uncharged triprotic acid, and protonated ethylenediaminetetraacetic acid, EDTA, a doubly charged hexaprotic acid are represented herein as HB, H_2B^{2+}, H_3B and H_6Y^{2+}, respectively. Stepwise deprotonation reactions are written as follows:

$$HB \underset{\phantom{K_{a1}}}{\overset{K_{a1}}{\rightleftharpoons}} H^+ + B^-$$

$$H_2B^{2+} \overset{K_{a1}}{\rightleftharpoons} H^+ + HB^+ \overset{K_{a2}}{\rightleftharpoons} H^+ + B$$

$$H_3B \overset{K_{a1}}{\rightleftharpoons} H^+ + H_2B^- \overset{K_{a2}}{\rightleftharpoons} H^+ + HB^{2-} \overset{K_{a3}}{\rightleftharpoons} H^+ + B^{3-}$$

$$H_6Y^{2+} \overset{K_{a1}}{\rightleftharpoons} H^+ + H_5Y^+ \overset{K_{a2}}{\rightleftharpoons} H^+ + H_4Y \overset{K_{a3}}{\rightleftharpoons} H^+ + H_3Y^- \overset{K_{a4}}{\rightleftharpoons} H_2Y^{2-} \overset{K_{a5}}{\rightleftharpoons} H^+ + HY^{3-} \overset{K_{a6}}{\rightleftharpoons} H^+ + Y^{4-}$$

You may find it informative to do Exercise 3.1.

EXERCISE 3.1
Complete the reaction sequence below for the stepwise deprotonation of a singly charged tetraprotic acid such as nitrilotriacetic acid, H_4B^+.

$$H_4B^+ \overset{K_{a1}}{\rightleftharpoons} H^+ + \cdots \overset{K_{a2}}{\rightleftharpoons} H^+ + \cdots \overset{K_{a3}}{\rightleftharpoons} H^+ + \cdots \overset{K_{a3}}{\rightleftharpoons} H^+ + \cdots$$

Hint: One of the products is H_2B^-.

As done by Ricci,[1] numbers of hydrogen ions lost in successive steps of deprotonation reactions are represented herein by the symbol, i. For example, the stepwise deprotonation reactions for a triprotic weak acid such as phosphoric acid correspond to losses of $i = 0-3$ protons. Similarly, the stepwise deprotonation reactions for a hexaprotic acid such as protonated EDTA correspond to losses of $i = 0-6$ protons.

The charge, z_i, on each form of an acid is the charge, z, on the parent acid minus the number, i, of protons the acid has lost to form the base, i.e., $z_i = z - i$. For example, for the zeroth, second, and sixth steps in the stepwise deprotonation of protonated EDTA, H_6Y^{2+}, charges on the forms are $z_0 = z - i = 2 - 0 = 2$, i.e., H_6Y^{2+}, $z_2 = z_0 - i = 2 - 2 = 0$, i.e., H_4B, and $z_6 = z - i = 2 - 6 = -4$, i.e., Y^{4-}. These relationships will be important when we talk about charge-balance equations and effects of ionic strength in later chapters.

Each pair of acids and bases differing by one proton is called a conjugate acid-base pair or more simply herein, an acid-base pair.

STEPWISE DEPROTONATION CONSTANTS

Quantities, $K_{a1}-K_{a6}$, are called stepwise deprotonation constants. Equations for stepwise deprotonation constants for charged and uncharged monoprotic, diprotic, and triprotic acids are summarized in Table 3.1.

TABLE 3.1
Forms of Deprotonation Constant Equations for Monoprotic, Diprotic, and Triprotic Acids

Monoprotic	Diprotic		Triprotic		
Uncharged Acids					
$K_a = \dfrac{[H^+][B^-]}{[HB]}$	$K_{a1} = \dfrac{[H^+][HB^-]}{[H_2B]}$	$K_{a2} = \dfrac{[H^+][B^{2-}]}{[HB^-]}$	$K_{a1} = \dfrac{[H^+][H_2B^-]}{[H_3B]}$	$K_{a2} = \dfrac{[H^+][HB^{2-}]}{[H_2B^-]}$	$K_{a3} = \dfrac{[H^+][B^{3-}]}{[HB^{2-}]}$
Singly Charged Acids					
$K_a = \dfrac{[H^+][B]}{[HB^+]}$	$K_{a1} = \dfrac{[H^+][HB]}{[H_2B^+]}$	$K_{a2} = \dfrac{[H^+][B^-]}{[HB]}$	$K_{a1} = \dfrac{[H^+][H_2B^+]}{[H_3B^{2+}]}$	$K_{a2} = \dfrac{[H^+][HB]}{[H_2B^+]}$	$K_{a3} = \dfrac{[H^+][B^-]}{[HB]}$
Doubly Charged Hexaprotic Acid, e.g., Protonated EDTA, H_6Y^{2+}					
$K_{a1} = \dfrac{[H^+][H_5Y^+]}{[H_6Y^{2+}]}$	$K_{a2} = \dfrac{[H^+][H_4Y]}{[H_5Y^+]}$	$K_{a3} = \dfrac{[H^+][H_3Y^-]}{[H_4Y]}$	$K_{a4} = \dfrac{[H^+][H_2Y^{2-}]}{[H_3Y^-]}$	$K_{a5} = \dfrac{[H^+][HY^{3-}]}{[H_2Y^{2-}]}$	$K_{a6} = \dfrac{[H^+][Y^{4-}]}{[HY^{3-}]}$

Reactions and equations for stepwise deprotonation constants for acids with other charges and other numbers, n, of acidic hydrogens are written by analogy.

pH's AND pK$_a$'s

pH's and pK$_a$'s are logarithmic functions of hydrogen ion concentrations and deprotonation constants, or more specifically, negative logarithms of hydrogen ion concentrations and deprotonation constants.

Relationships between the pH and hydrogen ion concentration are as shown in Equations 3.3a and 3.3b:

$$pH = -\log\left[H^+\right] \qquad (3.3a) \qquad\qquad \left[H^+\right] = 10^{-pH} \qquad (3.3b)$$

Similarly, relationships between pK$_a$'s and K$_a$'s are shown in Equations 3.3c and 3.3d:

$$pK_a = -\log K_a \qquad (3.3c) \qquad\qquad K_a = 10^{-pK_a} \qquad (3.3d)$$

Given H^+ or pH or K_a or pK$_a$, you can always calculate the other.

CUMULATIVE DEPROTONATION CONSTANTS

Some equations we will discuss later in this and subsequent chapters involve products of stepwise deprotonation constants. For example, equations for fractions of protonated EDTA in different forms involve several products of deprotonation constants, e.g., K_{a1}, $K_{a1}K_{a2}$, $K_{a1}K_{a2}K_{a3}$, ..., $K_{a1}K_{a2}K_{a2}K_{a4}K_{a5}K_{a6}$. As you can imagine, the resulting equations can be rather messy. An easy way to simplify the equations is to replace products of stepwise constants with cumulative constants.

Symbols, κ_i, are used for cumulative constants herein. The cumulative deprotonation constant, κ_i, for any step, i, in a set of deprotonation reactions is the product of all constants up to and including that step. As examples, cumulative deprotonation constants for the three steps in the deprotonation of any triprotic acid such as phosphoric acid are

$$\kappa_1 = K_{a1} \qquad \kappa_2 = K_{a1}K_{a2} \qquad \kappa_3 = K_{a1}K_{a2}K_{a3}$$

Analogous equations apply for acids with fewer or more acidic hydrogens.

EXERCISE 3.2

Given stepwise deprotonation constants for phosphoric acid, $K_{a1} = 7.1 \times 10^{-3}$, $K_{a2} = 6.3 \times 10^{-8}$, and $K_{a3} = 7.1 \times 10^{-13}$, show that cumulative constants are $\kappa_1 = 7.1 \times 10^{-3}$, $\kappa_2 = 4.4_7 \times 10^{-10}$, and $\kappa_3 = 3.1_8 \times 10^{-22}$.

For any acid with n acidic hydrogens, it is relatively easy to show that all cumulative constants beyond κ_n are zero. As examples, κ_2–$\kappa_6 = 0$ for a monoprotic acid, κ_3–$\kappa_6 = 0$ for a diprotic acid, κ_4–$\kappa_6 = 0$ for a triprotic acid, etc.

As mentioned earlier, cumulative deprotonation constants are used herein to simplify a variety of otherwise unwieldy equations.

For the record, pK$_a$'s are additive meaning that pK$_a$'s are sums of pK$_a$'s. For example, given pK$_{a1} = 2.15$, pK$_{a2} = 7.20$, and pK$_{a3} = 12.15$ for phosphoric acid, pK$_a$'s are pK$_{a1} = 2.15$, pK$_{a2} = 2.15 + 7.20 = 9.35$, and pK$_{a3} = $ pK$_{a1} + $ pK$_{a2} + $ pK$_3 = 2.15 + 7.20 + 12.15 = 21.50$.

A Systematic Set of Symbols Based on Numbers of Protons Lost

We will talk in this and subsequent chapters about a variety of weak acids ranging from uncharged monoprotic acids to doubly charged hexaprotic acids. As examples, acetic acid is an uncharged monoprotic acid, HB, protonated histidine is a singly charged triprotic acid, H_3B^+ and fully protonated EDTA is a doubly charged, hexaprotic acid, H_6Y^{2+}. It will be convenient to have a systematic set of symbols to represent quantities associated with different forms of this wide range of acids.

Numbers of protons lost during the stepwise deprotonation of any acid have been used[1-4] and are used herein to associate different quantities such as diluted concentrations and concentration fractions with the forms of acids they represent.

Numbers of protons lost during the stepwise deprotonation of weak acids are represented herein by the letter, i. For example, the stepwise deprotonation reaction below for phosphoric acid is annotated to identify numbers, i, of hydrogen ions lost by fully protonated phosphoric acid for successive steps in the deprotonation reactions.

$$\overbrace{H_3PO_4}^{i=0} + \overset{-H^+}{\underset{}{\rightleftharpoons}} + \overbrace{H_2PO_4^-}^{i=1} \overset{-H^+}{\underset{}{\rightleftharpoons}} + \overbrace{HPO_4^{2-}}^{i=2} \overset{-H^+}{\underset{}{\rightleftharpoons}} + \overbrace{PO_4^{3-}}^{i=3}$$

Numbers of hydrogen ions lost vary from $i = 0$ for fully protonated phosphoric acid to $i = 3$ for phosphate ion. For any acid of the form, H_nB, numbers of protons lost will vary from $i = 0$ for the fully protonated acid to $i = n$ for the fully deprotonated basic form of the acid.

As shown in Table 3.2, this i notation is used herein to identify a wide range of quantities associated with acid-base equilibria.

For any acid, H_nB, there are $n + 1$ of each quantity. Regarding deprotonation constants and cumulative deprotonation constants, $K_{a0} = 1$ and $\kappa_0 = 1$.

As an example of the systematic symbols, consider a solution containing 0.10 M phosphoric acid, H_3PO_4, 0.0750 M dihydrogen phosphate, $H_2PO_4^{2-}$, 0.050 M monohydrogen phosphate, HPO_4^{2-}, and 0.0250 M phosphate ion, PO_4^{3-}. These concentrations are represented as shown below using both conventional and the systematic symbols for comparison purposes.

$$C_{H_3PO_4} \equiv C_0 = 0.10 \text{ M}, \quad C_{H_2PO_4^-} \equiv C_1 = 0.075 \text{ M},$$

$$C_{HPO_4^{2-}} \equiv C_2 = 0.050 \text{ M} \quad \text{and} \quad C_{PO_4^{3-}} \equiv C_3 = 0.0250 \text{ M}$$

Here and throughout, the symbol, $\equiv$, means "defined as." In other words, $C_{H_3PO_4}$ is defined as C_0, $C_{H_2PO_4^-}$ is defined as C_1, etc.

TABLE 3.2
A Systematic Set of Symbols for Weak Acids Based on Numbers, i, of Protons Lost during Stepwise Deprotonation

Number, i, of Protons Lost	i	0	1	2	3	4	5	6
Undiluted concentrations	C_i^0	C_0^0	C_1^0	C_2^0	C_3^0	C_4^0	C_5^0	C_6^0
Diluted concentrations	C_i	C_0	C_1	C_2	C_3	C_4	C_5	C_6
Volumes	V_i	V_0	V_1	V_2	V_3	V_4	V_5	V_6
Charges	z_i	z_0	z_1	z_2	z_3	z_4	z_5	z_6
Concentration fractions	α_i	α_0	α_1	α_2	α_3	α_4	α_5	α_6
Activity coefficients	f_i	f_0	f_1	f_2	f_3	f_4	f_5	f_6
Deprotonation constants	K_{ai}	K_{a0}	K_{a1}	K_{a2}	K_{a3}	K_{a4}	K_{a5}	K_{a6}
Cumulative constants	κ_i	κ_0	κ_1	κ_2	κ_3	κ_4	κ_5	κ_6

On the one hand, conventional symbols identify the form of phosphoric acid each concentration represents. On the other hand, the systematic symbols are much simpler and more general in the sense that the same symbols apply for all acids with equal numbers, n, of acidic hydrogens. This symbolism will be particularly helpful when we talk later about some very complex equations associated with mixtures containing any or all forms of up to three hexaprotic acids, H_6B.

Concentration symbols are discussed in more detail here; other symbols are discussed in more detail at appropriate points in subsequent parts of this and other chapters.

SYMBOLS FOR UNDILUTED, DILUTED, AND EQUILIBRIUM CONCENTRATIONS

When solutions are mixed as is the case in acid-base titrations, it is necessary to account for dilution in the calculation process. To do this, it is important to distinguish among concentrations before mixing, i.e., *undiluted concentrations*, concentrations after mixing, i.e., diluted concentrations, and concentrations after equilibration reactions are complete.

Volumes of different forms of weak acids used to prepare solutions are represented by symbols, V_i, in which the subscript, i, identifies the form of the acid. Undiluted concentrations are represented by symbols, C_i^0, in which the subscripted i represents the form of the acid and the superscripted 0 identifies this as an initial or undiluted concentration. Diluted concentrations, i.e., concentrations after dilution but before equilibration reactions begin are represented by symbols, C_i, in which the subscripted i identifies the form of the acid of interest. Equilibrium concentrations are represented using conventional square-bracketed symbols, e.g., $[H_{n-i}B^z]$ in which n − i represents the number of hydrogens remaining and z is the charge.

Example 3.1 illustrates an example of the symbolism.

Example 3.1 Calculate the diluted dihydrogenphosphate ion, $H_2PO_4^-$, concentration before any reaction occurs in a solution after adding V_{NaOH} = 10.0 mL of 0.10 M sodium hydroxide to V_1 = 25.0 mL of a solution containing C_1^0 = 0.075 M dihydrogenphosphate.

$$C_1 = \frac{V_1 C_1^0}{V_1 + V_{NaOH}} = \frac{(25.0)(0.0750)}{25.0 + 10.0} = 0.0536 \text{ M}$$

As illustrated by this example, the diluted concentration of a reactant represents the concentration after effects of dilution are taken into account but before any reaction occurs.

As noted in Table 3.2, equilibrium concentrations are represented by square-bracketed symbols. A program to be discussed in a later chapter was used to show that equilibrium concentrations of the hydrogen ion and the four forms of phosphate at equilibrium for the situation in Example 3.1 are $[H^+] = 5.5 \times 10^{-8}$ M, $[H_3PO_4] = 1.9 \times 10^{-7}$ M, $[H_2B^-] = 0.0250$ M, $[HB^{2-}] = 0.0286$ M, and $[B^{3-}] = 3.7 \times 10^{-7}$ M in which B^{3-} represents PO_4^{3-}.

SALTS OF WEAK ACIDS

It is assumed throughout this discussion that solutions are prepared using sodium salts of negatively charged bases such as the phosphate ion and chloride salts of charged acids such as NH_4^+. As examples, the sodium salts of the three forms of a triply charged base such as phosphate ion, PO_4^{3-}, are represented as follows:

$$NaH_2B, \quad Na_2HB, \quad \text{and} \quad Na_3B$$

in which B^{3-} is the phosphate ion.

The chloride salts of the different forms of a doubly charged acid such as protonated ethylenediamine are represented as follows:

$$H_2BCl_2 \quad \text{and} \quad HBCl^-$$

in which B is ethylenediamine, i.e., $H_2NCH_2CH_2NH_2 \equiv B$.

It is also assumed that all salts dissociate completely.

Sodium Salts of an Uncharged Acid

Let's consider the sodium salts of the different ionic forms of phosphoric acid, i.e., NaH_2B, Na_2HB and Na_3B. Reactions of the salts are as follows:

$$NaH_2B \rightarrow Na^+ + H_2B^-, \quad Na_2HB \rightarrow 2Na^+ + HB^{2-} \quad \text{and} \quad Na_3B \rightarrow 3Na^+ + B^{3-}$$

in which B^{3-} is PO_4^{3-} and the unidirectional left-to-right arrows imply that the salts dissociate completely.

Assuming that the salts dissociate completely, the solution behaves as if the H_2B^-, HB^{2-}, and B^{3-} ions were dissolved in the solution. Therefore, concentrations of the sodium salts can be represented using Ricci's i notation discussed earlier, i.e.,

$$C_{NaH_2B} \equiv C_1; \quad C_{Na_2HB} \equiv C_2; \quad C_{Na_3B} \equiv C_3$$

Given that sodium ions are not involved in subsequent reactions, sodium ion concentrations from the individual reactions are equal to the concentrations of the salts multiplied by numbers of sodium ions per molecule, i.e.,

$$\left[Na^+\right]_{NaH_2B} = C_1; \quad \left[Na^+\right]_{Na_2HB} = 2C_2; \quad \left[Na^+\right]_{Na_3B} = 3C_3$$

> **Example 3.2 Calculate the sodium ion concentration in a solution prepared to contain 0.10 M monosodium phosphate, 0.050 M disodium phosphate, and 0.00 M sodium phosphate.**
>
> $$\left[Na^+\right] = C_1 + 2C_2 + 3C_3 = 0.10 + 2(0.050) + 3(0.00) = 0.20 \text{ M}$$

You may wonder why I included the concentration of sodium phosphate in Example 3.2 even though it is zero. The reason is that, by including a term for the concentration in the mass-balance equation even though it is zero, we obtain an equation that is applicable to other situations for which the concentration may not be zero.

Chloride Salts of a Charged Acid

Doubly and singly protonated forms of ethylenediamine, $HH_2NCH_2CH_2NH_2H^{2+}$ and $H_2NCH_2CH_2NH_2H^+$ are examples of doubly and singly charged acids. Generic formulae for the chloride salts of the acids are H_2BCl_2 and $HBCl$. Dissociation reactions of the salts are

$$H_2BCl_2 \rightarrow H_2B^{2+} + 2Cl^-; \quad HBCl \rightarrow HB^+ + Cl^-$$

Assuming that the salts dissociate completely, the solution behaves as if the H_2B^{2+} and HB^+ ions were dissolved in the solution. Therefore, concentrations of the chloride salts can be represented using Ricci's i notation discussed in connection with Table 3.2, i.e.,

$$C_{H_2BCl_2} \equiv C_0; \quad C_{HBCl} \equiv C_1$$

These relationships are used with mass-balance equations to correlate equilibrium concentrations of ions that don't undergo additional reactions with concentrations of reactants from which they were produced.

MASS-BALANCE EQUATIONS

Mass-balance equations are based on Lavoisier's law of the conservation of mass, meaning that mass is neither created nor destroyed. Assuming that there are no side reactions such as precipitation or vaporization of reactants or products, it follows that the sum of initial concentrations of reactants must be equal to the sum of equilibrium concentrations of the final forms of those reactants. There are usually two or more mass-balance equations for each situation.

SITUATIONS INVOLVING AN UNCHARGED ACID

Example 3.3 is used to illustrate the basis for mass-balance equations for a moderately complex situation involving phosphoric acid, a triprotic uncharged acid, H_3B.

> **Example 3.3 Given a solution containing concentrations equal to or larger than zero of HCl, NaOH, phosphoric acid and the sodium salts of the charged forms of phosphoric acid, write mass-balance equations for Cl⁻, Na⁺, and phosphate.**

Let's start with the chloride ion because it is the simplest.

Chloride Ion Concentration, [Cl⁻]

There is only one source of chloride ion, namely dissociation of hydrochloric acid, i.e.,

$$HCl \rightarrow H^+ + Cl^-$$

Assuming that HCl dissociates completely and that there are no side reactions involving the chloride ion, the chloride ion concentration will be equal to the diluted HCl concentration. This is expressed in Equation 3.4a:

$$\left[Cl^-\right] = C_{HCl} \tag{3.4a}$$

Sodium Ion Concentration, [Na⁺]

There are four sources of sodium ion, namely sodium hydroxide and the sodium salts of the three charged forms of phosphate, NaH_2B, Na_2HB, and Na_3B. Assuming that NaOH and the three sodium salts dissociate completely, reactions are as follows:

$$NaOH \rightarrow Na^+ + OH^-$$

$$NaH_2B \rightarrow Na^+ + H_2B^-; \quad Na_2HB \rightarrow 2Na^+ + HB^{2-}; \quad Na_3B \rightarrow 3Na^+ + B^{3-}$$

Assuming that there are no side reactions involving the sodium ion, the sodium ion concentration is the weighted sum of the diluted concentrations. This is expressed in Equation 3.4b:

$$\left[Na^+\right] = C_{NaOH} + C_1 + 2C_2 + 3C_3 \tag{3.4b}$$

in which the coefficients, 1, 2, and 3, associated with C_1, C_2, and C_3 account for the number of sodium ions per molecule of each salt.

Phosphate

Four forms of phosphoric acid, H_3B, NaH_2B, Na_2HB, and Na_3B, are included in the solution in Example 3.3. Representing the phosphate ion by B^{3-}, the stepwise equilibration reactions for the four forms are as follows:

$$H_3B + \underset{\rightleftharpoons}{\overset{-H^+}{\rightleftharpoons}} + H_2B^- \underset{\rightleftharpoons}{\overset{-H^+}{\rightleftharpoons}} + HB^{2-} \underset{\rightleftharpoons}{\overset{-H^+}{\rightleftharpoons}} + B^{3-}$$

Assuming no reactions other than the equilibration reactions, total diluted and equilibrium concentrations of all forms of the acid must be equal to the sum of the concentrations of the four forms included in the solution (Equation 3.4c):

$$C_T = C_0 + C_1 + C_2 + C_3 = [H_3B] + [H_2B^-] + [HB^{2-}] + [B^{3-}] \tag{3.4c}$$

in which C_T is the total concentration of all forms of phosphate and other symbols are as defined earlier.

Given that the different forms of phosphoric acid are involved in equilibration reactions, all that we can say about the equilibrium concentrations without doing equilibrium calculations is that the sum of the concentrations at equilibrium is the same as the total of all forms of phosphate added to the solution. Procedures used to do the equilibrium calculations are discussed in a later chapter.

EXERCISE 3.3

Given a solution containing 0.10 M concentrations each of HCl and NaOH and 0.025 M each of phosphoric acid and the three sodium salts of the charged forms of phosphoric acid, show that equilibrium concentrations of the chloride and sodium ions and the total phosphate concentration are as given below.

$$[Cl^-] = 0.10 \text{ M}, \quad [Na^+] = 0.25 \text{ M}, \quad \text{and} \quad C_T = 0.10 \text{ M}.$$

On the one hand, the use of the i notation and generic symbol, B^{3-}, for phosphate may mask some of the chemistry involved. On the other hand, from a computational point of view, equations written using the i notation and generic symbol for phosphate ion are the same for any uncharged triptotic acid. Moreover, they are easily modified to represent any uncharged or charged acid with fewer or more acidic hydrogens.

SITUATIONS INVOLVING A CHARGED ACID

Example 3.4 extends the discussion to a charged acid, protonated lysine, a doubly charged triprotic acid.

Example 3.4 Given a solution containing concentrations equal to or larger than zero of HCl, NaOH, the chloride salts of the positive forms of protonated lysine, and the sodium salt of the negatively charged form of lysine, write mass-balance equations for Cl^-, Na^+, and lysine. *Note*: Fully protonated lysine is a doubly charged triprotic acid, H_3B^{2+}.

Given that you likely are less familiar with charged acids than uncharged acids, it may help your understanding of subsequent discussion if we start with the stepwise deprotonation reactions of protonated glycine, a doubly charged triprotic acid.

$$\overbrace{H_3B^{2+}}^{i=0} + \underset{\rightleftharpoons}{-H^+} + \overbrace{H_2B^+}^{i=1} \underset{\rightleftharpoons}{-H^+} + \overbrace{HB}^{i=2} \underset{\rightleftharpoons}{-H^+} + \overbrace{B^-}^{i=3}$$

The i values are used to associate diluted concentrations with the appropriate forms of the acid.

Let's start with the chloride ion.

Chloride Ion Concentration, [Cl⁻]

Reactions involving the chloride ion include dissociations of HCl and the chloride salts of lysine, i.e.,

$$HCl \rightarrow H^+ + Cl^-, \quad H_3BCl_2 \rightarrow H_3B^{2+} + 2Cl^-, \quad H_2BCl \rightarrow H_2B^+ + Cl^-$$

The mass-balance equation is expressed as Equation 3.4d:

$$\left[Cl^-\right] = C_{HCl} + 2C_0 + C_1 \tag{3.4d}$$

Sodium Ion Concentration, [Na⁺]

Reactions involving the sodium ion include dissociation of sodium hydroxide and the sodium salt of the deprotonated lysine, i.e.,

$$NaOH \rightarrow Na^+ + OH^-, \quad NaB \rightarrow Na^+ + B^-$$

The sodium ion concentration is the sum of the two diluted concentrations, as expressed in Equation 3.4e:

$$\left[Na^+\right] = C_{NaOH} + C_3 \tag{3.4e}$$

Forms of Lysine

Stepwise deprotonation reactions for lysine are repeated here for your convenience.

$$\overbrace{H_3B^{2+}}^{i=0} + \underset{\rightleftharpoons}{-H^+} + \overbrace{H_2B^+}^{i=1} \underset{\rightleftharpoons}{-H^+} + \overbrace{HB}^{i=2} \underset{\rightleftharpoons}{-H^+} + \overbrace{B^-}^{i=3}$$

Assuming that there are no reactions other than the equilibration reactions, total diluted and equilibrium concentrations of all forms of the acid must be equal to the sum of the concentrations of the four forms included in the solution. This is expressed in Equation 3.4f:

$$C_T = C_0 + C_1 + C_2 + C_3 = \left[H_3B^{2+}\right] + \left[H_2B^+\right] + [HB] + \left[B^-\right] \tag{3.4f}$$

in which C_T is the total concentration of all forms of lysine, B^- is fully deprotonated lysine and other symbols are as defined earlier. Except for charges on the different forms of lysine, the mass-balance equations for lysine are the same as for phosphoric acid.

Exercises 3.4A and 3.4B extend the foregoing concepts to a different type of situation.

EXERCISE 3.4A

Given a solution containing HCl, NaOH, and the sodium salts, NaX and NaY, of two weak acids, HX and HY, write dissociation reactions and use them to explain the rationale for the following equations for the chloride ion and sodium ion concentrations, i.e., $[Cl^-] = C_{HCl}$ and $[Na^+] = C_{NaOH} + C_{1X} + C_{1Y}$.

Note: Subscripted X and Y represent the acids, HX and HY.

EXERCISE 3.4B

Show that the chloride ion and sodium ion concentrations in a solution containing 0.023 M HCl, 0.012 M NaOH, 0.014 M sodium acetate, and 0.041 M sodium formate are $[Cl^-] = 0.023$ M and $[Na^+] = 0.067$ M.

Similar relationships are used in the next chapter to adapt Ricci's systematic approach[1] to solutions containing two or more monoprotic weak acid-base pairs.

CONCENTRATION FRACTIONS, EQUILIBRIUM CONCENTRATIONS, AND DISTRIBUTION PLOTS

In his 1952 text, J. E. Ricci[1] showed, and we will show later that, whatever the starting concentrations of different forms of an acid, fractions of the total concentration of an acid in different forms at equilibrium depend on the hydrogen ion concentration and deprotonation constant(s). He also showed that the equilibrium concentration of each form of an acid is the fraction in that form times the total concentration of all forms of the acid.

Ricci used symbols of the form, α_i, to represent fractions of weak acids in forms that differ by $i = 0 - n$ protons from the parent acid. As examples, fractions of a triprotic weak acid in forms that have lost $i = 0, 1, 2$, and 3 protons, respectively, are represented by symbols, $\alpha_0, \alpha_1, \alpha_2$, and α_3.

Let's talk first about a typical application of fractions for a triprotic acid after which we will talk about the basis for and some special features and applications of the fractions.

AN APPLICATION OF CONCENTRATION FRACTIONS

One application of concentration fractions for weak acids is to calculate equilibrium concentrations of different forms of an acid. For example, equilibrium concentrations of forms of a triprotic acid, H_3B, in forms that have lost 0–3 protons are expressed as products of the fractions, α_i, of the acid in different forms times the total concentration, C_T, of all forms of the acid. For example, equations for phosphoric acid concentrations in different forms are as follows:

$$[H_3PO_4] = \alpha_0 C_T \quad [H_2PO_4^-] = \alpha_1 C_T \quad [HPO_4^{2-}] = \alpha_2 C_T \quad [PO_4^{3-}] = \alpha_3 C_T$$

in which $\alpha_0 - \alpha_3$ are fractions in forms of the acid that have lost 0–3 protons and C_T is the total concentration of all forms of phosphoric acid in the solution.

It will be shown later that fractions of a triprotic acid in forms that have lost $i = 0 - 3$ protons depend on the hydrogen ion concentration and cumulative deprotonation constants as follows (Equations 3.5a–3.5d):

$$\alpha_0 = \frac{[H^+]^3}{D_3} \quad (3.5a) \qquad\qquad \alpha_1 = \frac{\kappa_1 [H^+]^2}{D_3} \quad (3.5b)$$

$$\alpha_2 = \frac{\kappa_2 [H^+]}{D_3} \quad (3.5c) \qquad\qquad \alpha_3 = \frac{\kappa_3}{D_3} \quad (3.5d)$$

in which $[H^+]$ is the hydrogen ion concentration κ_1–κ_3 are cumulative deprotonation constants and D_3 is

$$D_3 = \left[H^+\right]^3 + \kappa_1\left[H^+\right]^2 + \kappa_2\left[H^+\right] + \kappa_3 \qquad (3.5e)$$

Bases for these equations are discussed shortly. For now, let's solve an example to illustrate a typical application. I will guide you through the procedure in Example 3.5 and trust you to do the substitutions and calculations.

Example 3.5 A spreadsheet program to be described in a later chapter was used to show that the hydrogen ion concentration in a solution containing $C_0 = 0.10$ M phosphoric acid, H_3PO_4, and $C_2 = 0.050$ M disodium phosphate, Na_2HPO_4, is $[H^+] = 3.22 \times 10^{-3}$ M. Calculate fractions and concentrations of the four forms of phosphoric acid in the solution at equilibrium. For phosphoric acid: $\kappa_1 = 7.1 \times 10^{-3}$, $\kappa_2 = 4.4_7 \times 10^{-10}$, and $\kappa_3 = 3.1_8 \times 10^{-22}$.

Step 1: Calculate the total concentration of all forms of phosphate.

$$C_T = C_0 + C_1 + C_2 + C_3 = 0.10 + 0.00 + 0.050 + 0.00 = 0.15\,M$$

Step 2: Calculate the denominator term, D_3, for the triprotic acid.

$$D_3 = [H^+]^3 + \kappa_1[H^+]^2 + \kappa_2[H^+] + \kappa_3 = 1.07 \times 10^{-7}$$

Step 3: Calculate the concentration fractions.

$$\alpha_0 = \frac{\left[H^+\right]^3}{D_3} = 0.312,\ \alpha_1 = \frac{\kappa_1\left[H^+\right]^2}{D_3} = 0.688,\ \alpha_2 = \frac{\kappa_2\left[H^+\right]}{D_3} = 1.4 \times 10^{-5},\ \alpha_3 = \frac{\kappa_3}{D_3} = 3.0 \times 10^{-15}$$

Step 4: Calculate equilibrium concentrations of the different forms of phosphoric acid.

$$[H_3PO_4] = \alpha_0 C_T = 0.0468\,M, \quad [H_2PO_4^-] = \alpha_1 C_T = 0.103\,M,$$

$$[HPO_4^{2-}] = \alpha_2 C_T = 2.0 \times 10^{-6}\,M, \quad [PO_4^{3-}] = \alpha_3 C_T = 4.5 \times 10^{-16}\,M$$

For those more comfortable with percentages than fractions, the fractions, $\alpha_0 = 0.312$, $\alpha_1 = 0.688$ mean that 31.2% of the total phosphate is present as phosphoric acid and 68.8% is present as dihydrogen phosphate. Fractions $\alpha_2 = 1.4 \times 10^{-5}$ and $\alpha_3 = 3.0 \times 10^{-15}$ mean that very small percentages of the total concentration are present as monohydrogen phosphate and phosphate ion. In other words, virtually all of the monohydrogen phosphate from disodium phosphate has been converted to dihydrogen phosphate by reaction with phosphoric acid. The phosphoric acid concentration has decreased from 0.10 M to 0.047 M and the dihydrogen phosphate has increased from zero to 0.103 M.

Now that we have discussed one way that fractions can be used, let's talk about the basis for and some special features of the fractions.

BASIS FOR CONCENTRATION FRACTIONS OF WEAK ACIDS

I will talk here about the basis for fractions of monoprotic and triprotic weak acids in different forms and then show you how systematic patterns in equations for these acids can be used to extend the equations to other situations.

Basis for Fractions for Monoprotic Weak Acids

Example 3.6 will guide you through a procedure used to develop equations for the fractions, α_0 and α_1, for a monoprotic acid-base pair. I suggest that you work through the process for each fraction independently.

> **Example 3.6 Derive equations for fractions, α_0 and α_1, of a monoprotic acid-base pair in fully protonated and fully deprotonated forms.**
>
> Fraction $\alpha_{HB} \equiv \alpha_0$ Fraction $\alpha_B \equiv \alpha_1$
>
> *Step 1:* Define each fraction in terms of equilibrium concentrations.
>
> $$\alpha_{HB} \equiv \alpha_0 = \frac{[HB]}{C_T} = \frac{[HB]}{[HB]+[B^-]} \qquad\qquad \alpha_B \equiv \alpha_1 = \frac{[B^-]}{C_T} = \frac{[B^-]}{[HB]+[B^-]}$$
>
> *Step 2:* Write the equation for the deprotonation constant.
>
> $$K_a = \frac{[H^+][B^-]}{[HB]}$$
>
> *Step 3:* Rearrange the deprotonation constant to express the concentration of the other form of the acid as a multiple of the form for which the fraction is being derived.
>
> $$[B^-] = \left(\frac{K_a}{[H^+]}\right)[\mathbf{HB}] \qquad\qquad [HB] = \left(\frac{[H^+]}{K_a}\right)[\mathbf{B^-}]$$
>
> *Step 4:* Substitute equations for [B⁻] and [HB] into the equations in Step 1 and cancel like terms.
>
> $$\alpha_0 = \frac{[HB]}{[HB]+[B^-]} = \frac{[\cancel{HB}]}{[\cancel{HB}]+\left(\dfrac{K_a}{[H^+]}\right)[\cancel{HB}]} \qquad\qquad \alpha_1 = \frac{[\cancel{B^-}]}{\left(\dfrac{[H^+]}{K_a}\right)[\cancel{B^-}]+[\cancel{B^-}]}$$
>
> *Step 5:* Multiply right sides of equations for α_0 and α_1 by [H⁺]/[H⁺] and K_a/K_a, respectively, and simplify into Equations 3.6a and 3.6b:
>
> $$\alpha_0 = \frac{[H^+]}{[H^+]+K_a} \qquad (3.6a) \qquad\qquad \alpha_1 = \frac{K_a}{[H^+]+K_a} \qquad (3.6b)$$

Two features of these fractions that are common to fractions for any acid with any number, n, of acidic hydrogens are that (a) denominator terms are the same for both fractions and (b) the numerator term for each successive fraction is the same as each successive denominator term. This means that, given the denominator term, D_n, for any acid with any number of acidic hydrogens, one can write equations for successive fractions, α_0–α_n, as successive denominator terms divided by the denominator.

I will show you how this works for a triprotic acid.

Basis for Fractional Equations for a Triprotic Acid

Example 3.7 illustrates a process used to derive an equation for the fraction, α_0, of a triprotic acid in the fully protonated form. Remember, having developed the equation for one fraction, we can use the denominator term to write equations for all the fractions.

Example 3.7 Develop an equation for the fraction of an uncharged triprotic weak acid in the form that has lost zero protons, i.e., the fraction in the form, H_3B.

Step 1: Write the fraction in the fully protonated form as the equilibrium concentration, $[H_3B]$, divided by the sum of equilibrium concentrations of all forms of the acid.

$$\alpha_0 = \frac{[H_3B]}{C_T} = \frac{[H_3B]}{[H_3B] + [H_2B^-] + [HB^{2-}] + [B^{3-}]}$$

Step 2: Write equations for the stepwise deprotonation constants.

$$K_{a1} = \frac{[H^+][H_2B^-]}{[H_3B]}; \quad K_{a2} = \frac{[H^+][HB^{2-}]}{[H_2B^-]}; \quad K_{a3} = \frac{[H^+][B^{3-}]}{[HB^{3-}]}$$

Step 3: Rearrange the deprotonation-constant equations to express concentrations of H_2B^-, HB^{2-}, and B^{3-} in terms of the concentration of H_3B.

$$[H_2B^-] = \frac{K_{a1}}{[H^+]}[H_3B]; \quad [HB^{2-}] = \frac{K_{a1}K_{a2}}{[H^+]^2}[H_3B]; \quad [B^{3-}] = \frac{K_{a1}K_{a2}K_{a3}}{[H^+]^3}[H_3B]$$

Step 4: Substitute these equations into the equation in Step 1 and cancel like terms.

$$\alpha_0 = \frac{[\cancel{H_3B}]}{\left(1 + \dfrac{K_{a1}}{[H^+]} + \dfrac{K_{a1}K_{a2}}{[H^+]^2} + \dfrac{K_{a1}K_{a2}K_{a3}}{[H^+]^3}\right)[\cancel{H_3B}]}$$

Step 5: Multiply the top and bottom of the resulting equation by $[H^+]^3/[H^+]^3$.

$$\alpha_0 = \frac{[H^+]^3}{[H^+]^3 + K_{a1}[H^+]^2 + K_{a1}K_{a2}[H^+] + K_{a1}K_{a2}K_{a3}} = \frac{[H^+]^3}{[H^+]^3 + \kappa_1[H^+]^2 + \kappa_2[H^+] + \kappa_3} \qquad (3.7a)$$

The process in this example applies for any fraction of any acid. The first step for any form of any acid is to express the fraction in terms of equilibrium concentrations of all forms of the acid. The second step is to write equations for the stepwise deprotonation constants. The third step is to rearrange and combine deprotonation constant equations into forms that express concentrations of other forms of the acid as a proportional factor times the concentration of the form for which the fractional equation is being developed. The fourth and fifth steps involve substitutions and simplifications.

On the one hand, this process can be used to derive equations for fractions of any forms of any acid with any number of acidic hydrogens. On the other hand, I will show you how systematic patterns in Equation 3.7a can be used to write equations for other forms of any triprotic acid as well as any acid with any number of acidic hydrogens.

Fractions for Other Forms of Triprotic Acids

From Equation 3.7a, the denominator term for any acid with n = 3 acidic protons is as follows (Equation 3.7b):

$$D_3 = [H^+]^3 + \kappa_1[H^+]^2 + \kappa_2[H^+] + \kappa_3 \qquad (3.7b)$$

Given that successive fractions are successive denominator terms divided by the same denominator, it follows that the four fractions for any triprotic weak acid are as follows:

$$\alpha_0 = \frac{\left[H^+\right]^3}{D_3} \quad \text{(3.8a)}$$

$$\alpha_1 = \frac{\kappa_1\left[H^+\right]^2}{D_3} \quad \text{(3.8b)}$$

$$\alpha_2 = \frac{\kappa_2\left[H^+\right]}{D_3} \quad \text{(3.8c)}$$

$$\alpha_3 = \frac{\kappa_3}{D_3} \quad \text{(3.8d)}$$

in which all symbols are as defined earlier.

I suggest that you use Exercise 3.5 to test your understanding of the process in Example 3.7 and to confirm that the numerator and denominator terms for α_1 are the same as given above.

EXERCISE 3.5

(A) Confirm equations below for the fraction, α_1, of a triprotic weak acid that has lost one proton.

$$\alpha_1 = \frac{\left[H_2B^-\right]}{C_T} = \frac{\left[H_2B^-\right]}{\left[H_3B\right] + \left[H_2B^-\right] + \left[HB^{2-}\right] + \left[B^{3-}\right]}$$

$$\left[H_3B\right] = \frac{\left[H^+\right]}{K_{a1}}\left[H_2B^-\right], \quad \left[HB^{2-}\right] = \frac{K_{a1}}{\left[H^+\right]}\left[H_2B^-\right], \quad \left[B^{3-}\right] = \frac{K_{a2}K_{a3}}{\left[H^+\right]^2}\left[H_2B^-\right] \quad \text{and}$$

$$\alpha_1 = \frac{K_{a1}\left[H^+\right]^2}{\left[H^+\right]^3 + K_{a1}\left[H^+\right]^2 + K_{a1}K_{a2}\left[H^+\right] + K_{a1}K_{a2}K_{a3}} = \frac{\kappa_1\left[H^+\right]^2}{\left[H^+\right]^3 + \kappa_1\left[H^+\right]^2 + \kappa_2\left[H^+\right] + \kappa_3}$$

(B) Use a similar process to confirm equations for α_2 and α_3.

In summary, the equation for any fraction of any acid with n acidic hydrogens can be used to write equations for all fractions of all acids with the same number of acidic hydrogens.

Procedures similar to that described above can be used to develop equations for fractions of acids with different numbers of acidic hydrogens. However, as shown below, patterns in the equations for a triprotic acid can be used to adapt equations for fractions of triprotic acids to analogous fractions for acids with different numbers of acidic hydrogens.

Extensions to Acids with Different Numbers of Acidic Hydrogens

One of the attractive features of Ricci's systematic approach is the ease with which patterns in equations developed for an acid with one number of acidic hydrogens can be used to extend the equations to acids with different numbers of acidic hydrogens. For example, patterns in the denominator term for a triprotic acid can be used to write equations for acids with fewer and more acidic hydrogens.

The denominator term for a triprotic acid, H_3B, can be written in the following specific and general forms

$$D_3 = \left[H^+\right]^3 + \kappa_1\left[H^+\right]^2 + \kappa_2\left[H^+\right] + \kappa_3\left[H^+\right]^{3-3}$$

$$D_n = \left[H^+\right]^n + \kappa_1\left[H^+\right]^{n-1} + \kappa_2\left[H^+\right]^{n-2} + \kappa_3\left[H^+\right]^{n-3}$$

(3.9a)

in which n = 3 and powers of the hydrogen ion concentration are n − i = 3 − i. It follows that Equation 3.9a can be adapted to any acid with any number, n, of acidic hydrogens as follows:

$$D_n = \left[H^+\right]^n + \kappa_1 \left[H^+\right]^{n-i} + \kappa_2 \left[H^+\right]^{n-i} + \kappa_3 \left[H^+\right]^{n-i} \cdots \kappa_n \left[H^+\right]^{n-n} \tag{3.9b}$$

in which n is the number of acidic hydrogens and i varies from 0 to n, i.e., $0 \leq i \leq n$.

Equation 3.9b can be simplified to the following more compact form:

$$D_n = \sum_{i=0}^{n} \kappa_i \left[H^+\right]^{n-i} \tag{3.9c}$$

This latter Equation 3.9c can be used to write denominator terms for any acid with any number of acidic hydrogens.

EXERCISE 3.6

Use Equation 3.9b or Equation 3.9c to confirm the following denominator terms, D_n, for monoprotic, diprotic, and hexaprotic acids.

$$D_1 = \left[H^+\right] + \kappa_1; \quad D_2 = \left[H^+\right]^2 + \kappa_1 \left[H^+\right] + \kappa_2 \quad \text{and}$$

$$D_6 = \left[H^+\right]^6 + \kappa_1 \left[H^+\right]^5 + \kappa_2 \left[H^+\right]^4 + \kappa_3 \left[H^+\right]^3 + \kappa_4 \left[H^+\right]^2 + \kappa_5 \left[H^+\right] + \kappa_6$$

I will leave it to you to use the denominator terms to write equations for fractions for the various forms of each acid as well as denominator terms for triprotic and tetraprotic acids. I am not aware of any pentaprotic acids.

Equation 3.9c not only puts the equation for dominator terms in a more compact form but also can be used to simplify programs used to solve the equations. For example, consider an Excel program to calculate the denominator term for acids with up to 6 acidic hydrogens using parameters in cells summarized below.

$$1 \leq n \leq 6: C42 \quad [H^+]: I82 \quad 0 \leq i \leq 6: C86 - I86 \quad \kappa_0 - \kappa_n: B106 - H106$$

The equation for the denominator term using Equation 3.9b is

$$= B106 * I82 \wedge (C42 - C86) + C\$106 * I82 \wedge (C42 - D86) + D\$106 * I82 \wedge (C42 - E86)$$

$$+ E\$106 * I82 \wedge (C42 - F86) + F\$106 * I82 \wedge (C42 - G86)$$

$$+ G\$106 * I82 \wedge (C42 - H86) + H\$106 * I82 \wedge (C42 - I86)$$

The equation using Equation 3.9c with κ_0 set to 1 in B106 is

$$= SUM\left((B106 : H106) * I82 \wedge (C42 - C86 : I86)\right)$$

There is less opportunity for error in the latter form of the equation.

Some of my programs are written using the format in Equation 3.9b, others are written using the more compact form in Equation 3.9c. Frankly, I would have saved myself a lot of time and grief if

I had recognized the simplifications inherent in the more compact equation sooner. In any event, for reasons described below, either implementation of the denominator equation will give correct denominators for acids with up to 6 acidic hydrogens if all constants beyond the last constant for each acid, K_{an}, are set to zero. For example, all constants beyond K_{a3} would be set to zero for a triprotic acid.

Alternative Forms of the Fractional Equations

I will continue to use fractional equations in the general forms in Equations 3.8a–3.8d throughout this and subsequent chapters. However, you can use Exercise 3.7 to confirm alternate forms of the equations that you might find more appealing for some applications, particularly calculations using hand-held calculators.

EXERCISE 3.7

(A) Using Equations 3.5a–3.5d, show that the four fractions for any triprotic acid are related as shown in Equations 3.10a–3.10d:

$$\alpha_0 = \frac{\left[H^+\right]^3}{D_3} \qquad (3.10a)$$

$$\alpha_1 = \frac{\kappa_1}{\left[H^+\right]^1}\alpha_0 \qquad (3.10b)$$

$$\alpha_2 = \frac{\kappa_2}{\left[H^+\right]^2}\alpha_0 \qquad (3.10c)$$

$$\alpha_3 = \frac{\kappa_3}{\left[H^+\right]^3}\alpha_0 \qquad (3.10d)$$

Hint: Divide each fraction, α_1–α_3, by α_0 and rearrange the result into the corresponding form.

(B) Confirm that the following general equation can be used to write each of the foregoing equations.

$$\alpha_{i>0} = \frac{\kappa_i}{\left[H^+\right]^i}\alpha_0 \qquad (3.10e)$$

(C) Confirm that fractions calculated using these equations for conditions as in Example 3.5 are the same as calculated in that example.

Having calculated α_0 for any acid, H_nB or H_nB^{z+}, each successive fraction can be calculated as a proportionality factor, $\kappa_i/[H^+]^i$, times α_0. Given that $\kappa_i = 0$ for $i > n$, all fractions beyond α_n are zero, i.e., $\alpha_{i>n} = 0$.

I haven't used this feature in my programs. However, you may find it useful in future applications.

A Computational Simplification

Another useful feature of these fractional equations is based on a feature of cumulative constants. *For any acid, H_nB, all stepwise deprotonation constants beyond K_{an} are zero and, therefore, all cumulative constants beyond κ_n are zero.* So, how is this useful?

Let's suppose we want to write a program or programs to calculate concentration fractions for acids with 1 to 6 acidic hydrogens. One way we could do this is to write a different program for each group of acids with the same number of acidic hydrogens. Another more efficient way is to write a single program to do calculations for acids with numbers of acidic hydrogens from 1–6.

To do this, write the equation for the denominator term for a hexaprotic acid in the following form (Equation 3.11):

$$D_n = \left[H^+\right]^n + \kappa_1\left[H^+\right]^{n-1} + \kappa_2\left[H^+\right]^{n-2} + \kappa_3\left[H^+\right]^{n-3} + \kappa_4\left[H^+\right]^{n-4} + \kappa_5\left[H^+\right]^{n-5} + \kappa_6 \qquad (3.11)$$

Then, (a) tell the program the number, n, of acidic hydrogens in each acid of interest and (b) *set all deprotonation constants beyond* K_{an} *to zero.* The net effect is that the program will calculate values of fractions from $i = 0$ to $i = n$ and set all fractions beyond $i = n$ to zero. I will help you write a spreadsheet program to do this.

SPREADSHEET PROGRAM TO CALCULATE CONCENTRATION FRACTIONS

A template for a spreadsheet program to calculate concentration fractions for acids with one to six acidic hydrogens is included as *Template_ABFractions*, on the programs website http://crcpress.com/9781138367227 for this text. To help you confirm that you have entered formulae correctly when the template is complete, the template is set up to calculate fractions of phosphoric acid in different forms for a hydrogen ion concentration of $[H^+] = 1.0 \times 10^{-3}$ M. Numerical values of fractions you should obtain are summarized in the last row of Table 3.3.

To complete the program, open the template and type formulae from Table 3.3 below into the indicated cells and press Ctrl/Shift/Enter simultaneously after entering each formula.

As noted above, numerical values of fractions you should see for phosphoric acid with $[H^+] = 1.0 \times 10^{-3}$ M are summarized in the last row of the table.

I suggest that you use Exercise 3.8 to confirm that that the program is functioning properly and that you know how to use it.

EXERCISE 3.8

Confirm concentration fractions for protonated EDTA,[a] a hexaprotic acid, i.e., $n = 6$, at a hydrogen ion concentration of $[H^+] = 1.0 \times 10^{-10}$ M.

K_{a1}	K_{a2}	K_{a3}	K_{a4}	K_{a5}	K_{a6}	$[H^+]$
1.0	3.2E-2	1.0E-2	2.2E-3	6.9E-7	5.8E-11	1.0E-10
α_0	α_1	α_2	α_3	α_4	α_5	α_6
1.30E-38	1.30E-28	4.17E-20	4.17E-12	9.17E-05	6.33E-01	3.67E-01

[a] EDTA is an acronym for ethylenediaminetetraacetic acid, H_6B^{2+}.

TABLE 3.3 Cell Contents for a Program to Calculate Concentration Fractions for Acids with One through Six Acidic Hydrogens

E15	E16	E17	E18	E19
=C15	=C15*C16	=C15*C16*C17	=C15*C16*C17*C18	=C15*C16*C17*C18*C19
E20	=C15*C16*C17*C18*C19*C20			
E27	=C21^C14+E15*C21^(C14-1)+E16*C21^(C14-2)+E17*C21^(C14-3)+E18*C21^(C14-4) +E19*C21^(C14-5)+E20			
B25		C25	D25	E25
=(C21^C14)/E27		=E15*C21^(C14-1)/E27	=E16*C21^(C14-2)/E27	=E17*C21^(C14-3)/E27
B27		C27	D27	
=E18*C21^(C14-4)/E27		=E19*C21^(C14-5)/E27	=E20/E27	
Quantities You Should See When Your Program Is Complete				

α_0	α_1	α_2	α_3	α_4	α_5	α_6	D_n
B25	C25	D25	E25	B27	C27	D27	E27
1.23E-01	8.77E-01	5.54E-05	2.49E-14	0.00E+00	0.00E+00	0.00E+00	8.11E-09

Concentration fractions calculated as described above can be used in a variety of ways, one of which is to help us visualize how hydrogen ion concentrations influence distributions of weak acids among their different forms. Plots used to do this are commonly called *distribution plots.*

DISTRIBUTION PLOTS

Distribution plots are plots of concentration fractions vs. pH. Distribution plots can be used to visualize how pH influences fractions of weak acids among their different forms.

A program to calculate and plot concentration fractions vs. pH for acids with up to six acidic hydrogens is included on Sheet 2 of the template, *Template_ABFractions*, described above.

To use the program, enter the number of acidic hydrogens and deprotonation constants for the acid of interest in Cells C14–C20, being careful to *set all constants beyond K_{an} to zero*, and observe the distribution plots in the figure adjacent to the data entry box.

Distribution Plots for Monoprotic Acid-Base Pairs

As discussed earlier, distribution plots for weak acids are plots of concentration fractions vs. pH. Figure 3.2 is the distribution plot for protonated imidazole, $K_a = 1.0 \times 10^{-7}$. To interpret such plots, remember that low pH corresponds to high H^+ concentrations. For example, for pH's from 0 to 14 as in Figure 3.2, the hydrogen ion concentration decreases from 1.0 M at pH = 0.0 to 10^{-14} M at pH = 14.

The solid and dotted plots in Figure 3.2 correspond to α_0 and α_1, respectively. At low pH, high $[H^+]$, virtually all of the imidazole is in the fully protonated form, i.e., HB^+. As the pH is increased, the hydrogen ion concentration will be decreased and more and more of the protonated imidazole will be deprotonated decreasing α_0 and increasing α_1. For pH's above about 10.0, virtually all of the protonated imidazole will be converted to imidazole, i.e., the basic form, B.

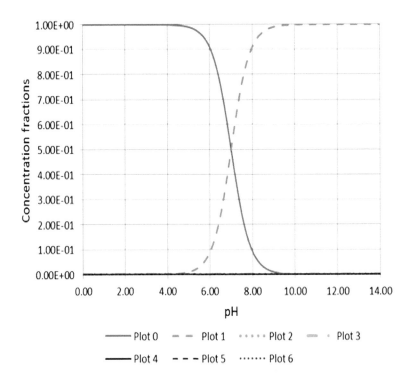

FIGURE 3.2   Distribution plots for protonated imidazole, $K_a = 1.0 \times 10^{-7}$.

Plots 0 and 1: Fractions, α_0 and α_1.

Notice that the intersection point for the two plots corresponds to $\alpha_0 = \alpha_1 = 0.5$ and pH = $pK_a = 7.0$. This isn't an accident. It results from the relationship between K_a, H^+ and concentrations of the acid-base pair. Given that $[HB] = [B^-]$ when $\alpha_0 = \alpha_1 = 0.5$, it follows that $[H^+] = K_a$ and $pK_a = pH$.

For a sufficiently wide pH range, distribution plots for all monoprotic acid-base pairs exhibit analogous sigmoid shapes. The larger the deprotonation constant, the closer the intersection point will be to pH ≤ 0; the smaller the deprotonation constant, the closer the intersection point will be to pH ≤ 14. However, plots for acids with very large and very small deprotonation constants will not show regions of complete deprotonation and complete protonation for the pH range included in Figure 3.2.

For example, a program on Sheet 5 of a weak acids program to be discussed in Chapter 5 was used to plot distribution plots for dichloroacetic acid/dichloroacetate ion, $pK_a = 1.3$, protonated imidazole/imidazole, $pK_a = 7.0$, and protonated diethylamine/diethylamine, $pK_a = 10.9$, in Figure 3.3.

Whereas both complete sigmoid plots are observed for the protonated imidazole/imidazole pair and the protonated diethlamine/diethylamine pair, the pH range plotted in the figure is too narrow to include complete sigmoid plots for dichloroacetic acid.

Distribution Plot for a Triprotic Acid

Figure 3.4 is the distribution plot for phosphoric acid.

The three pairs of plots from left to right represent fractions of the total concentration of phosphoric acid in forms of that have lost i = 0, 1, 2, and 3 protons, i.e., in forms H_3PO_4, $H_2PO_4^-$, HPO_4^{2-},

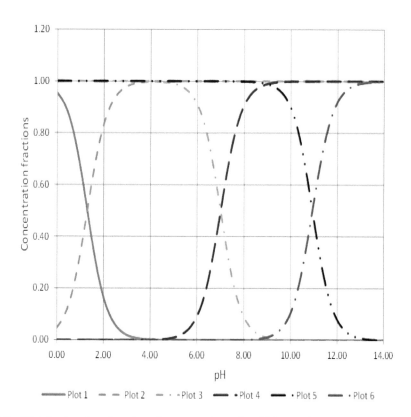

FIGURE 3.3 Distribution plots for selected monoprotic weak acid-base pairs.

Plots 1 and 2: Acidic and basic fractions of dichloroacetic acid, $pK_a = 1.3$.
Plots 3 and 4: Acidic and basic fractions of protonated imidazole, $pK_a = 7.0$.
Plots 5 and 6: Acidic and basic fractions of protonated diethylamine, $pK_a = 10.9$.

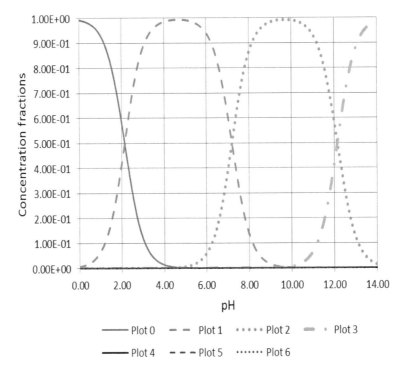

FIGURE 3.4 Distribution plot for phosphoric acid, $pK_{a1} = 2.15$, $pK_{a2} = 7.20$, and $pK_{a3} = 12.35$.

Plots 0 – 3: Fractions $\alpha_0 - \alpha_3$.

and PO_4^{3-}. At pH = 1, i.e., $[H^+] = 0.10$ M, virtually all the phosphate is present as H_3PO_4. As the pH is increased, percentages present as H_3PO_4 decrease and percentages present as $H_2PO_4^-$ increase and reach a maximum near pH = 5. As the pH is increased above 5, percentages present as $H_2PO_4^-$ begin to decrease and percentages present as HPO_4^{2-} increase to a maximum near pH = 10. As the pH is increased above 10, percentages present as HPO_4^{2-} decrease and the percentages present as PO_4^{3-} increase until virtually all the phosphate is present in the fully deprotonated form near pH = 14.

Fractional distributions for phosphoric acid are well separated because there are large differences among deprotonation constants. Distribution plots for acids with smaller differences among deprotonation constants will not be so well separated.

For example, Figure 3.5 is the distribution plot for citric acid, $K_{a1} = 7.4 \times 10^{-4}$, $K_{a2} = 1.7 \times 10^{-5}$ and $K_{a3} = 4.0 \times 10^{-7}$. The descending and ascending sigmoid plots represent fractions, α_0 and α_3, corresponding to fully protonated citric acid and fully deprotonated citrate, respectively. The intermediate peaks represent fractions, α_1 and α_2, for fractions of citric acid in the forms, H_2B^- and HB^{2-}. Differences between deprotonation constants are so small that the second deprotonation step begins before the first is complete and the third step begins before the second is complete. Therefore, neither α_1 nor α_2 is equal to 1.0 before the next deprotonation step begins. The result is a distribution plot with two sigmoid-shaped plots and two peaks corresponding to α_1 and α_2.

Any acid, H_nB, with n acidic hydrogen ions will exhibit descending and ascending sigmoid plots at low and high pH and plots with n – 2 peaks in between. Peaks will be broad as in Figure 3.4 if deprotonation constants are separated sufficiently or narrow as in Figure 3.5 if deprotonation constants are not well separated. Also, by analogy with the crossing points for monoprotic acid-base pairs, the crossing point for each successive pair corresponds to the pK_i for that pair. Fractions at

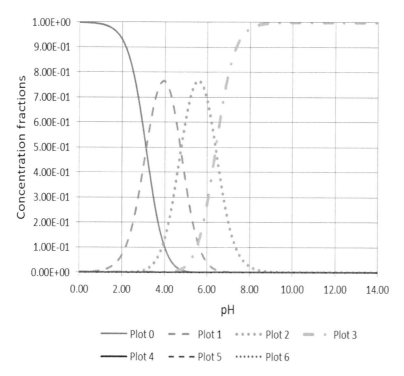

FIGURE 3.5 Distribution plot for citric acid, $pK_{a1} = 3.13$, $pK_{a2} = 4.76$, and $pK_{a3} = 6.40$.

Plots 0 – 3: Fractions $\alpha_0 - \alpha_3$.

crossing points may or may not be equal to 0.5 because some of the total concentration may be in other forms.

In summary, distribution plots are a convenient way to visualize how pH influences fractions of weak acids in different forms.

For the record, both programs described in this chapter are included in a more general weak acids program to be described in a later chapter.

SUMMARY

This chapter includes discussions of a range of topics and concepts relevant to a systematic approach to acid-base equilibria. Regarding strong acids, it is noted that concentrations of anionic parts of strong acids, e.g., Cl^-, and cationic parts of strong bases, e.g., Na^+, are equal to the diluted concentrations of the strong acids and strong bases when dissociation reactions are complete.

Equilibration reactions and constants for weak acids are discussed stating with the autoprotolysis of water and continuing with reactions and constants for acids with n = 1 to n = 6 acidic hydrogens, i.e., monoprotic acids to hexaprotic acids. The discussion includes reactions and constants for uncharged and uncharged acids. For polyprotic acids, it is noted that equations that involve products of deprotonation constants can be simplified by replacing products of constants with cumulative constants. For example, the product of constants for a hexaprotic acid, $K_{a1}K_{a2}...K_{a6}$, is replaced by a cumulative constant, κ_6.

It is noted that the stepwise deprotonation of an n-protic acid, H_nB, involves the loss one proton per step and that the number of protons lost by the fully protonation acid varies from i = 0 for the fully protonated acid to i = n for the fully deprotonated acid. As described by J. E. Ricci,[1] cumulative numbers, i, of protons lost are used to describe a systematic set of symbols that unifies symbolism for a wide range of weak acids and bases. As examples, undiluted concentrations, diluted concentrations and activity coefficients of forms of acids that have lost $i = 0 - n$ protons are represented by symbols, C_i^0, C_i, and f_i, respectively.

The i symbolism is used to correlate equilibrium concentrations of sodium ion and chloride ions form salts of the ions of uncharged and charged ions with diluted concentrations of salts that produce the ions. For example, the sodium ion concentration produced by dissociation of disodium phosphate, Na_2HPO_4, is represented by the symbol, $2C_2$, because the salt dissociates completely producing, HPO_4^{2-}, the form of phosphoric acid that has lost 2 protons. Using the same symbolism, dissociation of the chloride salt of a doubly charged triprotic acid, H_3BCl_2, produces a chloride ion concentration, $[Cl^-] = 2C_0$.

Mass-balance equations are used to account for all sources of all reactants. Mass-balance equations are based on Lavoisier's law of the conservation of mass, meaning that mass is neither created nor destroyed. Whereas forms of reactants change, the total mass of the different parts of reactants remains unchanged. As an example, when a solution prepared to contain concentrations, C_0, C_1, C_2 and C_3, of phosphoric acid, monosodium phosphate, disodium phosphate, and trisodium phosphate, respectively reaches equilibrium, the sodium ion concentration will be $[Na^+] = C_1 + 2C_2 + 3C_3$ and the total phosphate concentration will be $C_T = C_0 + C_1 + C_2 + C_3$. Analogous equations apply for chloride salts of charged acids.

Total concentrations, C_T, of different forms of weak acids based on mass-balance equations are used with concentration fractions, α_0, α_1, $\alpha_2...\alpha_n$, to calculate equilibrium concentrations of different forms of weak acids. For example, the equilibrium concentration of a form of an uncharged weak acid that has lost i protons is $[H_{n-i}B^{i-}] = \alpha_i C_T$. As shown by Ricci and by derivations in this chapter, concentration fractions are moderately complex functions of deprotonation constants and the hydrogen ion concentration.

Finally, it is shown that plots of concentration fractions vs. pH, called distribution plots, can be used to visualize how pH and the hydrogen ion concentration influence distributions of total concentrations of weak acids among different forms. The larger the deprotonation constant, the smaller the pH, higher the H^+ concentration, at which the deprotonation reaction corresponding to that deprotonation constant will approach completion.

REFERENCES

1. J. E. Ricci, *Hydrogen Ion Concentration, New Concepts in a Systematic Treatment*, Princeton U. P.: Princeton, NJ, 1952.
2. H. Freiser, *Concepts & Calculations in Analytical Chemistry, A Spreadsheet Approach*, CRC Press: Boca Raton, FL, 1992.
3. R. de Levie, *Principles of Quantitative Chemical Analysis*, McGraw-Hill: New York, NY, 1997.
4. Guido Frison, Alberto Calatroni, www.academia.edu/518142/ Some Notes on pH Computation Methods and Theories, 1–29.

4 A Systematic Approach to Monoprotic Acid-Base Equilibria without and with Correction for Ionic Strength

In a text published in 1952, J. E. Ricci[1] described a systematic approach to acid-base equilibria. This chapter describes adaptations of Ricci's systematic approach to monoprotic acid-base equilibria. One advantage of the systematic approach is that it can be used to avoid the types of errors associated with classical approximations. A second advantage is that it is easily adapted to situations involving two or more weak acid-base pairs with one or more acidic hydrogens each. A third advantage is that systematic patterns in equations based on the systematic approach make it easy to extend equations developed for one situation to equations for both simpler and more complex situations.

My purpose in this chapter is to show you how the systematic approach can be adapted to monoprotic acid-base equilibria. To begin, I will help you develop an exact equation for the hydrogen ion concentration in solutions containing concentrations equal to or larger than zero of HCl, NaOH, and up to three weak acid-base pairs. I will then help you write a spreadsheet program to solve the equation using activity-based constants. Finally, I will discuss some applications, including procedures used to correct for effects of ionic strength.

Whereas many texts focus on pH, I will focus on the hydrogen concentrations for three reasons. *First*, most quantities in which chemists are interested, including deprotonation constants, weak-acid/weak-base concentrations, concentration fractions, reaction rates, enzyme activity, etc. depend more directly on the hydrogen ion concentration than on pH. Second, given the hydrogen ion concentration, it is very easy to calculate pH when it is needed to display large ranges of H^+ concentrations in small spaces such as in titration curves, distribution plots and H^+ dependencies of quantities such as rate constants, enzyme activity, etc. Finally, decisions based on pH errors can be misleading because, given the logarithmic relationship between pH and the H^+ concentration, pH errors tend to be misleadingly small relative to concentration errors.

Regarding the last point, a 5% error in a pH of 5.0 corresponds to a pH of 5.25. These pH's correspond to hydrogen ion concentrations of 1.0×10^{-5} M and 5.6×10^{-6} M. This corresponds to an error of $100(5.6 \times 10^{-6}/5 \times 10^{-5} - 1) = 44\%$ in the hydrogen ion concentration. The error in the hydrogen ion concentration is about eight times larger than the pH error. This can be very misleading if, as is the case for most chemical processes, the process in which you are interested depends more directly on the hydrogen ion concentration than on pH.

SYSTEMATIC APPROACH TO MONOPROTIC ACID-BASE EQUILIBRIA

The process called a *systematic approach to acid-base equilibria* involves combinations of the autoprotolysis constant for water, mass-balance equations, concentration fraction equations and either charge-balance or proton-balance equations to obtain exact equations for the hydrogen ion concentration. Several authors[2,3,4] have described applications of the systematic approach to acid-base equilibria. These authors used graphical procedures to augment adaptations of the systematic approach to approximate calculations. Anyone interested in graphical solutions to the equations should consult these texts.[2,3,4] This discussion emphasizes iterative solutions of the exact equations using spreadsheet programs.

As noted above, two options for adapting the systematic approach to acid-base equilibria involve proton-balance and charge-balance equations.[2,3] I chose to use charge-balance equations primarily because they are more general than proton balance equations in the sense that they can be adapted to different types of ionic equilibria.

A Preview of Acid-Base Titration Curves

Figure 4.1A includes titration curves for a strong monoprotic acid, HCl, titrated with a strong monobasic base, NaOH, and a strong monobasic base, NaOH, titrated with a strong monoprotic acid, HCl.

Figure 4.1B includes titration curves for mixtures of weak acids with strong base and weak bases with strong acid, both without and with correction for ionic strength.

The ascending plots are titration curves for the titration with 0.10 M sodium hydroxide of 25.0 mL of a solution containing 0.10 M each of dichloroacetic acid, $K_a = 0.050$, acetic acid, $K_a = 1.8 \times 10^{-5}$ M, and ammonium ion, $K_a = 5.7 \times 10^{-10}$. As you can see, a reasonably sharp break is obtained for each of the three acids.

The descending plots are titration curves for titrations with 0.10 M HCl of 25.0 mL of a solution containing 0.10 M each of the conjugate bases of the three acids, i.e., dichloroacetate, acetate, and ammonia. The two breaks in the titration curves are for ammonia and acetate ion. The dichloroacetate ion is too weak a base to be titrated in aqueous solution. For example, the protonation constant for dichloroacetate is the autoprotolysis constant for water divided by the deprotonation constant for

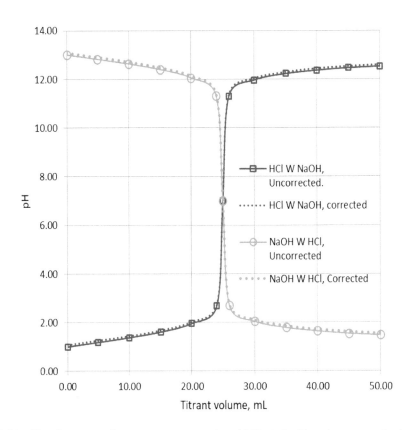

FIGURE 4.1A Titration curves for a strong monoprotic acid titrated with a strong monobasic base and a strong monoprotic base titrated with a strong monoprotic acid without and with correction for ionic strength.

Ascending plots: 25.0 mL of 0.10 M HCl titrated with 0.10 M NaOH.
Descending plots: 25.0 mL of 0.10 M NaOH titrated with 0.10 M HCl.

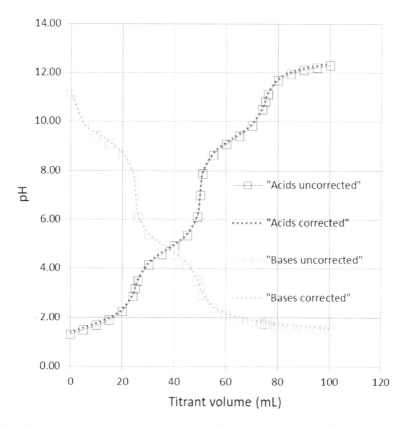

FIGURE 4.1B Titration curves for mixtures of monoprotic weak acids titrated with strong base and mixtures of monobasic weak bases titrated with strong acid without and with correction for ionic strength.

Ascending plots: titration with 0.10 M sodium hydroxide of 25.0 mL of a solution containing 0.10 M each of dichloroacetic acid, $K_a = 0.050$, acetic acid, $K_a = 1.8 \times 10^{-5}$ M, and ammonium ion, $K_a = 5.7 \times 10^{-10}$.
Descending plots: titration with 0.10 M HCl of 25.0 mL of a solution containing 0.10 M each of the conjugate bases of the three acids, i.e., dichloroacetate, acetate, and ammonia.

dichloroacetic acid, i.e., $K_b = K_w/K_a = 1.0 \times 10^{-14}/0.050 = 2.0 \times 10^{-13}$. This shows that dichloroacetate is much too weak a base to be titrated with a strong acid in water.

Templates for the first step in each of the four sets of titrations are in Cells A64–I72, A79–I87, K64–S72 and K79–S87, respectively, on Sheet 2 of the monoprotic acid/base program. Detailed procedures for using such templates are discussed later. For now, my primary goal is to help you understand the basis for the calculations used to obtain these results.

Generic Symbols for Acids and Bases

We will talk about a program to do calculations for solutions containing a strong acid such as hydrochloric acid, a strong base such as sodium hydroxide, and up to three monoprotic *charged or uncharged* weak acids and their conjugate bases. This necessarily requires some unconventional symbols for the three weak acid-base pairs. I chose to use X^-, Y^-, and Z^- as symbols for conjugate bases of uncharged acids such as acetic acid and X, Y, and Z as symbols for conjugate bases of charged acids such as the ammonium ion.

As examples, reactions for conjugate acid-base pairs of acetic acid and ammonium ion using more conventional symbols are compared below for the same reactions written in terms of symbols, X^- and X, as substitutes for Ac^- and NH_3.

Uncharged Acids	Salts of the Acids	Salts of Charged Acids	Charged Acids
$HAc \overset{K_{ax}}{\rightleftharpoons} H^+ + Ac^-$	$NaAc \rightarrow Na^+ + Ac^-$	$\overbrace{HNH_3Cl}^{NH_4Cl} \rightarrow HNH_3^+ + Cl^-$	$\overbrace{HNH_3^+}^{NH_4^+} \overset{K_{ax}}{\rightleftharpoons} H^+ + NH_3$
$HX \overset{K_{ax}}{\rightleftharpoons} H^+ + X^-$	$NaX \rightarrow Na^+ + X^-$	$HXCl \rightarrow HX^+ + Cl^-$	$HX^+ \overset{K_{ax}}{\rightleftharpoons} H^+ + X$

Analogous comparisons for conjugate bases represented by symbols Y and Z can be written by replacing X with Y or Z.

A CHARGE-BALANCE EQUATION

As a starting point, I will develop the charge-balance equation for solutions containing sodium hydroxide, hydrochloric acid, acetic acid, sodium acetate, ammonia, and ammonium chloride. Having developed the charge-balance equation for this situation, I will then show you how we can generalize the equation by adapting it to uncharged and charged generic acids, HX and HY^+ and their conjugate bases. Having done that, I will show you how the equation can be extended to a third charged or uncharged weak acid, HZ of HZ^+, and its conjugate base. I will then explain how the equation and a program based on it can be adapted to situations ranging from pure water to solutions containing a strong acid, a strong base, and up to three weak acid-base pairs.

AN EXAMPLE

Example 4.1 is used to illustrate an application of the systematic approach to solutions containing concentrations equal to or larger than zero of HCl, NaOH, acetic acid, sodium acetate, ammonium chloride, and ammonia.

Example 4.1 Derive a charge-balance equation for the hydrogen ion concentration that can be used with *either activity-based or concentration-based constants* in solutions containing *concentrations equal to or larger than zero* of HCl, NaOH, acetic acid, sodium acetate, ammonium chloride, and ammonia.

Step 1: Write the reactions that produce ions and underline **one each** of the different ions.

$$H_2O \rightleftharpoons \underline{H^+} + \underline{OH^-} \; ; HCl \rightarrow H^+ + \underline{Cl^-} \; ; NaOH \rightarrow \underline{Na^+} + OH^- \; ;$$

$$NaAc \rightarrow Na^+ + \underline{Ac^-}; NH_4Cl \rightarrow \underline{NH_4^+} + Cl^-; HAc \rightleftharpoons H^+ + Ac^-; NH_4^+ \rightleftharpoons H^+ + NH_3$$

Step 2: Write an *ionic form* of the charge-balance equation, CBE, by *equating the algebraic sum of products of signed charges times equilibrium concentrations of **one each** of ions to zero.*

$$CBE = \left[H^+\right] - \left[OH^-\right] + \left[Na^+\right] - \left[Cl^-\right] - \left[Ac^-\right] + \left[NH_4^+\right] = 0$$

Step 3: Write equations for the ions using the *autoprotolysis-constant, mass-balance equations, and fractional equations* discussed in the background chapter.

$$\left[OH^-\right] = \frac{K_w}{\left[H^+\right]} \qquad \left[Cl^-\right] = C_{HCl} + C_{NH_4Cl} \qquad \left[Na^+\right] = C_{NaOH} + C_{NaAc}$$

$$\left[Ac^-\right] = \alpha_{1HAc}\left(C_{HAc} + C_{Ac}\right) \qquad \left[NH_4^+\right] = \alpha_{0NH_4}\left(C_{NH_4} + C_{NH_3}\right)$$

Step 4: Write a *fractional form* of the charge-balance equation by substituting these equations into the ionic equation and rearranging the result into the form below.

$$CBE = \left[H^+\right] - \frac{K_w}{\left[H^+\right]} + \overbrace{C_{NaOH} - C_{HCl}}^{NaOH/HCl\ term} + \overbrace{C_{NaAc} - \alpha_{1HAc}\left(C_{HAc} + C_{Ac}\right)}^{HAc/Ac\ term}$$

$$- \overbrace{C_{NH_4} + \alpha_{0NH_4}\left(C_{NH_4} + C_{NH_3}\right)}^{NH_4/NH_3\ term} = 0$$

Step 5: Write an *expanded fractional form* of the CBE by replacing concentration fractions and total concentrations with equations from the background chapter for α_{Ac}, and α_{NH_4}.

$$CBE = \left[H^+\right] - \frac{K_w}{\left[H^+\right]} + \overbrace{C_{NaOH} - C_{HCl}}^{NaOH/HCl\ term} + \overbrace{C_{Ac} - \frac{K_{aHAc}}{\left[H^+\right] + K_{aHAc}}\left(C_{HAc} + C_{Ac}\right)}^{Weak-acid\ term:\ HAc/Ac\ term}$$

$$- \overbrace{C_{NH_4} + \frac{\left[H^+\right]}{\left[H^+\right] + K_{aNH_4}}\left(C_{NH_4} + C_{NH_3}\right)}^{Weak-acid\ term:\ NH_4^+/NH_3\ term} = 0$$

(4.1a)

Step 6: Write a generalized form of this equation by defining acetic acid and acetate ion as HX and X$^-$ and ammonium ion and ammonia as HY$^+$ and Y, respectively, and using Ricci's i notation for numbers of protons lost.

$$CBE = \left[H^+\right] - \overbrace{\frac{K_w}{\left[H^+\right]}}^{Water\ term} + \overbrace{C_{NaOH} - C_{HCl}}^{NaOH/HCl\ term} + \overbrace{C_{1x} - \frac{K_{ax}}{\left[H^+\right] + K_{ax}}\left(C_{0x} + C_{1x}\right)}^{Weak-acid\ term:\ HX/X\ term}$$

(4.1b)

$$- \overbrace{C_{0y} + \frac{\left[H^+\right]}{\left[H^+\right] + K_{ay}}\left(C_{0y} + C_{1y}\right)}^{Weak-acid\ term:\ HY^+/Y\ term} = 0$$

Symbols used in this example are summarized below:

C_{HCl} and C_{NaOH}: Diluted concentrations of HCl and NaOH.

C_{HAc}, C_{Ac}, C_{NH_4Cl}, C_{NH_3}: Diluted concentrations of acetic acid, sodium acetate, ammonium chloride, and ammonia, respectively.

α_{1HAc}, α_{0NH_4}: Fractions of acetic acid and ammonium ion in forms that have lost 1 and 0 protons, respectively.

C_{0x}, C_{1x}, C_{0y}, C_{1y}: Diluted concentrations of forms of Acids X and Y that have lost $i = 0$ and $i = 1$ protons, respectively. (Acids X and Y are acetic acid and ammonium ion in this example.)

K_{aHAc}, K_{aNH_4}: Deprotonation constants for acetic acid and the ammonium ion.

K_{ax}, K_{ay}: Deprotonation constants for acids X and Y, e.g., acetic acid and ammonium ion.

Regarding Steps 1 and 2, each ionic concentration represents the *combined concentrations from all sources of each ion* and therefore is included only once in the charge-balance equation. Mass-balance equations are used to account for different sources of the ions.

Also regarding Step 2, the charge-balance equation is usually written by equating the sum of positive charges to the sum of negative charges. While working with these equations, I discovered that *the charge-balance equation written by equating the algebraic sum of positive and negative charges to zero as done in the example is negative for all hydrogen ion concentrations less than the equilibrium concentration and positive for all H⁺ concentrations larger than the equilibrium concentration. This feature is used later as a convergence criterion for an iterative procedure to solve the charge-balance equation for the H⁺ concentration.*

Regarding Step 3, the autoprotolysis-constant equation, mass-balance equations, and concentration-fraction equations are used to account for different sources of the ions. All these equations are discussed in the background chapter and are not discussed further here.

The *fractional form* of the charge-balance equation in Step 4 is obtained by substituting equations from Step 3 into the ionic form of the equation in Step 2. The computational form of the equation in Step 5 is obtained by replacing terms, α_{Ac}, and α_{NH_3}, for the fractions with equations in terms of the hydrogen ion concentration and deprotonation constants. The equations are

$$\alpha_{1Ac} = \frac{K_{aAc}}{\left[H^+\right] + K_{aAc}} \quad \text{and} \quad \alpha_{0NH_4} = \frac{\left[H^+\right]}{\left[H^+\right] + K_{aNH_4}}$$

in which α_{1HAc} is the fraction of acetic acid and acetate in the form that has lost one proton, i.e., acetate; α_{0NH_4} is the fraction of the ammonium ion and ammonia in the form that has lost zero protons, i.e., NH_4^+.

Step 6 shows how the equation written in terms of more conventional symbols for acetate and ammonia, Ac^- and NH_3, can be written in terms of the generic symbols, X^- and Y, for the same reactants. Subscripted x's and y's are used here and elsewhere to differentiate among quantities related to the two acid-base pairs. Subscripted z's are used for a third acid, HZ or HZ^+, and its conjugate base, Z^- or Z.

A SIMPLIFYING FEATURE OF THE CHARGE-BALANCE EQUATION

The forms of the weak acid terms in Equation 4.1b for the uncharged and charged acids, HX and HY^+, appear to be different. However, as shown in Table 4.1, the form of the equation for the charged acid can be converted to that of the uncharged acid.

Don't believe me? Let's do a little algebra with the weak-acid term for the charged acid to show that *the general form of the term for the charged acid is in fact equivalent to the form of the term for the uncharged acid, HX.*

As a starting point, let's set the sum of fractions for the charged acid equal to 1, i.e., $\alpha_{0y} + \alpha_{1y} = 1$. Now, let's rewrite this sum of fractions in the form, $\alpha_{1y} = 1 - \alpha_{0y}$. Now comes the fun part. Let's replace α_{0y} and C_{Ty} in Equation 4.1a with $\alpha_{0y} = 1 - \alpha_{1y}$ and $C_{Ty} = C_{0y} + C_{1y}$ and do some arithmetic with the result. Here we go.

$$-C_{0y} + \alpha_{0y}C_{Ty} = -C_{0y} + \left(1 - \alpha_{1y}\right)C_{Ty} = -C_{0y} + \overbrace{C_{0y} + C_{1y}}^{C_{Ty}} - \alpha_{1y}C_{Ty} = C_{1y} - \alpha_{1y}C_{Ty}$$

TABLE 4.1

Weak Acid Terms for Uncharged and Charged Weak Acids and Their Conjugate Bases

Uncharged Acid, HX	Charged Acid, HY^+
$C_{1x} - \dfrac{K_{ax}}{\left[H^+\right] + K_{ax}}\left(C_{0x} + C_{1x}\right)$	$-C_{1y} + \dfrac{\left[H^+\right]}{\left[H^+\right] + K_{ay}}\left(C_{0y} + C_{1y}\right) = C_{1y} - \dfrac{K_{ay}}{\left[H^+\right] + K_{ay}}\left(C_{0y} + C_{1y}\right)$

It follows that the term for the charged weak acid is the same as that for the uncharged acid. It also follows that **the weak-acid terms for uncharged and charged acids have the same forms**. This means that Equation 4.1b can be written as follows:

$$
\text{CBE} = \overbrace{\left[\text{H}^+\right] - \frac{K_w}{\left[\text{H}^+\right]}}^{\text{Water term}} + \overbrace{C_{\text{NaOH}} - C_{\text{HCl}}}^{\text{NaOH/HCl term}} + \overbrace{C_{1x} - \frac{K_{ax}}{\left[\text{H}^+\right] + K_{ax}}\left(C_{0x} + C_{1x}\right)}^{\text{Weak-acid term: HX/X term}}
$$

$$
+ \overbrace{C_{1y} - \frac{K_{ay}}{\left[\text{H}^+\right] + K_{ay}}\left(C_{0y} + C_{1y}\right)}^{\text{Weak-acid term: HY}^+\text{/Y term}} = 0
$$

(4.1c)

All symbols are as defined in Example 4.1.

This simplifies our lives in the sense that *it isn't necessary to differentiate between uncharged and charged acids until we begin to discuss effects of ionic strength*. We will show in the next chapter that the same simplification applies for polyprotic acids and bases.

EXTENSION OF EQUATION 4.1B TO INCLUDE THREE WEAK ACID-BASE PAIRS

A procedure similar to that described for two weak acids could be used to develop equations for solutions containing three or more weak acid-base pairs. However, that isn't necessary. One of the attractive features of Ricci's systematic approach is the ease with which *patterns in equations developed for one situation can be used to adapt the equations to other simpler and more complex situations*.

For example, Equation 4.1c is annotated to emphasize the fact that the equation consists of four sets of *additive terms*, i.e., one set of terms each for water, the strong acid/base pair, and each of the weak acid/base pairs. *This additive feature applies for any number of weak acid-base pairs.* Therefore, we can extend Equations 4.1a and 4.1b to include a third weak acid-base pair, e.g., HZ and NaZ or HZ$^+$ and Z by adding the terms in Table 4.2.

A complete equation for situations that include HCl, NaOH, and three charged or uncharged weak acid-base pairs can be written by adding these terms to Equation 4.1c. I will trust you to write the fractional form of the equation for situations that include three weak acid-base pairs. The computational form of the equation is summarized in Table 4.3.

The equation is annotated to identify the origin of each set of terms. It is more important that you understand the basis for the equation than that you try to remember it. However, if you remember the first three sets of terms, then you can extend the equation to additional weak acid-base pairs by repeating the third set of terms for each additional acid-base pair.

SOME USEFUL FEATURES OF EQUATION 4.3

The equation for two weak acid-base pairs was extended to an equation for three acid-base pairs by adding terms for the third pair, HZ and Z or HZ$^+$ and Z. Similarly, the equation can be adapted to simpler situations by setting appropriate concentrations to zero.

TABLE 4.2

Terms Used to Extend Equations 4.1a and 4.1b to Include a Third Acid-Base Pair, HZ and Z$^-$ or HZ$^+$ and Z

Fractional Term		Computational Term	
$C_{1z} - \alpha_{1z}\left(C_{0z} + C_{1z}\right)$	(4.2a)	$C_{1z} - \dfrac{K_{az}}{\left[\text{H}^+\right] + K_{az}}\left(C_{0z} + C_{1z}\right)$	(4.2b)

TABLE 4.3

Computational Form of the Charge-Balance Equation for Solutions Containing HCL, NaOH, and Three Weak Acid-Base Pairs

$$\left[H^+\right] - \overbrace{\frac{K_w}{\left[H^+\right]}}^{\left[OH^-\right]} + \overbrace{C_{NaOH} - C_{HCl}}^{NaOH, HCl} + \overbrace{C_{1x} - \frac{K_{ax}}{\left[H^+\right] + K_{ax}}\left(C_{0x} + C_{1x}\right)}^{Acid-base\ pair,\ HX,\ X} \cdots continue\ below$$

$$\cdots continue \cdots + \overbrace{C_{1y} - \frac{K_{ay}}{\left[H^+\right] + K_{ay}}\left(C_{0y} + C_{1y}\right)}^{Acid-base\ pair,\ HY,\ Y} + \overbrace{C_{1z} - \frac{K_{az}}{\left[H^+\right] + K_{az}}\left(C_{0z} + C_{1z}\right)}^{Acid-base\ pair,\ HZ,\ Z} = 0 \qquad (4.3)$$

C_{HCl} and C_{NaOH}: Diluted concentrations of HCl and NaOH.

C_{0x}, C_{1x}, C_{0y}, C_{1y}, C_{0z}, C_{1z}: Diluted concentrations of forms of acids X, Y, and Z that have lost $i = 0$ and 1 protons, respectively.

K_{ax}, K_{ay}, K_{az}: Deprotonation constants for acids X, Y, and Z.

For example, Equation 4.3 can be adapted to situations involving two weak acid-base pairs by setting $C_{0z} = C_{1z} = 0$. Similarly, the equation can be adapted to still simpler situations by setting appropriate concentrations to zero. In the extreme, the equation can be adapted to pure water by setting all concentrations to the right of the "water" term to zero.

EXERCISE 4.1

(A) Show that Equation 4.3 can be converted to the equation for two weak acid-base pairs by setting concentrations of one weak acid-base pair to zero, e.g., $C_{HZ} = C_Z = 0$. (B) Show that Equation 4.3 can be converted to the equation for one weak acid-base pair by setting concentrations of two weak acid-base pairs to zero, e.g., $C_{HZ} = C_Z = C_{HY} = C_Y = 0$. (C) Show that Equation 4.3 can be converted to the equation for pure water, i.e., $[H^+] = K_w^{1/2} = 10^{-7}$ M, by setting all reactant concentrations to zero, e.g., $C_{HCl} = C_{NaOH} = C_{HX} = C_X = C_{HY} = C_H = C_{HZ} = C_Z = 0$.

The foregoing features of the charge-balance equation can be used as the basis for a program to solve problems ranging from pure water to mixtures of concentrations equal to or larger than zero of HCl, NaOH, and three weak acid-base pairs of uncharged or charged acids. I will describe a spreadsheet program to do this.

SPREADSHEET PROGRAM FOR CALCULATIONS WITHOUT CORRECTION FOR IONIC STRENGTH

The spreadsheet program is based on a feature of the charge-balance equation discovered during the preparation of these chapters on acid-base equilibria. I discovered early in this process that *charge-balance equations written by equating algebraic sums of contributions of positive and negative ions to the charge to zero are negative for hydrogen ion concentrations less than the equilibrium concentration and positive for hydrogen ion concentrations larger than the equilibrium concentration.* This feature is used as a *convergence criterion* for iterative programs to solve Equation 4.3.

AN EXAMPLE

Let's use Example 4.2 to talk about some features of the program.

Example 4.2 Calculate hydrogen ion concentrations and pH's for solutions prepared by adding V_b = (A) 0.0 mL, (B) 5.0 mL, (C) 10.0 mL, (D) 15.0 mL, ..., 100 mL of 0.10 M NaOH in 5.0 mL increments to V_a = 25.0 mL of a solution containing 0.10 M each of dichloroacetic acid, acetic acid, and the ammonium ion.

Deprotonation constants: dichloroacetic acid, 0.050, acetic acid, 1.8×10^{-5}, ammonium ion, 5.7×10^{-10}.

This example illustrates calculations associated with titrations of mixtures of thee weak acids with a strong base, NaOH.

OPENING AND PREPARING THE PROGRAM

The program to solve Example 4.2 and others like it is on Sheet 2 of the program, *Monoprotic_Systematic*, on the programs website http://crcpress.com/9781138367227. To use the program, open it into a fresh Excel file and select Sheet 2. To confirm that iterative settings are as intended, select File/Options/Formulas and confirm the settings, (a) Workbook calculations, Automatic; (b) *Enable iterative calculations, check*; (c) *Maximum Iterations*, 2,000; and (d) *Maximum Change*, 1E-14.

TEMPLATES

Templates containing information to be copied and pasted into the input sections of some programs are included with the relevant programs. For example, a template in Cells A94–I102 of Sheet 2 of the monoprotic weak acids program can be used to restore the program to calculations for pure water. Other templates on that sheet can be used to illustrate applications to other combinations of acids and bases.

The templates are intended to serve two purposes. First, they are intended to help you solve some specific problems. Second, and perhaps more importantly, they are intended to serve as guides as to how you can adapt each program to situations other than those represented by the templates.

One reviewer suggested that each template should be on a separate page on which all calculations associated with the template can be done. To show how this would be done, I have included templates for four different situations on Sheets 7–10 of the monoprotic weak acids program. However, my preferred approach and the approach emphasized throughout the text is to include all templates relevant to each sheet of a program on the sheet to which they apply. Each template on a particular sheet is copied and pasted into the input section of the program on that sheet using the Paste (P) option. When all calculations relevant to each template are complete then the program can be closed or calculations related to another template can be done on the same sheet. This greatly reduces the number of sheets with which a user must become familiar.

TEMPLATE FOR PART A OF EXAMPLE 4.2

A template for Part A of Example 4.2 is in Cells K64–S72 on Sheet 2 of the monoprotic acid-base program. To use the template, copy it and paste it into Cells A22–I30 of the program using the Paste (P) option. Table 4.4 is a copy of the parts of the program in Rows 23–38 for Part A of the example with B24 set to 0.

Numbers in all cells correspond to control parameters in Row 23 and input concentrations, volumes, constants, etc. in Rows 24–30 before the problem is solved, i.e., with B24 set to 0.

To solve Part A of the example, i.e., V_b = 0, set B24 to 1 and wait for I36 to change from No to Yes. Hydrogen ion concentrations in E38 and E41 should be 5.01E-2 and 5.85E-2, respectively, and the pH's in F38 and F41 should be 1.30 and 1.23, respectively. To solve other parts of the example, set B24 to 0, set H30 to the desired volume, e.g., 5, 10 15, ... 95, 100, set B24 to 1, wait for I36 to

TABLE 4.4

Initial Setup and First Few Lines of a Program to Do Calculations for Part A of Example 4.2

A	B	C	D	E	F	G	H	I
Row 23	Control parameters							
RSet/Calc	0	Sig. Figs.	3	$C_{H,Initl.}$	1.00E-15	Dil., D, or Undil., U?		D
Rows 26, 27: Diluted, D, or undiluted, U, concentrations. (D or U in Cell I24)								
$C_{HCl}^{\,0}$	$C_{NaOH}^{\,0}$	$C_{0x}^{\,0}$	$C_{1x}^{\,0}$	$C_{0y}^{\,0}$	$C_{1y}^{\,0}$	$C_{0z}^{\,0}$	$C_{1z}^{\,0}$	
0.00E+00	1.00E-01	1.00E-01	0.00E+00	1.00E-01	0.00E+00	1.00E-01	0.00E+00	
Row 28	Deprotonation constants, charges on the acids, 0 or 1, and volumes							
K_{ax}	K_{ay}	K_{az}	z_{0x}	z_{0y}	z_{0z}	V_a	V_b	K_W
5.00E-02	1.80E-05	5.70E-10	0	0	1	25.00	0.00	1.00E-14
DON'T CHANGE ANYTHING BEYOND THIS ROW								
Row 32	Diluted concentrations							
C_{HCl}	C_{NaOH}	C_{0x}	C_{1x}	C_{0y}	C_{1y}	C_{0z}	C_{1z}	
0.00E+00	0.00E+00	1.00E-01	0.00E+00	1.00E-01	0.00E+00	1.00E-01	0.00E+00	
Row 35	Iterative calculations without correction for ionic strength							
Row 36	Low resolution step		High resolution step		Iterations complete?			No
Row 37	CBE_{LoRes}	$C_{H,LowRes}$	CBE_{HiRes}	[H⁺]	pH	Number of iterations		0
Row 38	-1.0E+01	1.00E-15	-1.0E+02	1.00E-16	16.00			

$C_{HCl}^{\,0}$ and $C_{NaOH}^{\,0}$: Initial HCl and NaOH concentrations.

$C_{0x}^{\,0}$, $C_{1x}^{\,0}$, $C_{0y}^{\,0}$, $C_{1y}^{\,0}$, $C_{0z}^{\,0}$, $C_{1z}^{\,0}$: Initial concentrations of forms of acids X, Y, and Z that have lost i = 0 and 1 protons, respectively.

K_{ax}, K_{ay}, and K_{az}: Deprotonation (dissociation) constants for acids X, Y, and Z.

$z_{0,x}$, $z_{0,y}$, z_{0z}: Charges on fully protonated forms of acids X, Y, and Z.

V_a: Volume of the strong acid, e.g., HCl; V_b: Volume of the strong base, e.g., NaOH.

C_{HCl}, C_{NaOH}: Diluted HCl and NaOH concentrations.

C_{0x}, C_{1x}, C_{0y}, C_{1y}, C_{0z}, C_{1z}: Diluted concentrations of forms of acids X, Y, and Z that have lost i = 0 and 1 protons, respectively.

CBE_{LoRes}, CBE_{HiRes}: Charge-balance equations for the low resolution and high resolution steps.

$C_{H,LoRes}$: Trial hydrogen ion concentrations for the low resolution step.

[H⁺] and pH: Equilibrium H⁺ concentration and pH.

change to Yes, and record hydrogen ion concentrations and pH's. pH's I obtained for NaOH volumes from 0 to 100 mL without and with correction for ionic strength are in Cells AD33–AE100. Results for pH vs. the NaOH volume are plotted as the ascending plots in the figure in Cells AC4–AI25. As you can see, there is very little difference between plots of results without and with correction for ionic strength.

A Two-Step Procedure to Resolve Hydrogen Ion Concentrations

Calculations such as those described above are complicated by the fact that hydrogen ion concentrations for situations of interest can vary over very large ranges, e.g., 10^{-14} M for 1.0 M NaOH to 1.0 M for 1.0 M HCl. To obtain acceptable resolution of the H⁺ concentration within a reasonable time, the iterative process is implemented in two stages in the spreadsheet program.

The first stage is a *low resolution step* in which the program finds the upper and lower limits of a 10-fold range of concentrations that includes the equilibrium concentration, e.g., 10^{-3} to 10^{-2} M. The second stage is a *high resolution step* in which the program resolves the H⁺ concentration to a number of user-selected significant figures in D24. The user can set the number of significant figures in D24 to any desired value. However, three significant figures is a good compromise between limitations imposed by uncertainties in deprotonation constants and convergence time.

Low Resolution Step

The *low resolution step* is implemented in Cells B38 and C38. The low resolution step starts with a trial value of the H^+ concentration equal to 10^{-15} M in F24 and increases it in 10-fold steps until the sign of charge-balance equation in B38 changes from $-$ to $+$. The $-$ to $+$ change causes this part of the program to stop with the H^+ concentration, e.g., 10^{-1} M, that caused the $-$ to $+$ change in Cell C38. Given that charge-balance equation should be zero at equilibrium, it follows that the equilibrium H^+ concentration is between the start and end of the $-$ to $+$ change, i.e., 1.0×10^{-2} M and 1.0×10^{-1} M in this case.

High Resolution Step

With D24 set to 3 significant figures, the high resolution step starts with 10% of the upper limit concentration from E38 in E41 and increases it by amounts equal to $10^{-3}(0.1$ M$) = 10^{-4}$ M until the sign of the charge-balance equation in D38 changes from $-$ to $+$. The $-$ to $+$ change causes this part of the program to stop with the H^+ concentration that produced the $-$ to $+$ change, i.e., 5.01E-2 M, in E38. It follows that the H^+ concentration has been resolved to three significant figures, i.e., 5.01E-2.

The same principles apply for other H^+ concentrations. Whatever the hydrogen ion concentration, the low resolution step will stop with the upper limit of a 10-fold range that includes the equilibrium concentration in C38 and the high resolution step will resolve concentrations in that range to the number of significant figures in D24. Moreover, the same general approach is used for polyprotic acids and bases discussed in a later chapter.

With D24 set to 3, percentage errors in hydrogen ion concentrations resulting solely from the iterative process will vary from 1% at the lower end of each 10-fold range 0.1% at the upper limit of each 10-fold range. These errors are well within uncertainties associated with constants used in the calculations.

Let's talk about contents of some cells in the program.

CONTENTS OF SELECTED CELLS IN THE MONOPROTIC WEAK-ACID PROGRAM

Table 4.5 summarizes contents of selected cells in the spreadsheet program on Sheet 2 of the overall program, *Monoprotic_Systematic*.

Cell I24 is used to determine if concentrations in Row 27 are treated as diluted concentrations or undiluted concentrations. If I24 is set to D, then the program treats concentrations in Row 27 as diluted concentrations. If I24 is set to U as for Example 4.2, the program calculates diluted concentrations using volumes, V_a and V_b, of acids and bases in Cells G30 and H30 and concentrations in Row 27.

Algorithms in C38 and E38 control increments in the hydrogen ion concentration used to find the H^+ concentration that satisfies the charge-balance equation. As shown in bold type, the algorithm in C38 increases trial values of the hydrogen ion concentration in 10-fold increments until the sign

---

TABLE 4.5

Contents of Selected Cells for the Part of the Spreadsheet Program Used to Solve the Charge-Balance Equation for the Hydrogen Ion Concentration without Correction for Ionic Strength

A34 =IF($I24="D",A27,$G30*A27/($G30+$H30))   B34 =IF($I24="D",B27,$H30*B27/($G30+$H30))

C38 =IF(B24=0,F24,IF(B38<0,**10*C38**,C38))   E38 =IF(B38<0,0.1*C38,IF(D38<0,**E38+10^-D24*C38**,E38))

B38 =C38-I30/C38+(B34-A34)+(D34+F34+H34)-((A30/(C38+A30))*(C34+D34)+(B30/(C38+B30))*(E34+F34)
   +(C30/(C38+C30))*(G34+H34))

D38 =E38-I30/E38+(B34-A34)+(D34+F34+H34)-((A30/(E38+A30))*(C34+D34)+(B30/(E38+B30))*(E34+F34)
   +(C30/(E38+C30))*(G34+H34))

of the charge-balance equation changes from − to +. This gives the upper limit of a 10-fold range of hydrogen ion concentrations that includes the equilibrium concentration.

Algorithms in D38 and E38 resolve the hydrogen ion concentration within the 10-fold range to the number of significant figures in D24. It does this by increasing trial concentrations by increments equal to 10^{-D24} times the upper-limit concentration found in the preceding step. For example, with the significant figures in D24 set to 3 and the upper limit concentration in C38 equal to 0.10 M, the hydrogen ion concentration is resolved to $\Delta[H^+] = 10^{-3}(0.10\ M) = 10^{-4}\ M$. If the upper limit concentration in C38 were equal to $1.0 \times 10^{-9}\ M$, the hydrogen ion concentration in E38 would be resolved to $\Delta[H^+] = 10^{-3}(1 \times 10^{-9}\ M) = 10^{-12}\ M$. Analogous degrees of resolution apply to other situations.

Cells B38 and D38 contain Excel versions of the charge-balance equations for the low resolution and high resolution steps described above. An annotated version of the equation is included below.

$$\overbrace{\left[H^+\right]}^{C38,E38} - \overbrace{\frac{K_w}{\left[H^+\right]}}^{I30} + \overbrace{C_{NaOH} - C_{HCl}}^{B34-A34} + \overbrace{C_{1x}}^{A30} - \overbrace{\frac{K_{ax}}{\left[H^+\right]+K_{ax}}}^{A30}\left(\overbrace{C_{0x}}^{C34} + \overbrace{C_{1x}}^{D34}\right) \cdots continue\ below$$

$$(4.3)$$

$$\cdots continue \cdots C_{1y} - \frac{K_{ay}}{\left[H^+\right]+K_{ay}}\left(C_{0y} + C_{1y}\right) + C_{1z} - \frac{K_{az}}{\left[H^+\right]+K_{az}}\left(C_{0z} + C_{1z}\right) = 0$$

I have correlated some of the terms in the equation with cells in the program. I will trust you to correlate other terms with other cell numbers. Hopefully, this will give you a somewhat better insight into the inner workings of the program.

When you have done calculations for volumes in H30 from $V_b = 0.0$ mL to 100 mL at 5.0 mL increments, you will have sufficient information to plot titration curves, albeit curves that are poorly resolved near equivalence points. You can obtain more fully resolved plots by doing calculations using 1-mL increments just before and just after each of the three equivalence points at 25, 50, and 75 mL. Either way, I will describe an extension of the program on Sheet 4 later that is designed to calculate and plot titration curves for much smaller increments in titrant volumes.

You can use results in Exercise 4.2A to confirm that you are using the program correctly.

EXERCISE 4.2A
Use the program to confirm hydrogen ion concentrations for the NaOH volumes below for Example 4.2.

Volume, V_b	0	25	50	75	85
[H⁺], Example 4.2	5.01E-2	6.65E-4	1.02E-7	1.54E-11	1.10E-12

TITRATIONS OF MIXTURES OF WEAK BASES WITH A STRONG ACID

Example 4.3 involves titration of mixtures of weak bases with hydrochloric acid.

Example 4.3 Calculate hydrogen ion concentrations and pH's for solutions prepared by adding V_a = (A) 0.0 mL, (B) 5.0 mL, (C) 10.0 mL, (D) 15.0, ..., 100 mL of 0.10 M HCl in 5.0 mL increments to $V_b = 25.0$ mL of a solution containing 0.10 M each of dichloroacetate, acetate, and ammonia.

Deprotonation constants: dichloroacetic acid, $K_{ax} = 0.050$, acetate ion, $K_{ay} = 1.8 \times 10^{-5}$, and ammonia, $K_{az} = 5.7 \times 10^{-10}$.

A template for the input part of the monoprotic acids program for Part A of this example is included in Cells K79–S87 on Sheet 2 of the monoprotic weak-acid program. You can copy the template and paste it into Cells A22–I30 if you wish. However, I suggest that you will learn more by entering information for the example into the input section of the program and using the template to check your entries.

In any event, having completed the input section for Part A of this example, you should proceed as described earlier for Example 4.2 except that you vary V_a rather than V_b.

You can use Exercise 4.2B to confirm that you are using the program correctly.

EXERCISE 4.2B

Use the program to confirm hydrogen ion concentrations for the HCl volumes below for Example 4.3 without correction for ionic strength.

Volume, V_a	0	25	50	75	85
[H^+], Example 4.3	7.60E-12	1.02E-7	5.93E-5	1.84E-2	2.44E-2

More complete values of pH vs. HCl volumes without and with correction for ionic strength are in Cells AF33–AG59. Plots of pH vs HCl volume are in the figure in Cells AC4–AI25 as well as in Figure 4.1B discussed earlier.

CORRECTION FOR IONIC STRENGTH

The part of the spreadsheet program to account for effects of ionic strength is included in Rows 40–50, Columns A–I on Sheet 2 of the program, *Monoprotic_Systematic*.

Except for Cells D30–F30, input information is the same as that described above for calculations without correction for ionic strength. Cells D30–F30 are used for charges, z_{0x}, z_{0y}, and z_{0z}, on the fully protonated acids. As examples, charges on dichloroacetic acid, acetic acid, and ammonium ion are 0, 0, and 1, respectively.

I'll show you how to solve an example and then describe the basis for the part of the program that accounts for effects of ionic strength.

Example 4.4 Compare the hydrogen ion concentration and pH calculated without and with correction for ionic strength in a solution containing 0.10 M sodium hydroxide and 0.10 M protonated hydroxylamine, $K_a = 1.3 \times 10^{-6}$.

Set I24 to D telling the program that calculations are to be done for diluted solutions. Then set B24, A27–H27, and A30–F30 to zero. Then set B27 and C27 to 0.10 each, A30 to the value of the deprotonation constant, 1.3E-6, and D30 to 1 to tell the program that protonated hydroxylamine is a singly charged acid. Then set B24 to 1 and wait for I36 to change to Yes.

As shown in Cells K27 and L27, hydrogen ion concentrations calculated without and with correction for ionic strength, 3.61×10^{-10} M and 4.46×10^{-10} M, respectively, differ by 19% and corresponding pH's, 9.44 and 9.35, respectively, differ by about 1%.

The smaller difference for pH's results from the logarithmic relationship between concentration and pH. *For monoprotic acids and bases, ionic strength will seldom have effects on pH's larger than a few percent but often have much larger effects on hydrogen ion concentrations.*

BASIS FOR CALCULATIONS TO CORRECT FOR IONIC STRENGTH

Effects of ionic strength can be taken into account by replacing *activity-based constants*, commonly called *thermodynamic constants*, with *concentration-based constants*, commonly called *conditional constants*. After describing the basis for procedures used to calculate concentration-based constants, I will talk about how the program in Table 4.4 is extended to account for those effects.

Calculation Options and Some Related Symbols

A three-step process is used to correct hydrogen ion concentrations and pH for effects of ionic strength. The first step is to use reactant concentrations with tabulated values of *activity-based constants*, more commonly called *thermodynamic constants*, to calculate hydrogen ion concentrations and pH. Hydrogen ion concentrations and pH's calculated in this way are represented herein by the symbols, $[H^+]$ and $pH = -\log [H^+]$.

Concentrations calculated in this way are used to calculate the ionic strength that in turn is used to calculate activity coefficients that in turn are used to calculate values of deprotonation constants and concentration fractions corrected for effects of ionic strength. The program described herein uses an iterative process to repeat these calculations many times to account for dependencies of several quantities, including equilibrium concentrations, deprotonation constants and concentration fractions on each other. Quantities calculated in this way are represented by subscripted μ's to indicate that they are adjusted for effects of ionic strength. As examples, the hydrogen ion concentration and pH obtained in this step are represented herein by $[H^+]_\mu$ and $pH_\mu = -\log [H^+]_\mu$.

The third step is to use activity coefficients calculated in the foregoing option to calculate hydrogen ion activity and the activity-based pH. Symbols and equations used herein for hydrogen ion activity and activity-based pH are $\{H^+\} = f_H[H^+]_\mu$ and $p\{H\} = -\log \{H^+\} = -\log(f_H[H^+]_\mu)$.

Quantities used in this process are described below.

Activity Coefficients

Activity-based constants are represented herein by the usual symbols, K_{ax}, K_{ay}, K_{az}, and K_w for acids X, Y, Z, and water. Concentration-based constants are represented herein by the symbols, $K_{ax\mu}$, $K_{ay\mu}$, $K_{az\mu}$, and $K_{w\mu}$, respectively. The subscripted μs indicate that the constants include effects of ionic strength. Activity coefficients are used to convert activity-based constants to concentration-based constants and ionic strength is used to calculate activity coefficients.

Activity coefficients, f_i, are related to charges, z_i, on ions and ionic strength as follows:

$$f_i = 10 \wedge \left(\frac{-0.51 z_i^2 \sqrt{\mu}}{1 + 3.3 D \sqrt{\mu}} \right) \qquad (4.4)$$

in which f_i is the activity coefficient, z_i is the charge on an ion, μ_i is the ionic strength, and D_i is the hydrated diameter[5].

Given that hydrated diameters are not known for all ions associated with monoprotic weak acids and that activity coefficients are relatively insensitive to changes in hydrated diameters, an average value of $D_{Av} \cong 0.5$ nm is used for all ions except hydrogen ions and hydroxide ions. Hydrated diameters for the H^+ and OH^- ions are $D_H = 0.9$ nm and $D_{OH} = 0.3$ nm[5]. *Activity coefficients for hydrogen ions and hydroxide ions are represented by symbols f_H and f_{OH}, respectively; activity coefficients for all other ions are represented by f_{Av} based on an average value of the hydrated diameter, $D_{Av} = 0.5$. Errors resulting from this approximation are likely significantly less than errors that result when effects of ionic strength are ignored.*

Ionic Strength

Ionic strength, μ, is one half the sum of products of charges squared times concentrations of the ions. Given that monoprotic acids and bases involve singly charged ions, *the ionic strength of a solution containing monoprotic acids and bases and their salts is one half the sum of the concentrations of positive and negative ions.* By setting $z_i = 0$ in Equation 4.4 for uncharged reactants, you can show that uncharged reactants and products do not influence ionic strengths in the range for which this equation is valid.

As an example, consider a solution containing 0.035 hydrochloric acid and 0.025 M sodium chloride. Reactions and concentrations are

$$\overset{0.035}{\overbrace{HCl}} \rightarrow \overset{0.035}{\overbrace{H^+}} + \overset{0.035}{\overbrace{Cl^-}} \qquad \overset{0.025}{\overbrace{NaCl}} \rightarrow \overset{0.025}{\overbrace{Na^+}} + \overset{0.025}{\overbrace{Cl^-}}$$

It follows that the ionic strength is

$$\mu = 0.5\left(\left[H^+\right] + \left[Na^+\right] + \left[Cl^-\right]\right) = 0.5(0.035 + 0.025 + 0.035 + 0.025) = 0.060 \text{ M}$$

Converting Activity Based, or Thermodynamic, Constants to Concentration-Based, or Conditional, Constants

As is common practice, the equilibrium concentration of any reactant, X, is represented herein by the symbol, [X]. The activity of any reactant, X, is represented as the reactant in braces, i.e., {X}. The activity of any reactant, X, is the activity coefficient, f_X, times the equilibrium concentration, as expressed in Equation 4.5a:

$$\{X\} = f_x[X] \tag{4.5a}$$

For any reaction such as the following,

$$\overset{f_a}{\overbrace{A}} \rightleftharpoons \overset{f_b}{\overbrace{B}} + \overset{f_c}{\overbrace{C}}$$

the activity-based constant, K_a, and concentration-based constant, $K_{a\mu}$, are defined as follows:

$$K_a = \frac{\{B\}\{C\}}{\{A\}}; \quad K_{a\mu} = \frac{[B][C]}{[A]}$$

in which quantities in braces, {A}, {B}, and {C} are equilibrium activities and quantities in square brackets, [A], [B], and [C], are equilibrium concentrations.

A relationship between activity-based and concentration-based constants can be obtained by replacing activities in the activity-based constant with products of activity coefficients times concentrations. For example, the relationship between activity-based and concentration-based constants for the reaction above is expressed in Equation 4.5b:

$$K_a = \frac{\{B\}\{C\}}{\{A\}} = \frac{f_b[B]f_c[C]}{f_a[A]} = \frac{f_b[B]f_c[C]}{f_a[A]} = \frac{f_b f_c}{f_a}\frac{[B][C]}{[A]} = \frac{f_b f_c}{f_a}K_{a\mu} \tag{4.5b}$$

in which K_a is the activity-based constant, {A}, {B}, and {C} are activities of the reactant and products, f_a, f_b, and f_c are activity coefficients, and $K_{a\mu}$ is the concentration-based constant.

The first and last terms in the foregoing equation can be rearranged into the form

$$K_{a\mu} = \frac{f_a}{f_b f_c}K_a \tag{4.5c}$$

in which symbols are as described above.

Unless stated otherwise, tabulated deprotonation constants are activity-based constants. As shown in Equation 4.5c, concentration-based constants are calculated using activity coefficients and activity-based constants.

Now, let's apply these concepts to water, acetic acid, an uncharged acid, and the ammonium ion, a singly charged acid. Results are summarized below in Equations 4.5d–4.5f.

$$\overset{f_{H_2O}=1}{\overbrace{H_2O}} \overset{K_w}{\rightleftharpoons} \overset{f_H}{\overbrace{H^+}} + \overset{f_{OH}}{\overbrace{OH^-}} \quad K_{w,\mu} = \frac{f_{H_2O}}{f_H f_{OH}} \frac{[H^+][OH^-]}{[H_2O]} \overset{f_{H_2O}=1}{\cong} \frac{1}{f_H f_{OH}} K_w \tag{4.5d}$$

$$\overset{f_{HAc}=1}{\overbrace{HAc}} \overset{K_a}{\rightleftharpoons} \overset{f_H}{\overbrace{H^+}} + \overset{f_{Ac}}{\overbrace{Ac^-}} \quad K_{a,\mu} = \frac{f_{HAc}}{f_H f_{Ac}} \frac{[H^+][Ac^-]}{[HAc]} \overset{f_{HAc}=1}{\cong} \frac{1}{f_H f_{Ac}} K_a \tag{4.5e}$$

$$\overset{f_{NH4}}{\overbrace{NH_4^+}} \overset{K_a}{\rightleftharpoons} \overset{f_H}{\overbrace{H^+}} + \overset{f_{NH3}=1}{\overbrace{NH_3}} \quad K_{a\mu} = \frac{f_{NH4}}{f_H f_{NH3}} \frac{[H^+][NH_3]}{[NH_4^+]} \overset{f_{NH3}=1}{\cong} \frac{f_{NH4}}{f_H} K_a \tag{4.5f}$$

Constants defined in this way can be used to write equations for concentration fractions based exclusively on concentration-based constants and the hydrogen ion concentration corrected for effects of ionic strength. Equations 4.5g and 4.5h are as follows:

$$\alpha_{0\mu} = \frac{[H^+]_\mu}{[H^+]_\mu + K_{a,\mu}} \tag{4.5g}$$

$$\alpha_{1\mu} = \frac{K_{a,\mu}}{[H^+]_\mu + K_{a,\mu}} \tag{4.5h}$$

in which subscripted µs identify quantities corrected for ionic strength.

Overview of the Program

The program to be discussed is designed to do calculations for solutions containing concentrations equal to or larger than zero of sodium hydroxide, hydrochloric acid, and acid-base pairs of three sets of uncharged **or** charged acids, HX or HX$^+$, HY or HY$^+$, and HZ or HZ$^+$. To understand and implement the program, we need to develop equations for ionic strength and a charge-balance equation. To simplify our discussion, we will begin with solutions containing sodium hydroxide, hydrochloric acid, and acid-base pairs of **either** acetic acid, HAc, **or** ammonium ion, NH$_4^+$. Having developed the equation for this situation, we will show you how the equation can be extended to three sets of uncharged or charged acids.

Finally, given that *only one set of concentrations of each charged or uncharged acid-base pair will be included in each solution, the program is designed to select reactants and products associated with the uncharged or charged acid for each pair. For example, using acetic acid and the ammonium ion as examples of Acid X, the program will do calculations for acetic acid and acetate ion if z_{0x} in D30 is set to 0 and it will do calculations for ammonium ion and ammonia if z_{0X} in D30 is set to 1.*

Ionic Strength

The program to be discussed is designed to do calculations for solutions containing concentrations equal to or larger than zero of NaOH, HCl and acid-base pairs of three sets of uncharged or charged acids, HX or HX$^+$, HY or HY$^+$, and HZ or HZ$^+$.

We will talk first about the ionic strength in a solution containing sodium hydroxide, hydrochloric acid and **either** sodium acetate **or** ammonium chloride and then extend the equation to the salts of two additional sets of uncharged or charged acids. Salts of the acids are used because they produce all ions associated with each acid except the hydrogen ion that is accounted for independently. Example 4.5A summarizes a process used to develop an equation for the ionic strength for such situations.

Example 4.5A Develop an equation for the ionic strength in a solution containing diluted concentrations C_{HCl} of hydrochloric acid and C_{NaOH} of sodium hydroxide plus *either* C_{HAc} and C_{NaAc} of acetic acid and sodium acetate ion *or* C_{NH_4Cl} and C_{NH_3} of ammonium chloride and ammonia.

Step 1: Account for hydrogen ion and hydroxide ion concentrations *from all sources.*

$$0.5 * \left(\left[H^+ \right] + \left[OH^- \right] \right) = 0.5 * \left(\left[H^+ \right] + \frac{K_w}{\left[H^+ \right]} \right)$$

Step 2: Account for contributions from hydrochloric acid and sodium hydroxide by assuming that each dissociates completely.

$$NaOH \rightarrow Na^+ + OH^-; HCl \rightarrow H^+ + Cl^- \qquad 0.5 \left(\left[Na^+ \right]_{NaOH} + \left[Cl^- \right]_{HCl} \right) = 0.5 \left(C_{NaOH} + C_{HCl} \right)$$

Step 3: If sodium acetate is included, account for it by assuming that it dissociates completely as shown below.

$$NaAc \rightarrow Na^+ + Ac^- \qquad 0.5 \left(\left[Na^+ \right]_{NaAc} + \left[Ac^- \right] \right) = 0.5 \left(C_{NaAc} + \alpha_{Ac} \left(C_{HAc} + C_{NaAc} \right) \right)$$

Step 4: If ammonium chloride is included, account for it by assuming that it dissociates completely as shown below.

$$NH_4Cl \rightarrow Cl^- + NH_4^+ \qquad 0.5 \left(\left[Cl^- \right]_{NH_4Cl} + \left[NH_4^+ \right] \right) = 0.5 \left(C_{NH_4Cl} + \alpha_{NH_4} \left(C_{NH_4Cl} + C_{NH_3} \right) \right)$$

Step 5: Combine terms for sodium acetate and ammonium chloride.

$$0.5 \left(C_{NaAc} + \alpha_{Ac} \left(C_{HAc} + C_{NaAc} \right) \text{ or } C_{NH_4Cl} + \alpha_{NH_4} \left(C_{NH_4Cl} + C_{NH_3} \right) \right)$$

Step 6: Combine all the foregoing terms by using If/Then logic to choose between terms for sodium acetate or ammonium chloride.

$$\mu = 0.5 * \left\{ \left[H^+ \right] + \frac{K_{w\mu}}{\left[H^+ \right]} + \left(C_{NaOH} + C_{HCl} \right) + IF \left[z_{0HAc} = 0, C_{NaAc} + \alpha_{Ac\mu} \left(C_{HAc} + C_{NaAc} \right), \right. \right.$$

$$\left. \left. C_{NH_4Cl} + \alpha_{NH_4\mu} \left(C_{NH_4Cl} + C_{NH_3} \right) \right] \right\}$$

Note: In Excel syntax, the comma between terms, $(C_{HAc} + C_{NaAc})$ and , translates to "otherwise" or more explicitly, If $(z_{NH_4} = 1, C_{NH_4Cl} + \cdots)$.

Step 7: Rewrite the equation using Ricci's i notation and subscripted x's to generalize the equation for the generic acids, HX or HX$^+$, and the conjugate bases.

$$\mu = 0.5 * \left(\left[H^+ \right] + \frac{K_{w\mu}}{\left[H^+ \right]} + \left(C_{NaOH} + C_{HCl} \right) + IF \left(z_{0x} = 0, C_{1x} + \alpha_{1x\mu} \left(C_{0x} + C_{1x} \right), C_{0x} + \alpha_{0x\mu} \left(C_{0x} + C_{1x} \right) \right) \right)$$

Step 8: Add the terms below to extend the equation to two additional sets of uncharged or charged acids, HY or HY$^+$ and HZ or HZ$^+$, and their conjugate bases.

$$+ IF \left(z_{0y} = 0, C_{1y} + \alpha_{1y\mu} \left(C_{0y} + C_{1y} \right), C_{0y\mu} + \alpha_{0y\mu} \left(C_{0y} + C_{1y} \right) \right)$$

$$+ IF \left(z_{0z} = 0, C_{1z} + \alpha_{1z\mu} \left(C_{0z} + C_{1z} \right), C_{0z} + \alpha_{0z} \left(C_{0z} + C_{1z} \right) \right)$$

Step 9: Complete the equation by adding terms for two additional sets of charged or uncharged acids and their conjugate bases (see Equation 4.6).

$$\mu = 0.5 * \left(\left[H^+ \right] + \frac{K_{w\mu}}{\left[H^+ \right]} + \left(C_{NaOH} + C_{HCl} \right) + IF \left(z_{0x} = 0, C_{1x} + \alpha_{1x\mu} \left(C_{0x} + C_{1x} \right), \right. \right.$$

$$C_{0x} + \alpha_{0x\mu} \left(C_{0x} + C_{1x} \right) \right) + IF \left(z_{0y} = 0, C_{1y} + \alpha_{1y\mu} \left(C_{0y} + C_{1y} \right), C_{0y} + \alpha_{0y\mu} \left(C_{0y} + C_{1y} \right) \right) \quad (4.6)$$

$$\left. + IF \left(z_{0z} = 0, C_{1z} + \alpha_{1z\mu} \left(C_{0z} + C_{1z} \right), C_{0z} + \alpha_{0z\mu} \left(C_{0z} + C_{1z} \right) \right) \right)$$

As noted earlier, quantities with subscripted μs are adjusted for effects of ionic strength.

Step 1 shows how hydrogen ion and hydroxide ion concentrations from all sources are taken into account. Given that the hydrogen ion concentration is calculated by the iterative program, the hydroxide ion concentration is calculated as the autoprotolysis constant for water divided by the hydrogen ion concentration.

Assuming that sodium hydroxide and hydrochloric acid dissociate completely as shown in Step 2, the sodium ion and chloride ion concentrations are equal to the diluted NaOH and HCl concentrations.

Assuming that sodium acetate dissociates completely as shown in Step 3, the sodium ion concentration from the salt will be equal to the salt concentration and the acetate concentration will be equal to the fraction, α_{Ac}, times the sum of concentrations of acetic acid and sodium acetate.

Assuming that ammonium chloride dissociates completely as shown in Step 4, the chloride ion concentration will be equal to the salt concentration and the ammonium ion concentration will be equal to the fraction, α_{NH_4}, times the sum of concentrations of ammonium chloride and ammonia.

Step 5 is included to emphasize the fact that one term **or** the other is included for each situation.

Step 6 shows how charges on the fully protonated acids are used to choose the reactant concentrations used to calculate the ionic strength.

All the terms are combined in Step 7 using braces, {}, square brackets, [], and parentheses to help you identify which sets of terms are in which groups. As noted in the example, the comma between the $(C_{HAc} + C_{NaAc})$ and C_{NH_4} terms translates to "otherwise" in Excel syntax. In other words, the combined terms choose between contributions of the acetate ion for the acetic acid/acetate pair and the ammonium ion for the ammonium ion/ammonia pair.

Regarding Step 7, remember, in Ricci's i notation, subscripted 0's and 1's refer to forms of acids that have lost 0 and 1 protons, respectively. Subscripted x's refer to acids, HX or HX$^+$.

Steps 8 and 9 show how the equation for one acid, HX or HX$^+$, and its conjugate base can be extended to two additional sets uncharged or charged acids, HY or HY$^+$ and HZ or HZ$^+$ and their conjugate bases.

The complete equation is implemented in Cell B44 of the program, *Monoprotic_Systematic*.

Calculations used to account for effects of ionic strength are in Cells A40–I50. All the quantities in Cells A40–I50 depend on each other. If you change one you change all the others. This means that it isn't possible to describe the overall process in a simple set of sequential steps with a bow on top. However, we can talk about the process.

Let's begin with the hydrogen ion concentration calculated in Cell E50. As long as the part of the program without correction for ionic strength is searching for the upper limit of the 10-fold range of concentrations that includes the ionic strength, the hydrogen ion concentration in E41 is set to 1% of the upper limit concentration in C38. When the part of the program without correction for ionic strength finds the upper limit concentration, the part of the program in Cells A40–I50 begins to increase the hydrogen ion concentration in E50 by increments determined by the number of significant figures in D24.

As the hydrogen ion concentration in E50 is increased, it causes concentration fractions in B47–G47 to change. As these fractions change, concentrations of the hydrogen ion, the hydroxide ion, and different forms of the acids change. These changes cause the ionic strength in B44 to change. As the ionic strength in B44 changes, this causes activity coefficients in B50–I50 to change, Changes in activity coefficients produce changes in the ionic-strength-dependent constants in C44–F44. Changes in these constants produce changes in concentration fractions and all other quantities in these cells.

The sign of the charge-balance equation in D41 will be negative when the foregoing process begins. As the hydrogen ion concentration in E41 is increased, the foregoing processes cause the charge-balance equation to become less negative until eventually its sign changes from $-$ to $+$. The $-$ to $+$ sign change corresponds to a hydrogen ion concentration for which the charge-balance equation is very close to zero. The hydrogen ion concentration that causes the $-$ to $+$ change is accepted as the equilibrium concentration resolved to the number of significant figures in D24.

Concentrations and pH corrected for effects of ionic strength are summarized in Cells K26–R26. Corresponding fractions of different forms of the weak acids are summarized in Cells M34–R34. The hydrogen ion activity, $\{H^+\}$, and activity-based pH, $p\{H\}$, are in G41 and H41.

The ionic strength calculated in Cell B44 is used to calculate activity coefficients of the ions in Cells B50–I50 using the Debye and Hückel equation. Activity coefficients calculated in this way are used in Cells C44–F44 to calculate equilibrium constants adjusted for effects of ionic strength. Constants for the weak acids calculated in this way are used in Cells B47–G47 with the hydrogen ion concentration calculated in E41 to calculate fractions of the weak acids in different forms. Those fractions are used with diluted concentrations of different forms of the weak acids to solve the charge-balance equation in Cell D41.

The hydrogen ion concentration in E41 starts at 1% of the hydrogen ion concentration calculated in C38 without correction for ionic strength and is increased stepwise by amounts determined by the number of significant figures in Cell D24. For example, if D24 is set to 3, then the hydrogen ion concentration in C38 will be increased by 10^{-3} times the upper limit concentration.

Given that the reactions and ions are the same as those for Example 4.5A, I will use an abbreviated approach to develop the charge-balance equation.

Example 4.5B Develop a charge-balance equation to account for effects of ionic strength in a solution containing concentrations, C_{HCl} of hydrochloric acid and C_{NaOH} of sodium hydroxide plus C_{HAc} of acetic acid and C_{NaAc} of acetate ion *or* C_{NH_4Cl} of ammonium chloride and C_{NH_3} of ammonia and show how the equation can be adapted to charged or uncharged generic acids HX or HX$^+$, HY or HY$^+$, and HZ or HZ$^+$.

Step 1: Write reactions needed to identify all ions.

$$H_2O \overset{K_W}{\rightleftharpoons} H^+ + OH^-, \quad HCl \rightarrow H^+ + Cl^-, \quad NaOH \rightarrow Na^+ + OH^-$$

$$\overset{z_{0,HAc}=0}{\overbrace{NaAc}} \rightarrow Na^+ + Ac^- \quad or \quad \overset{z_{0NH4}=1}{\overbrace{NH_4Cl}} \rightarrow Cl^- + NH_4^+$$

Step 2: Write the ionic form of a charge-balance equation to account for effects of ionic strength for the solution described above.

$$CBE = \left(\left[H^+\right] - \left[OH^-\right] + \left[Na^+\right] - \left[Cl^-\right] + IF\left(z_{0HAc} = 0, \left[Na^+\right] - \left[Ac^-\right], -\left[Cl^-\right] + \left[NH_4^+\right]\right)\right)$$

Step 3: Write a fractional form of the equation.

$$CBE = \left[H^+\right] - \frac{K_w}{\left[H^+\right]} + \left(C_{NaOH} - C_{HCl}\right) + IF\left(z_{0HAc} = 0, C_{NaAc} - \alpha_{Ac}\left(C_{HAc} + C_{NaAc}\right),\right.$$

$$\left. -C_{NH_4Cl} + \alpha_{0NH_4}\left(C_{NH_4Cl} + C_{NH_3}\right)\right)$$

Step 4: Use Ricci's i notation to convert the foregoing equation to a form in terms of an uncharged acid, HX, such as acetic acid **or** a charged acid, HX$^+$, such as the ammonium ion.

$$CBE = \left[H^+\right] + \frac{K_w}{\left[H^+\right]} + \left(C_{NaOH} - C_{HCl}\right) + IF\left(z_{0x} = 0, C_{1x} - \alpha_{1x}\left(C_{0x} + C_{1x}\right),\right.$$

$$\left. -C_{0x} + \alpha_{0x}\left(C_{0x} + C_{1x}\right)\right)$$

(4.7a)

Step 5: Use the relationship, $\alpha_{0x} = 1 - \alpha_{1x}$, to convert Equation 4.7a to Equation 4.7b as was done earlier to convert Equation 4.1b to Equation 4.1c.

$$CBE = \left[H^+\right] + \frac{K_w}{\left[H^+\right]} + \left(C_{NaOH} - C_{HCl}\right) + C_{1x} - \alpha_{1x}\left(C_{0x} + C_{1x}\right)$$

(4.7b)

Step 6: Write terms showing how the foregoing equation can be extended to two additional pairs of uncharged or charged acids, HY or HY$^+$ and HZ or HZ$^+$, by changing subscripted X's to subscripted Y's and Z's.

$$C_{1y} - \alpha_{1y}\left(C_{0y} + C_{1y}\right) + C_{1z} - \alpha_{1z}\left(C_{0z} + C_{1z}\right)$$

I will trust you to combine the terms to obtain the complete equation.

Except for the need to account for different signs on the ions, the process is very similar to that used to develop an equation for the ionic strength and is not discussed further.

IMPLEMENTING THE EQUATIONS

Calculations used to implement the equations described above are in Rows 39–50 of Sheet 2 of the monoprotic acids program, *Monoprotic_Systematic*.

As shown in Step 2 of Example 4.5A, the complete equation for ionic strength is the sum of a term that is the same for all weak acids and one or more terms that are unique for each of three weak acid-base pairs. Charges, z_0, on the parent acids in Cells D30–F30 are used with If/Then logic in Cell B44 to choose between weak acid terms in Step 2 of the example. The charges are also used in Cells C44, D44, and E44 to choose between equations for concentration-based constants as in Step 4 of the example.

This part of the program starts when calculations without correction for ionic strength have identified the upper limit of the 10-fold range that includes the equilibrium concentration. Then, calculations in D41 and E41 start with a trial H^+ concentration equal to 1% of the upper limit concentration from C38 in E41. Then, with the number of significant figures in D24 set to 3, the program increases trial values of the hydrogen ion concentration in E41 by amounts equal to 0.1% of the upper limit concentration in E41 until the charge-balance equation in D41 changes from $-$ to $+$. The program then stops with the hydrogen ion concentration that caused the sign change in E41. This concentration is accepted as the equilibrium hydrogen ion concentration corrected for ionic strength, i.e., $[H^+]_\mu$.

Updated values of the ionic strength, activity coefficients, and concentration-based constants are calculated for each increment in the H^+ concentration. Otherwise, the operation of the program is similar to that described for calculations without correction for ionic strength. Ionic-strength corrected H^+ concentration and pH are in Cells E41 and F41. The ionic strength is in L29. The hydrogen ion activity, $\{H^+\}$, and activity-based pH, $p\{H\}$, are in G41 and H41.

SELECTED APPLICATIONS OF THE SPREADSHEET PROGRAM, MONOPROTIC_SYSTEMATIC

Some applications of the program are described below. In these and other applications (a) set B24 to zero to enter new information, (b) set all concentrations, deprotonation constants, and charges relevant to each problem to their intended values, 0.0 or otherwise, and (c) set B24 to 1 to start the iterative calculations.

EFFECTS OF IONIC STRENGTH ON TITRATION CURVES

Given the relatively small effects of ionic strength on pH's, it seems reasonable to ask if ionic strengths up to about 0.1 M have significant effects on shapes of titration curves for monoprotic acids and bases. The program on Sheet 2 of the monoprotic acids program was used to answer this question.

Situations for which titration curves are compared are summarized in Table 4.6. In each case, calculations were done for 25.0 mL volumes of solutions containing 0.10 M concentration of each acid or each base titrated with 0.10 M NaOH or 0.10 M HCl, respectively.

Titration curves without and with correction for ionic strength are included in Figures 4.1A and 4.1B. Although there are slight differences between plots without and with correction for ionic strength, differences are too small to influence interpretations of the results.

For the record, whereas the plot for weak acids shows a relatively sharp break for each of the three acids, the plot for weak bases has breaks only for ammonia and acetate.

Dichloroacetate, $K_a = 0.05 \Rightarrow K_b = K_w/K_a = 2 \times 10^{-13}$, is much too weak a base to be titrated in water.

TABLE 4.6

Single- and Three-Component Situations for Which Titration Curves without and with Correction for Ionic Strength Are Compared

Acids	HCl alone	Dichloroacetic acid + acetic acid + ammonium ion
Bases	NaOH alone	Ammonia + acetate + dichloroacetate

For dichloroacetic acid: $K_a = 0.050$, For acetic acid: $K_a = 1.8 \times 10^{-5}$, For ammonium ion: $K_a = 5.7 \times 10^{-10}$. For all bases: $K_b = K_w/K_a$.

TABLE 4.7
Results for Example 4.6

C_{HCl}	0.000	0.002	0.004	0.006	0.008	0.010	Change[a]
$[H^+]$	1.80E-5	1.88E-5	1.95E-5	2.03E-5	2.12E-5	2.20E-5	22%
pH	4.74	4.73	4.71	4.69	4.67	4.66	-1.7%

[a] Percentage changes caused by increasing the HCl concentration from 0.00 M to 0.010 M.

BUFFER PERFORMANCE

The purpose of an acid-base buffer is to reduce effects of added H^+ or OH^- concentrations on changes in the H^+ concentration and pH. As you probably know, acid-base buffers are mixtures of conjugate acid-base pairs.

Example 4.6 illustrates an application of the weak acids program to an acid-base buffer.

Example 4.6 How well does a buffer containing 0.10 M each of acetic acid and acetate ion control the hydrogen ion concentration and pH for solutions in which the HCl concentration is varied from 0.000 M to 0.010 M in 0.0020 M steps without changing the volume?

To answer this question, set I24 to D, C27 and D27 to 1.0E-01 each, A30 to 1.8E-5, and all other concentrations in Row 27 to 0. Then record the H^+ concentration and pH in Cells E38 and F38 for different HCl concentrations in Cell A27. Results I obtained are summarized in Table 4.7.

Whereas the H^+ concentration increases by 22% for the 0.010 M change in HCl concentration, the pH decreases by only 1.7%. The relevance of these numbers depends on how results are to be used. If the application depends more on pH than the hydrogen ion concentration as is the case when plotting titration curves, then pH errors are quite satisfactory. However, if your application depends more directly on the H^+ concentration than pH as is the case for many types of applications such as calculating concentration fractions, weak acid and weak base concentrations, and H^+-dependent rate constants, then the percentage change in the hydrogen ion concentration may be larger than you are willing to accept.

One way to improve buffer performance is to increase concentrations of the buffer components.

EXERCISE 4.3

To observe effects of buffer concentrations on the ability of the buffer to resist changes in hydrogen ion concentrations, do the calculations with the HCl concentration set at 0.010 M and acetic acid and sodium acetate concentrations each set at 0.10, 0.20, 0.30, and 0.40 M and compare hydrogen ion concentrations and percentage changes from the concentration, 1.8×10^{-5} M, with the HCl concentration equal to zero. Concentrations I obtained for the four acetic-acid/acetate concentrations are 2.2×10^{-5}, 1.99×10^{-5} M, 1.93×10^{-5} M, and 1.90×10^{-5}, corresponding to percentage changes from 1.8×10^{-5} of 22%, 10.6%, 7.2%, and 5.6% respectively.

Increases in buffer concentrations improve the ability of the buffers to resist changes in pH. We will have more to say about this in a later chapter.

DISTRIBUTION PLOTS

Distribution plots are plots of fractions of weak acids in different forms vs. pH. These plots help us visualize how pH, and by extension, hydrogen ion concentrations, influence forms of weak acids in different forms.

As a reminder, fractions, α_0 and α_1, of a monoprotic weak acid in forms that have lost $i = 0$ and $i = 1$ proton are as follows:

$$\alpha_0 = \frac{\left[H^+\right]}{\left[H^+\right] + K_a} \qquad \text{and} \qquad \alpha_1 = \frac{K_a}{\left[H^+\right] + K_a}$$

in which α_0 and α_1 are fractions in forms that have lost $i = 0$ and $i = 1$ protons, respectively, $[H^+]$ is the hydrogen ion concentration and K_a is the deprotonation constant.

A program on Sheet 3 of the program, *Monoprotic_Systematic*, is designed to calculate and plot concentration fractions vs. pH for monoprotic acids and bases. Figure 4.2 is a distribution plot for acetic acid. As you can see, virtually all of the acetic acid is in the fully protonated form for pH's up to about pH = 2.5. For pH's larger than about 2.5, the fraction in the protonated form decreases toward 0.0 and fractions in the deprotonated form increase toward 1.0.

The pH at which the plots cross corresponds to a point at which the two fractions are equal. As shown below, the crossing point corresponds to a point at which the hydrogen ion concentration is equal to the deprotonation constant or pH = pK_a.

$$\alpha_0 = \alpha_1 \quad \Rightarrow \quad \frac{\left[H^+\right]}{\left[H^+\right] + K_a} = \frac{K_a}{\left[H^+\right] + K_a} \quad \Rightarrow \quad \left[H^+\right] = K_a \quad \text{or} \quad pH = pK_a$$

As pH's increase above the crossing point the fraction, α_0, in the protonated form decreases toward 0.0 and the fraction, α_1, in the deprotonated form increases toward 1.0.

A program to plot distribution plots for three weak acids on one figure is included on Sheet 5 of the weak acids program.

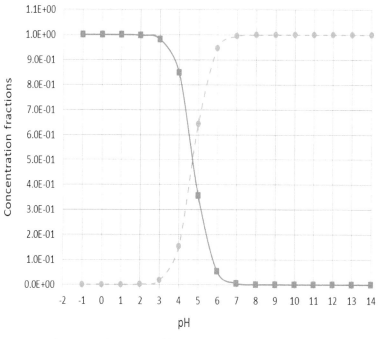

FIGURE 4.2 Distribution plots for acetic acid.

PROGRAM TO CALCULATE AND PLOT pH vs. TITRANT VOLUME

As shown earlier, the program on Sheet 2 of the monoprotic weak acids program can be used for point-by-point calculations of titration curves. There probably is some pedagogical value for doing point-by-point calculations as described earlier. However, a more efficient way to prepare and compare titration curves is described below.

An extension on Sheet 4 of the program, *Monoprotic_Systematic*, is designed to calculate and plot pH vs. titrant volume for multiple volumes simultaneously and to plot titration curves for up to six different sets of conditions on one figure. Let's begin with strong-acid/strong-base titrations.

EFFECTS OF CONCENTRATIONS ON STRONG-ACID/STRONG-BASE TITRATIONS

Example 4.7 illustrates applications of the program to strong-acid/strong-base titrations.

> **Example 4.7 Plot titration curves for the titration of 25.0 mL of (A) 0.20 M HCl with 0.20 M NaOH, (B) 0.020 M HCl with 0.020 M NaOH, (C) 0.0020 M HCl with 0.0020 M NaOH, (D) 0.20 M NaOH with 0.20 M HCl, (E) 0.020 M NaOH with 0.020 M HCl, and (F) 0.0020 M NaOH with 0.0020 M HCl.**

A template for Part A of this example is included in Cells A44–I52 on Sheet 4 of the monoprotic weak acids program. A scenario used to prepare the plots for the six parts of Example 4.7 is summarized in Table 4.8. All cells in Rows 27 and 30 except I30 and those identified below should be set to zero.

The value, 6.5, for I28 for Part F merits some comment. For any number, n, of plots up to six, the last number in I28 should be a number between the number of the last plot and the number of the next plot in the sequence. As examples, to plot 2, 3, or 5 plots on one figure, the last values of I28 should be 2.5, 3.5, or 5.5. This prevents the titration curve for the last set of conditions from being plotted twice on the figure.

If you forget to set I28 to the next larger number for any part, it will be necessary to repeat the process from that part to the end.

Figure 4.3 illustrates titration curves for the six situations in the example. I suggest that you confirm curve shapes in Figure 4.3 by working through the scenario yourself.

Ascending plots are for Parts A–C, i.e., titrations of different HCl concentrations with NaOH; descending plots are for Parts D–F, i.e., titrations of different NaOH concentrations with HCl. The plots show that the total pH change decreases by about two units for each 10-fold decrease in concentration.

TABLE 4.8
Scenario Used to Solve Example 4.7

Start by setting B24 to 0 and I28 to 0 and then to 1. Then follow the sequence below.							
Parts A–C							
Cell	B24	I23	G30	H30	A27	B27	B24	I28[a]
Part A	0	A	25.0	Ignore	0.20	0.20	0 → 1	1 → 2[a]
Part B	0	A	25.0	Ignore	0.02	0.02	0 → 1	2 → 3[a]
Part C	0	A	25.0	Ignore	0.0020	0.0020	0 → 1	3 → 4[a]
Parts D–F							
Cell	B24	I23	G30	H30	A27	B27	B24	I28
Part D	0	B	Ignore	25.0	0.20	0.20	0 → 1	4 → 5[a]
Part E	0	B	Ignore	25.0	0.02	0.02	0 → 1	5 → 6[a]
Part F	0	B	Ignore	25.0	0.0020	0.0020	0 → 1	6 → 6.5[b]

[a] Wait for I24 to change from No to Yes before setting I28 to the next larger number.
[b] For any number, n, plots on a figure, the last entry in I28 should be n.5, e.g., 2.5, 3.5, or 4.5 for 2, 3, or 4 plots.

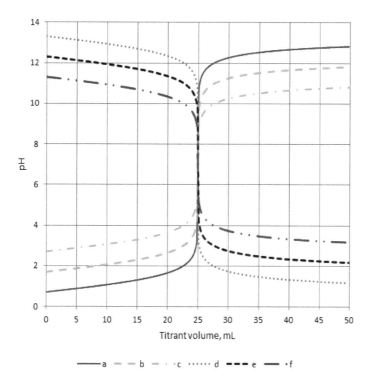

FIGURE 4.3 Effects of strong acid and strong base concentrations on shapes of titration curves.

Ascending plots: Titration curves for the titration of 25.0 mL of (a) 0.20 M HCl with 0.20 M NaOH, (b) 0.020 M HCl with 0.020 M NaOH, (c) 0.0020 M HCl with 0.0020 M NaOH.

Descending plots: (d) 0.20 M NaOH with 0.20 M HCl, (e) 0.020 M NaOH with 0.020 M HCl, and (f) 0.0020 M NaOH with 0.0020 M HCl.

Calculations used to prepare these plots are done in Cells U28–AH427 on Sheet 4. Volumes and pH's for each situation are organized in Cells AI28–AJ427 and plotted separately in Spreadsheet Figure 4.1, Cells K6–S31, for each set of conditions. Titration curves for one to six sets of conditions are plotted in Spreadsheet Figure 4.2, Cells K36–S59.

EFFECTS OF CONCENTRATION ON TITRATION CURVES FOR WEAK ACIDS AND WEAK BASES

Example 4.8 illustrates effects of weak-acid, weak-base, and titrant concentrations on shapes of titration curves.

> **Example 4.8 Illustrate effects of concentrations on shapes of titration curves for (A) titrations of 25.0 mL volumes of solutions containing (a) 0.10 M, (b) 0.010 M, (c) 0.0010 M acetic acid with NaOH concentrations equal to each acetic acid concentration and (B) titrations of 25.0 mL volumes of solutions containing (a) 0.10 M, (b) 0.010 M, (c) 0.0010 M ammonia with HCl concentrations equal to each ammonia concentration.**

A template for Part A of this example is included in Cells A60–I68 on Sheet 4 of the monoprotic weak acids program. The setup and order in which cells are changed for this example are summarized in Table 4.9. All cells in Rows 27 and 30 except I30 and those identified below should be set to zero.

Titration curves are plotted in Figure 4.4. As shown in the figure, the magnitude of the pH change and sharpness of the equivalence point decreases with decreasing acetic acid and ammonia concentrations.

Hopefully, this example will give you sufficient information to do Exercise 4.4.

TABLE 4.9
Scenario Used to Solve Example 4.8

Start by setting B24 to 0 and I28 to 0 and then to 1. Then follow the sequence below.							
Part A							
Cell	B24	A30	I23	G30	B27	C27	B24	I28[a]
Part a	0	1.8E-5	A	25.0	0.10	0.10	0 → 1	1 → 2[a]
Part b	0	1.8E-5	A	25.0	0.010	0.010	0 → 1	2 → 3[a]
Part c	0	1.8E-5	A	25.0	0.0010	0.0010	0 → 1	3 → 4[a]
Part B							
Cell	B24	A30	I23	H30	A27	D27	B24	I28
Part d	0	5.7E-10	B	25.0	0.10	0.10	0 → 1	4 → 5[a]
Part e	0	5.7E-10	B	25.0	0.010	0.010	0 → 1	5 → 6[a]
Part f	0	5.7E-10	B	25.0	0.0010	0.0010	0 → 1	6 → 6.5

[a] Change only after I24 changes from No to Yes.

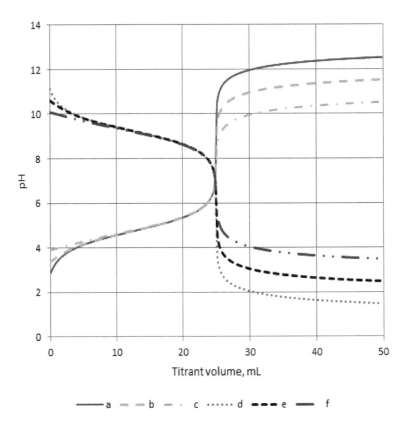

FIGURE 4.4 Effects of concentration on titration curves for acetic acid and ammonia.

Ascending plots: Titration curves for 25.0 mL each of (a) 0.10, (b) 0.010, and (c) 0.0010 M acetic acid, $K_a = 1.8 \times 10^{-5}$, with NaOH concentrations equal to each acetic acid concentration.

Descending plots: Titration curves for 25.0 mL each of (d) 0.10, (e) 0.010, and (f) 0.0010 M ammonia, $K_a = 5.7 \times 10^{-10}$, with HCl concentrations equal to each ammonia concentration.

EXERCISE 4.4

Illustrate effects of deprotonation constants on shapes of titration curves by confirming the plots in Figure 4.5 for (A) titrations with 0.10 M NaOH of 25.0 mL volumes of solutions containing (a) 0.10 M trichloroacetic acid, $K_a = 0.22$, (b) 0.10 M bromoacetic acid, $K_a = 1.3 \times 10^{-3}$, and (c) 0.10 M acetic acid, $K_a = 1.8 \times 10^{-5}$, and (B) titrations with 0.10 M HCl of 25.0 mL volumes of solutions containing (d) 0.10 M guanidine, $K_a = 2.9 \times 10^{-14}$, (e) 0.10 M butylamine, $K_a = 2.3 \times 10^{-11}$, and (f) 0.10 M 3-nitrophenolate ion, $K_a = 4.1 \times 10^{-9}$. (*Hints:* Use patterns in Tables 4.8 and 4.9 as guides. If you need it, a template for Part A, a is in Cells A75–I83.)

Figure 4.5 is the figure you should obtain for the exercise. Notice that the smaller the deprotonation constant, the smaller the range of pH's over which the break in the titration curve occurs.

RELATIONSHIPS BETWEEN SHAPES OF TITRATION CURVES AND CONCENTRATION FRACTIONS

As discussed in the background chapter, hydrogen ion concentrations can be used with deprotonation constants to calculate fractions of weak acid-base pairs in different forms. Fractions calculated in this way can be used with the total concentration, C_T, of each acid-base pair to calculate equilibrium concentrations of the acids and bases.

The part of the program in Cells M32–R35 on Sheet 2 is designed to calculate fractions of the acids in different forms. We can use these fractions to give us some insight into degrees of

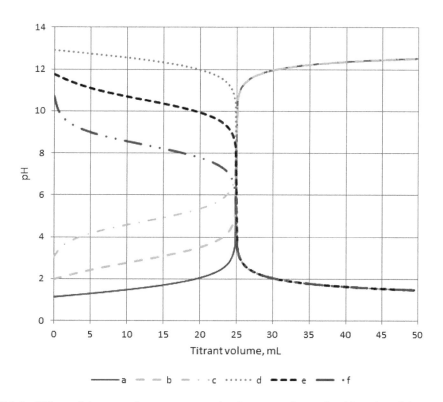

FIGURE 4.5 Effects of deprotonation constants on titration curves for weak acids and weak bases.

Ascending plots: 25.0 mL of 0.10 M each of (a) trichloroacetic acid, $K_a = 0.22$, (b) bromoacetic acid, $K_a = 1.3 \times 10^{-3}$, and (c) acetic acid, $K_a = 1.8 \times 10^{-5}$ with 0.10 M NaOH.

Descending plots: 25.0 mL of 0.10 M each of (d) guanidine, $K_a = 2.9 \times 10^{-14}$, (e) butylamine, $K_a = 2.3 \times 10^{-11}$, and (f) 3-nitrophenolate ion, $K_a = 4.1 \times 10^{-9}$, titrated with 0.10 M HCl.

completions of the various reactions at different points in titrations, particularly at equivalence points.

In a titration of a weak acid such as dichloroacetic acid with sodium hydroxide, it is desirable that all the acid be converted to the conjugate base at the equivalence point, i.e., that $\alpha_0 \rightarrow 0.0$ and that $\alpha_1 \rightarrow 1.0$. For example, let's compare fractions of dichloroacetic acid converted to dichloroacetate at the equivalence point for the titration of 25 mL 0.10 M dichloroacetic acid with 0.10 M sodium hydroxide without and with 0.10 M acetic acid.

Dichloroacetic Acid without Acetic Acid

Starting with conditions as in Table 4.4, set E27 = G27 = 0 and H30 = 25.0, then set B24 to 1, wait for I36 to change to Yes, and read the fractions α_{0X} and α_{1X} from Cells M33 and N33. You should read $\alpha_{0X} = 1.4 \times 10^{-6}$ and $\alpha_{1X} = 1.00$. This means that sodium hydroxide has converted virtually 100% of dichloroacetic acid to dichloroacetate, i.e.,

$$HB + NaOH \overset{100\%}{\rightarrow} Na^+ + B^- + H_2O$$

In other words, as is desirable in a titration, the reaction between the analyte, dichloreacetic acid, and the titrant, sodium hydroxide, is virtually 100% complete.

Dichloroacetic Acid with 0.10 M Acetic Acid

Starting with conditions described in the preceding paragraph, set B24 = 0 and E27 = 0.10. Then set B24 = 1 and wait for I36 to change to Yes. Then read the fractions α_{0X} and α_1 from Cells M33 and N33. Values I obtained are $\alpha_{0X} = 0.013$, $\alpha_{1X} = 0.987$ meaning that sodium hydroxide has converted only 98.7% of the dichloroacetic acid to dichloroacetate. This also means that some of the sodium hydroxide titrant will have reacted with acetic acid to convert part of it to acetate, i.e., $\alpha_{0Y} = 0.074$, $\alpha_{1Y} = 0.026$. What do these numbers mean? They mean that 98.7% of the sodium hydroxide titrant has reacted with dichloroacetic acid, Acid X, and that 2.6% has reacted with acetic acid, Acid Y. In other words, the inclusion of acetic acid interferes with the equivalence point for dichloroacetic acid.

Would you expect the percentage of trichloroacetic acid, $K_a = 0.22$, converted to the conjugate base to be larger or smaller than that for dichloroacetic acid? You can test your conclusion by rerunning that part of the example with A30 set to 0.22.

On the one hand, conventional titration curves of pH vs. titrant volume help us determine the suitability of a set of conditions for a satisfactory titration. On the other hand, these plots don't tell us much about reasons for the shapes we see. Plots of concentration fractions vs. titrant volume can help us fill that gap. A program in Columns AY–BB on Sheet 4 of the program is designed to calculate concentration fractions vs. titrant volume.

As an example, Figure 4.6 includes titration curves for a titration with 0.10 M NaOH of 25.0 mL of one solution containing 0.10 M each of dichloroacetic acid, $K_a = 0.050$, and protonated hydroxylamine, $K_a = 1.0 \times 10^{-7}$, and a second solution containing 0.10 M each of dichloroacetic acid and formic acid, $K_a = 1.8 \times 10^{-4}$. A template for the titration of the mixture of dichloroacetic acid and protonated hydroxylamine is included in Cells A90–I98 on Sheet 4 of the monoprotic acids program.

Whereas the titration curve for dichloroacetic acid and protonated hydroxylamine have relatively sharp breaks at the equivalence points for the two acids, the plot for dichloroacetic and formic acid shows a very slight break at the equivalence point for dichloroacetic acid, 25.0 mL, and a sharp break at the equivalence point for both acids, 50.0 mL. In other words, whereas it is possible to distinguish between dichloroacetic acid and protonated hydroxylamine titrimetrically, it isn't possible to differentiate between dichloroacetic acid and formic acid titrimetrically.

Reasons for these differences involve relative values of deprotonation constants. On the one hand, deprotonations constants for the dichloroacetic acid/protonated hydroxylamine mixture are 0.050 and 1.0×10^{-7}, respectively, corresponding to a ratio of constants of $0.05/1 \times 10^{-7} = 4.7 \times 10^5$.

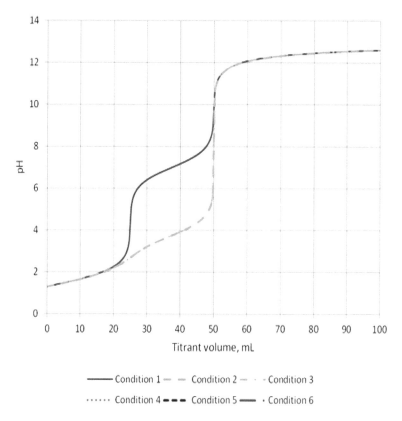

FIGURE 4.6 Titration curves for solutions containing Condition (1) 0.10 M each of dichloroacetic acid, $K_a = 0.050$, and protonated hydroxylamine, $K_a = 1.0 \times 10^{-7}$, and Condition (2) 0.10 M each of dichloroacetic acid, $K_a = 0.050$, and formic acid, $K_a = 1.8 \times 10^{-3}$, titrated with 0.10 M NaOH.

On the other hand, deprotonation constants for dichloroacetic acid and formic acid are 0.050 and 1.8×10^{-4}, respectively, corresponding to a ratio of $0.05/1.8 \times 10^{-4} = 28$. The result is that titration curves for the two acids are not resolved from each other. In general, a ratio of constants of about 1000 to 1 is needed to ensure good resolution of equivalence points for two weak acids. Analogous considerations apply for titrations of weak bases.

RELATIONSHIPS BETWEEN SHAPES OF TITRATION CURVES AND CONCENTRATION FRACTIONS

A program in Columns AY–BB on Sheet 4 of the monoprotic weak acids program is designed to calculate fractions of up to three weak acids converted to basic forms by titration with a strong base or fractions of weak bases converted to acidic forms by titration with a strong acid. Plots of fractions vs. titrant volume can give us insight into reasons for shapes of titration curves such as those in Figure 4.6.

Figure 4.7 includes plots of fractions of dichloroacetic acid, protonated hydroxylamine and formic acid converted to dichloroacetate, hydroxylamine, and formate during the titration of 25.0-mL volumes each of one solution containing 0.10 M dichloroacetic acid and 0.10 M protonated hydroxylamine and another solution containing 0.10 M dichloroacetic acid and 0.10 M formic acid.

The reaction of each acid with the titrant, sodium hydroxide, is as follows:

$$HB + NaOH \rightarrow Na^+ + B^- + H_2O$$

For satisfactory titrations, it is desirable that sodium hydroxide react completely with each acid. The fraction of each acid converted to the basic form is a direct measure of the completeness of the reaction.

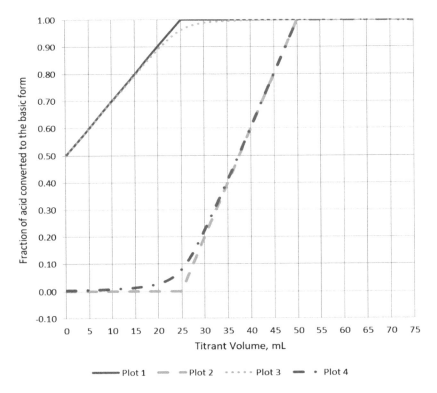

FIGURE 4.7 Fractions of acids converted to basic forms by titration with 0.10 M sodium hydroxide of 25.0-mL volumes of two solutions, one containing 0.10 M concentrations of dichloroacetic acid and protonated hydroxylamine, the other containing 0.10 M concentrations of dichloroacetic acid and formic acid.

Plots 1 and 2: Fractions converted to dichloroacetate and hydroxylamine, respectively.

Plot 3 and 4: Fractions converted to dichloroacetate and formate, respectively.

Let's talk first about Plots 1 and 2 for the solution containing dichloroacetic acetic acid and protonated hydroxylamine. Plot 1 for dichloroacetic acid shows that the fraction of dichloroacetic acid converted to dichloroacetate increases to 1.0 before Plot 2 for the fraction of protonated hydroxylamine converted to hydroxylamine begins to increase. In other words, virtually all of the dichloroacetic acid reacts with the sodium hydroxide titrant before any significant amount of the protonated hydroxylamine begins to react. This is the reason two distinct breaks are observed in the titration curve.

Looking next at Plots 3 and 4 for dichloroacetic acid and formic acid, you will observe that Plot 4 for the fraction of formic acid converted to formate ion begins to increase before Plot 3 for the fraction of dichloroacetic reacting with sodium hydroxide has reached 1.0, i.e., before all of the dichloroacetic acid has reacted.

This is the reason the titration curve for the solution containing dichloroacetic acid and formic acid has only one break corresponding to the titration of both acids.

On the one hand, conventional titration curves of pH vs. titrant volume help us determine the suitability of a set of conditions for a satisfactory titration. On the other hand, these plots don't tell us much about reasons for the shapes we see. Plots of concentration fractions vs. titrant volume can help fill that gap.

SUMMARY

This chapter describes a systematic approach to monoprotic acid-base equilibria. The systematic approach is used to develop ionic, fractional, and computational forms of a charge-balance equation for hydrogen ion concentrations in solutions containing concentrations equal to or larger than zero of HCl, NaOH, and up to three weak acid-base pairs. Having written an ionic form of the charge-balance equation, CBE, mass-balance, and concentration-fraction equations are used to express the charge-balance equation in terms of the H^+ concentration and known quantities. Those quantities include the autoprotolysis constant for water, deprotonation constants, and diluted concentrations of reactants.

Iterative programs to solve the CBE are based on an observation that the equation is less than zero for hydrogen ion concentrations less than the equilibrium concentration and positive for H^+ concentrations larger than zero. A spreadsheet program is implemented by starting with a very small H^+ concentration and increasing the concentration by known amounts until the sign of the charge-balance equation changes from minus to plus. The concentration that causes the − to + change is the equilibrium concentration resolved to a number of significant figures selected by the user, usually three.

Equations and procedures for converting activity-based constants to concentration-based constants are described. The spreadsheet program can be used with either activity-based or concentration-based constants. The spreadsheet program is adapted to several applications, including titration curves, buffer performance and distribution plots.

REFERENCES

1. J. E. Ricci, *Hydrogen ion concentration, New concepts in a Systematic Treatment*, Princeton U. P., Princeton, NJ, 1952.
2. R. de Levie, *Principles of Quantitative Chemical Analysis*, McGraw-Hill: New York, NY, 1997.
3. H. Freiser, *Concepts & Calculations in Analytical Chemistry, A Spreadsheet Approach*, CRC Press: Boca Raton, FL, 1992.
4. H. A. Laitinen, W. E. Harris, *Chemical Analysis*, Second Edition, McGraw-Hill: New York, NY, 1975.
5. J. Kielland, *J. Am. Chem. Soc.*, **1937**, *59*, 1675–1678. (Search the internet for "Individual Activity Coefficients of Ions in Aqueous Solutions" to find tabulated values of hydrated diameters for several ions.)

5 A Systematic Approach to Monoprotic, Diprotic, and Polyprotic Acid-Base Equilibria

This chapter focuses on equilibrium calculations for acids with up to six acidic hydrogens. My overall goal is to help you understand the basis for equations and a spreadsheet program to calculate concentrations of all reactants and products in solutions containing concentrations equal to or larger than zero of a monoprotic strong acid, a monoprotic strong base, and different forms of three uncharged, singly charged, or doubly charged weak acids with up to six acidic protons each without and with correction for ionic strength. The first step for all situations is to calculate the hydrogen ion concentration. Subsequent steps are to use the hydrogen ion concentration to calculate concentration fractions and to use the concentration fractions to calculate concentrations of different forms of the weak acids. Of the several options available to do this,[1] I chose a systematic approach described initially by J. E. Ricci[2] and extended by others.[3–6]

Two options considered as the basis for the systematic approach are proton-balance equations and charge-balance equations. Proton-balance equations are based on the fact that protons lost and gained during an acid/base reaction must be the same. Charge-balance equations are based on the fact that the net charge on any solution at equilibrium must be zero. I chose to use charge-balance equations herein and elsewhere because charge-balance equations are more generally applicable than proton-balance equations in the sense that they apply for all types of ionic reactions.

The treatment of the systematic approach in this chapter differs from earlier treatments in two major ways. Other authors[3–6] used graphically assisted approximation procedures. The treatment herein uses a feature of the charge-balance equation as the basis for an iterative spreadsheet program to calculate hydrogen ion concentrations directly without any need to use graphical results or approximations. The resulting program allows the user to control the number of significant figures to which the hydrogen ion concentration is resolved by entering the desired number, e.g., 2, 3, 4, in a cell in the program. For example, with the number of significant figures set to 3, the program will resolve the hydrogen ion concentration to three significant figures with convergence times from about 1 to 30 seconds for most situations.

I will start by introducing the charge-balance equation and program used to do the calculations. I will then describe the basis for the equation and program without correction for ionic strength and then extend the discussion to calculations with correction for ionic strength. I will assume throughout the discussion that you are familiar with or will review background information in Chapter 3.

PREVIEW OF EQUATIONS FOR THE HYDROGEN ION CONCENTRATION AND IONIC STRENGTH AND A PROGRAM TO SOLVE THEM

We will show later that the computational form of the charge-balance equation for a solution containing HCl, NaOH, and different forms of a polyprotic acid, H_nB, can be written in the following compact form (Equation 5.1a).

$$\overbrace{\left[H^+\right] - \frac{K_w}{\left[H^+\right]} + \left(C_{NaOH} - C_{HCl}\right)}^{\text{Independent term}} + \sum_{i=0}^{n} iC_i - \overbrace{\left(\frac{\sum_{i=0}^{n} i\kappa_i \left[H^+\right]^{n-i}}{\sum_{i=0}^{n} \kappa_i \left[H^+\right]^{n-i}}\right) \left(\sum_{i=0}^{n} C_i\right)}^{\text{Weak-acid term}} = 0 \qquad (5.1a)$$

Symbols C_{NaOH} and C_{HCl} are diluted concentrations of NaOH and HCl, n is the number of acidic hydrogens in the weak acid, i is the number of protons lost by each form of the acid, C_i is the diluted concentration of each form of the acid that has lost i protons, κ_i are stepwise cumulative constants with $\kappa_0 = 1$, and Σs are the conventional symbols for summations. Summations are from i = 0 to the number, n, of acidic hydrogens.

We will show later that this equation applies for both charged and uncharged acids.

An equation for solutions containing different forms of two or more weak acids is obtained by adding one or more weak acid terms. An iterative spreadsheet program to solve the equation for solutions containing HCl, NaOH, and any or all forms of up to three weak acids, Acid X, Acid Y, and Acid Z, with up to six acidic hydrogens each is included on the programs website http://crcpress.com/9781138367227 for this text.

Exercise 5.1 illustrates an application of the equation to a specific situation.

EXERCISE 5.1

Use Equation 5.1a to show that the charge-balance equation for a solution containing HCl, NaOH, and different forms of an uncharged triprotic acid is

$$\left[H^+\right] - \frac{K_w}{\left[H^+\right]} + \left(C_{NaOH} - C_{HCl}\right) + C_1 + 2C_2 + 3C_3 - \left(\frac{\kappa_1\left[H^+\right]^2 + 2\kappa_2\left[H^+\right] + 3\kappa_3}{\left[H^+\right]^3 + \kappa_1\left[H^+\right]^2 + \kappa_2\left[H^+\right] + \kappa_3}\right)$$
$$\times \left(C_0 + C_1 + C_2 + C_3\right) = 0$$

Hint: Do the summations for i's from 0 to 3.

As reminders from Chapter 3, the denominator term for fractions of a triprotic acid in different forms is

$$\left[H^+\right]^3 + \kappa_1\left[H^+\right]^2 + \kappa_2\left[H^+\right] + \kappa_3$$

Numerator terms for successive fractions, $\alpha_0 - \alpha_3$, are successive denominator terms.

Equation 5.1a is obtained by equating differences between charges contributed by positive and negative ions to zero. I discovered during the preparation of this text that charge-balance equations for acids and bases written by equating differences between positive and negative charges to zero are negative for hydrogen ion concentrations less than the equilibrium concentration and positive for concentrations larger than the equilibrium concentration.

This feature is used as the convergence criterion for an iterative spreadsheet program to solve the equation for the hydrogen ion concentration without and with correction for ionic strength.

It is also shown later that an equation for the ionic strength in a solution containing HCl, NaOH, and different forms of a weak acid with n acidic hydrogens and charge, z_0, on the fully protonated acid is expressed in Equation 5.1b:

$$\mu = 0.5\left\{\overbrace{\left[H^+\right] + \frac{K_w}{\left[H^+\right]} + \left(C_{HCl} + C_{NaOH}\right)}^{\text{Independent term}} + \overbrace{\sum_{i=0}^{n}\left(|z_0 - i|C_i\right) + \left[\sum_{i=0}^{n}\left(z_0 - i\right)^2 \alpha_i\right]C_T}^{\text{Weak-acid term}}\right\} \qquad (5.1b)$$

in which C_T is the total concentration of all forms of the weak acid, z_0 is the charge on the fully protonated acid, α_i is the fraction in each form that has lost i protons, and other symbols are as in

Equation 5.1a. The equation for a solution containing HCl, NaOH, and different forms of three weak acids can be written by adding one additional weak-acid term for each of two additional acids.

Exercise 5.2 illustrates an adaptation of this equation to different forms of a triprotic weak acid.

EXERCISE 5.2

Show that the equation for the ionic strength in a solution containing HCl, NaOH, and different forms of a singly charged triprotic acid is

$$\left[H^+\right]+\frac{K_w}{\left[H^+\right]}+\left(C_{NaOH}+C_{HCl}\right)+C_0+C_2+2C_3-\left(\frac{\kappa_0\left[H^+\right]^3+\kappa_2\left[H^+\right]+4\kappa_3}{\left[H^+\right]^3+\kappa_1\left[H^+\right]^2+\kappa_2\left[H^+\right]+\kappa_3}\right)$$
$$\times\left(C_0+C_1+C_2+C_3\right)=0$$

Hints: Do the summations for $z_0 = 1$, and i from 0 to 3. For $z_0 = 1$ and i = 1, $z_0 - i = 1 - 1 = 0$.

We will talk later about the basis for these equations. For now, let's talk about a program designed to use these equations to calculate equilibrium concentrations of all reactants and products without and with correction for ionic strength for situations ranging from pure water to solutions containing concentrations equal to or larger than zero of HCl, NaOH, and any or all forms of three uncharged, singly charged, or doubly charged weak acids with up to six acidic protons each.

The part of the program used to do these calculations is on Sheet 1 of the program, *Acid Base Equilibria*, included on the programs website http://crcpress.com/9781138367227 for this text. We will talk later about the basis for the program. For now, let's talk about an overview of the program and selected examples used to illustrate applications of the program as well as the validity of results obtained using it.

OVERVIEW OF THE PROGRAM ON SHEET 1 OF THE ACID-BASE PROGRAM

The part of the program on Sheet 1 is designed to do calculations without and with correction for ionic strength for solutions ranging from pure water to solutions containing diluted concentrations, C_{HCl} and C_{NaOH}, of hydrochloric acid and/or sodium hydroxide with diluted concentrations of any or all forms of up to three weak acids with up to six acidic hydrogens each.

To begin, open the program, *Acid Base Equilibria*, from the programs website http://crcpress.com/9781138367227 into an unused Excel spreadsheet and select Sheet 1. Selected parts of the program are described below.

INPUT SECTION

Table 5.1A includes the input section of the program with all concentrations and constants except K_w in H42 set to zero as well as a line of the program that is used to inform the user when iterative calculations are complete.

The initial hydrogen ion concentration in F29, i.e., $C_H = 1.0 \times 10^{-15}$ M, is the starting concentration for the iterative process used to resolve the equilibrium concentration. You shouldn't need to change this concentration. The number, 3, in I29 tells the program to resolve the hydrogen ion concentration to three significant figures. You can change this to smaller or larger numbers to get smaller or larger degrees of resolution. The larger the number of significant figures, the longer the convergence time. Resolution to three significant figures is a good compromise between convergence time and uncertainties associated with deprotonation constants.

Weak acids are identified in the program as Acids X, Y, and Z. It doesn't matter which acid is represented as Acid X, Y, or Z. What matters is that contents of all cells for each acid correspond to the proper acid. With this in mind, stepwise deprotonation constants for the acids are entered into Cells B32–G34. Diluted concentrations of different forms of the acids that have lost 0 – n protons

TABLE 5.1A

Input and Program Status Sections of the Acid-Base Program (Columns A–I, Left to Right, Rows 29–47, Top to Bottom)

A-28	B	C	D	E	F	G	H	I
Reset, 0./Start, 1		0		Initial C_H	1.00E-15	Significant figures		3
Row 30: Stepwise Deprotonation Constants								
Row 31	K_{a1}	K_{a2}	K_{a3}	K_{a4}	K_{a5}	K_{a6}		
Acid X	0.00E+00	0.00E+00	0.00E+00	0.00E+00	0.00E+00	0.00E+00		
Acid Y	0.00E+00	0.00E+00	0.00E+00	0.00E+00	0.00E+00	0.00E+00		
Acid Z	0.00E+00	0.00E+00	0.00E+00	0.00E+00	0.00E+00	0.00E+00		
Diluted Concentrations								
Row 36	C_0	C_1	C_2	C_3	C_4	C_5	C_6	C_{HCl}
Acid X	0.00E+00	0.00E+00	0.00E+00	0.00E+00	0.00E+00	0.00E+00	0.00E+00	0.00E+00
Acid Y	0.00E+00	0.00E+00	0.00E+00	0.00E+00	0.00E+00	0.00E+00	0.00E+00	C_{NaOH}
Acid Z	0.00E+00	0.00E+00	0.00E+00	0.00E+00	0.00E+00	0.00E+00	0.00E+00	0.00E+00
		Additional Weak-Acid Information						
		n	z_0	$K_a @ \mu = 0$?			$K_{w,0}$	
Row 42	Acid X	0	0	Yes			1.00E-14	
Row 43	Acid Y	0	0	Yes				
Row 44	Acid Z	0	0	Yes				
DO NOT CHANGE ANY QUANTITIES BEYOND THIS POINT								
Row 46, Program status		Waiting			Number of iterations			0

are entered into Cells B37–H39. Numbers, n, of acidic hydrogens and charges, z_0, on fully protonated acids are entered into Cells C42–D44.

Entries in Cells E42–E44 merit special discussion. Whereas most tabulated deprotonation constants correspond to zero ionic strength, some constants, such as those for ethylenediaminetetraacetic acid, EDTA, correspond to ionic strengths larger than zero. A Yes or No in Cells E42–E44 tells the program that the constants do or don't correspond to zero ionic strength. The program accounts for effects of ionic strength on constants corresponding the zero ion strength, i.e., activity-based (thermodynamic) constants. It doesn't account for effects of ionic strength on constants for ionic strengths other than zero, i.e., concentration-based (conditional) constants.

It is important that numerical values of all quantities not relevant to each situation be set to zero. Quantities in Cells F29, I29, and H42 are relevant to all situations.

STEPWISE PROCEDURE TO SOLVE PROBLEMS

The following is a stepwise procedure to solve any problem.

1. Set C29 to 0.
2. Enter values of constants, concentrations, numbers, n, of acidic hydrogens and charges, z_0, on fully protonated acids in appropriate cells.
3. Set all cells not relevant to each problem to zero.
4. Set Cells E42–E44 to "Yes" for constants at zero ionic strength and "No" for constants at ionic strength other than zero.
5. Set C29 to 1, wait for the iteration counter in I46 to stop changing and D46 to change to Solved. If the iteration counter stops changing before D46 changes to "Solved," press F9 one or more times to initiate additional iterations.
6. Read calculated results from output sections described below.

TABLE 5.1B

Part of the Output with Calculated Results for Pure Water

J27	K27	L27	M27	N27	O27	P27	Q27	R27
Calculated Quantities								
Ionic Strength (M)		1.00E-07	Be Skeptical of Results for Ionic Strengths Larger Than 0.1 M.					
Row 29	f_{Na}, f_{Cl}	1.00E+00	f_H	1.00E+00				
[H$^+$]	Uncorrected		1.00E-07	Corrected		1.00E-07	Error (%)	3.7E-02
[OH$^-$]	Uncorrected		9.99E-08	Corrected		1.00E-07	Error (%)	−1.1E-01
pH	Uncorrected		7.00	Corrected		7.00	Error (%)	−2.3E-03
Protons lost, i		0	1	2	3	4	5	6
Concentrations of Different Forms of Acids Uncorrected for Ionic Strength								
Row 35	Acid X	0.00E+00	0.00E+00	0.00E+00	0.00E+00	0.00E+00	0.00E+00	0.00E+00
Row 36	Acid Y	0.00E+00	0.00E+00	0.00E+00	0.00E+00	0.00E+00	0.00E+00	0.00E+00
Row 37	Acid Z	0.00E+00	0.00E+00	0.00E+00	0.00E+00	0.00E+00	0.00E+00	0.00E+00
Concentrations of Different Forms of Acids Corrected for Ionic Strength								
Row 39	Acid X	0.00E+00	0.00E+00	0.00E+00	0.00E+00	0.00E+00	0.00E+00	0.00E+00
Row 40	Acid Y	0.00E+00	0.00E+00	0.00E+00	0.00E+00	0.00E+00	0.00E+00	0.00E+00
Row 41	Acid Z	0.00E+00	0.00E+00	0.00E+00	0.00E+00	0.00E+00	0.00E+00	0.00E+00

SELECTED PARTS OF THE OUTPUT SECTIONS

Table 5.1B illustrates a part of the section of the program in which calculated concentrations and other quantities are summarized. Results are for pure water.

As expected, hydrogen ion and hydroxide concentrations in Cells M30, M31, P30, and P31 without and with correction for ionic strength are all 1.0×10^{-7} M and pH's in M30 and P30 are both 7.00.

Concentrations of forms of acids that have lost numbers of protons in Row 33 are in Rows 35–39. Concentrations in Cells L35–R37 are calculated without correction for ionic strength; concentrations in Rows L39–R41 are calculated with correction for ionic strength. Other quantities, including concentration fractions, activity-based deprotonation constants, and pK$_a$s are in Cells J47–R66. Contents of these cells will change to numerical values larger than zero when calculations are done for constants and concentrations larger than zero.

As shown in Table 5.1A and as stored on the program website http://crcpress.com/9781138367227, the program is set up to do calculations for pure water. To test the program for pure water, set C29 to 1 and wait until D46 changes to "solved" and read hydrogen ion concentrations without and with correction or ionic strength from Cells M30 and P30. The program should require about 910 iterations to calculate hydrogen ion concentrations equal to the expected value, [H$^+$] = 1.00×10^{-7} M. Set C29 to 0 to reset all cells to their reset conditions.

My natural inclination at this point is to describe how the program does the calculations. However, I believe it will serve you better to learn how to use the program to solve some examples first and then to learn how the program does what it does.

SOME EXAMPLES

Some example calculations are used here to illustrate types of situations to which the program can be applied as well as some procedures I used to evaluate accuracies of quantities calculated using the program.

TABLE 5.2

Deprotonation Constants for Selected Acids

Acetic Acid	Carbonic Acid, K_{a1}, and K_{a2}		Phosphoric Acid, K_{a1}, K_{a2}, and K_{a3}		
1.8×10^{-5}	$\mathbf{4.45 \times 10^{-7}}$	4.69×10^{-11}	7.11×10^{-3}	$\mathbf{6.32 \times 10^{-8}}$	4.5×10^{-13}
	Protonated EDTA, K_{a1}–K_{a6}				
1.0	3.2×10^{-2}	1.0×10^{-2}	2.2×10^{-3}	$\mathbf{6.9 \times 10^{-7}}$	5.8×10^{-11}

SITUATIONS WITHOUT CORRECTION FOR IONIC STRENGTH

As discussed in Chapter 1, conventional approximations can give accurate result for some situations but large errors for other situations. We will use several situations for which conventional approximations are expected to give accurate results to illustrate applications of the weak acids program as well as to test the accuracy of hydrogen ion concentrations calculated using it.

We will discuss situations involving acid-base pairs of acetic acid, carbonic acid, phosphoric acid, and ethylenediaminetetraacetic acid, EDTA, a doubly charged hexaprotic acid, H_6B^{2+}. Deprotonation constants for the acids are summarized in Table 5.2.

Notice that deprotonation constants in bold type differ from nearest-neighbor constants by 1000-fold or more. This means that approximate calculations for acid-base pairs corresponding to these constants can be done by treating them as monoprotic acids and bases.

Representing numbers of protons lost by i, each situation to be discussed involves 0.010 M hydrochloric acid plus concentrations, $C_i = 0.050$ M and $C_{i+1} = 0.060$ M, of the conjugate acid-base pair of the forms of an acid with deprotonation constant, K_{ai}. For example, for dihydrogen phosphate and monohydrogen phosphate, K_{a2} is associated with concentrations, C_1 and C_2, of forms of phosphoric acid that have lost i = 1 and i = 2 protons. This and other analogous combinations represent situations for which conventional approximations are expected to give hydrogen ion concentrations accurate to within 5%.

Assuming that hydrochloric acid reacts completely with the basic form of each acid-base pair and that sodium hydroxide reacts completely with the acidic form of each pair, hydrogen ion concentrations for the acids chosen for these calculations can be approximated with good accuracy as

$$\left[H^+ \right] \cong K_{a,i} \frac{(C_i + C_H)}{(C_{i+1} - C_H)} \qquad \left[H^+ \right] \cong K_{a,i} \frac{(C_i - C_{OH})}{(C_{i+1} + C_{OH})}$$

in which K_{ai} is the deprotonation constant for the form of the acid that has lost i protons, C_i and C_{I+1} are forms of each acid that have lost i and i + 1 protons, and C_H and C_{OH} are diluted HCl and NaOH concentrations.

Templates for situations containing 0.010 M hydrochloric acid with solutions containing concentrations of the four acids mentioned earlier are included in Cells T29–BF44 of the program on Sheet 1 of the weak acids program. To use each template, copy it and paste it into Cells A29–I44 on Sheet 1 of the weak acids program using the Paste (P) option. Then set C29 to 1, wait for Cell D46 to change from "Solving" to "Solved" and read approximate and "exact" hydrogen ion concentrations from I33 and M30.

Calculations for added HCl concentrations from $C_H = 0$ to 0.050 M can be done by changing I37 to the desired HCl concentration with I39 = 0. Calculations for added NaOH concentration from $C_{OH} = 0$ to 0.040 M can be done by setting I37 = 0 and I39 to the desired NaOH concentration.

TABLE 5.3

Comparisons of Hydrogen Ion Concentrations Calculated Using Approximation Procedures and an Exact Equation

	Monoprotic[a]	Diprotic[b]	Trprotic[c]	Hexaprotic[d]
Approximate [H+]	2.16×10^{-5} M	5.34×10^{-7} M	7.58×10^{-8} M	8.28×10^{-7} M
AB Program [H+]	2.16×10^{-5} M	5.35×10^{-7} M	7.59×10^{-8} M	8.28×10^{-7} M

Additions of $C_H = 0.010$ M HCl to:
[a] 0,050 M acetic acid plus 0.060 M acetate ion;
[b] 0.050 M carbonic acid plus 0.060 M bicarbonate ion;
[c] 0.050 M dihydrogen phosphate plus 0.060 M monohydrogen phosphate; and
[d] 0.050 M and 0.060 M of forms of EDTA that have lost i = 4 and i = 5 protons each.

Results I obtained for selected situations are summarized in Table 5.3.

As you can see, there is very good agreement between hydrogen ion concentrations calculated using the approximate and exact options. Given that conditions for the approximate calculations are selected to give accurate concentrations, these and several other analogous sets of results support the validity of the part of the weak acids program that does calculations without correction for ionic strength. A template to restore Sheet 1 of the weak acids program is included in Cells BH29–BP44.

CALCULATIONS WITH CORRECTION FOR IONIC STENGTH

Results calculated using the weak acids program are compared here with results reported by other authors.

Mononoprotic Weak Acid and the Conjugate Base of a Monoprotic Weak Acid

R. de Levie[4] used approximation procedures based on Davies' equation[7] to calculate the ionic strength, activity coefficient, and concentration-based (conditional) pK_a of a 0.10 M solution of a monoprotic weak acid with an activity-based deprotonation constant of $K_a = 1.00 \times 10^{-5}$ (Ref. 4, p. 414). A template used to adapt the weak acids program to this situation is included in Cells T50–AB65 on Sheet 1 of the acid-base program. This template can be used to calculate results for this situation by copying it and pasting it into Cells A29–I44.

de Levie used Davies's equation[7] to calculate activity coefficients and the weak acids program used the extended Debye/Huckle equation[8] to calculate activity coefficients. The ionic strength and $pK_{a,\mu}$ reported by de Levie are $\mu = 1.00 \times 10^{-3}$ and $pK_a = 4.97$; values calculated using the weak acids program are $\mu = 1.03 \times 10^{-3}$ and $pK_a = 4.97$.

de Levie also reported results for a 0.10 M solution of the conjugate base of the weak acid. The template described above can be adapted to this situation by changing B37 to 0.0 and C37 to 0.10. The ionic strengths and pK_a values calculated using the two options are the same, $\mu = 0.100$, $pK_{a,\mu} = 4.81$. Also, pH's are similar, 8.80 vs. 8.82.

Intermediate Form of a Triprotic Weak Acid

de Levie also reported results for a 0.10 M solution of disodium phosphate, Na_2HPO_4. A template used to adapt the weak acids program to this situation is included in Cells AD50–AL65 on Sheet 1 of the weak acids program. The template is used in the usual way. Results reported by de Levie and calculated using the weak acids program are summarized in Table 5.4.

The two options give identical values of the ionic strength and similar values of pH, concentration-based (conditional) pK_as and pK_w. Differences in pK values are all less than 3%.

TABLE 5.4

Comparison of Selected Quantities Using Different Approaches

	μ	pH_μ	$pK_{a2,\mu}$	$pK_{a3,\mu}$	$pK_{w,\mu}$
de Levie	0.30	9.09	6.59	11.36	13.34
AB program	0.30	8.99	6.65	11.51	13.71
Difference	<0.1%	−1.1%	0.9%	1.3%	2.8%

TABLE 5.5

Effects of Ionic Strength on Experimentally Determined[a] and Calculated pK_a Values for Acetic Acid

μ	0.02	0.03	0.06	0.11	0.21	0.51	1.01
$pK_{a,\mu}^a$	4.64	4.62	4.58	4.54	4.51	4.48	4.50
$pK_{a,\mu}^b$	4.64	4.63	4.59	4.56	4.52	4.47	4.43
Difference	0.2%	0.2%	0.2%	0.2%	0.2%	0.2%	1.6%

[a] Harned and Hickey, Table II, p. 1287, $pK_{a0} = 4.754$, $K_{a0} = 1.78 \times 10^{-5}$ @ $\mu = 0$.

[b] Weak acids program.

Deprotonation Constant for Acetic Acid

Harned and Hickey[9] have reported results for effects of sodium chloride on pK_a values for acetic acid. A template for calculations for $\mu = 0.020$ M using the weak acids program is included in Cells AN50–AV65. Having pasted this template into Cells A29–I44 using the Paste (P) option, calculations for this and other ionic strengths can be done by setting HCl concentrations in I37 to the desired ionic strengths. $pK_{a\mu}$ values are in Cell L74.

Harned and Hickey's results for ionic strengths from 0.020 to 1.01 M are included in Table 5.5 along with results calculated using the weak acids program.

Values of pK_a calculated using the weak acids program were calculated by using the HCl concentration in I37 to control the ionic strengths. This forced the program to go through the iterative process to calculate pK_a values. This provides a more valid test of the ability of the program to account for effects of ionic strength than control of the ionic strength using an inert salt such as potassium chloride.

In any event, the small differences between experimentally determined and calculated pK_as in Table 5.5 lend support for the validity of this part of the weak acids program.

The agreement for the larger ionic strengths is a bit surprising given that the Debye-Hückle equation isn't valid for ionic strengths larger than about 0.10 M.

MIXTURES OF UP TO THREE WEAK ACIDS AND THEIR CONJUGATE BASES

To this point we have limitized our attention to situations involving just one weak acid and/or a conjugate base. As mentioned earlier, the weak acids program is designed to do calculations for different forms of up to three weak acids with up to six acidic hydrogens each without and with correction for ionic strength. We will talk here about solutions containing different forms of ethylenediaminetetraacetic acid, EDTA, a doubly charged hexaprotic acid commonly represented as H_6Y^{2+}, ammonium ion, and ammonia plus carbonic acid and bicarbonate produced by dissolved carbon dioxide.

Let's talk first about the rationale for the calculations we will discuss.

As noted above, fully protonated EDTA is a doubly charged hexaprotic acid commonly represented by H_6Y^{2+}. Using this representation, the stepwise deprotonation of protonated EDTA is as follows:

$$H_6Y^{2+} \underset{-H^+}{\overset{K_{a1}}{\rightleftharpoons}} H_5Y^+ \underset{-H^+}{\overset{K_{a2}}{\rightleftharpoons}} H_4Y \underset{-H^+}{\overset{K_{a3}}{\rightleftharpoons}} H_3Y^- \underset{-H^+}{\overset{K_{a4}}{\rightleftharpoons}} H_2Y^{2-} \underset{-H^+}{\overset{K_{a5}}{\rightleftharpoons}} HY^{3-} \underset{-H^+}{\overset{K_{a6}}{\rightleftharpoons}} Y^{4-}$$

Solutions containing EDTA are commonly prepared using the disodium salt, $Na_2H_2Y \cdot 2H_2O$, which dissociates to give sodium ions and H_2Y^{2-}.

The fully deprotonated form of EDTA, Y^{4-}, is commonly used as a titrant for metal ions. For example, the reaction with Zn^{2+} is as follows:

$$Zn^{2+} + Y^{4-} \rightleftharpoons ZnY^{2-}$$

Therefore, it is important that a significant fraction, α_6, of EDTA be in the form that reacts with metal ions, i.e., in the fully deprotonated form, Y^{4-}. Let's calculate hydrogen ion concentration and fractions in the fully deprotonation forms for a solution containing $C_4 = 0.010$ M disodium dihydrogen EDTA without and with correction for ionic strength.

A template for this situation is in Cells T71–AB86 on Sheet 1 of the weak acids program. To use the template, copy it and paste it into Cells A29–I44 using the Paste (P) option. After you paste the template into the input section of the program, notice that Cell E42 is set to "No." This tells the program that the constants don't correspond to zero ionic strength. The program doesn't adjust these constants for effects of ionic strength but rather uses them as is for all calculations.

To solve the problem, set C29 to 1, wait for D46 to change to "Solved" and read the ionic strength, hydrogen ion concentrations, and fractions of EDTA in the fully deprotonated forms without and with correction for ionic strength from Cells L28, M30, P30, R49, and R53, respectively.

Results you should obtain are summarized below.

μ (L28)	$[H^+]_0$ (M30)	$[H^+]_\mu$ (P30)	$\alpha_{6,0}$ (R49)	$\alpha_{6,\mu}$ (R53)
0.0303 M	3.51×10^{-5} M	3.51×10^{-5} M	3.14×10^{-8}	3.14×10^{-8}

Note: Subscripted 0s and μs here and elsewhere represent quantities for $\mu = 0$ and $\mu > 0$, respectively.

You may (should) wonder why hydrogen ion concentrations ostensibly without and with correction for ionic strengths are the same. Given that deprotonation constants for EDTA correspond to an ionic strength of 0.10 M, calculations without and with correction for ionic strength correspond to an ionic strength of $\mu = 0.10$ M.

More importantly, fractions, α_6, of EDTA in the fully deprotonated form, Y^{4-}, that reacts with metal ions are very small, i.e., about 3.1×10^{-8}. The small fractions mean that a very small percentage of EDTA will be in the form that reacts with metal ions. So, what can we do to increase these fractions? We can buffer the solution at a smaller hydrogen ion concentration, larger pH, to force the EDTA reaction further to the right. For reasons to be discussed in a later chapter, it is common practice to use ammonium ion/ammonia buffers near pH = 10 for EDTA titrations.

A template for a buffer solution containing 0.0175 M ammonium chloride and 0.10 M ammonia is in Cells AD71–AL86 as Acid Y. I suggest that you use the template to calculate the hydrogen ion concentration of the buffer. Selected results I obtained using the template are summarized below.

μ (L28)	$[H^+]_0$ (M30)	$[H^+]_\mu$ (P30)
0.0176 M	1.01×10^{-10}	9.91×10^{-11}

In the absence of any other acidic or basic reactants, the buffer controls the hydrogen ion concentration near the desired value of 1.0×10^{-10}, pH = 10.0. What happens if we combine the buffer with 0.010 M EDTA?

A template for a solution containing 0.0175 M ammonium chloride, 0.10 M ammonia, and $C_{4,Y} = 0.010$ M of H_2Y^{2-} is included in Cells AN71–AV86. Selected results I obtained using the template are summarized below.

μ (L28)	$[H^+]_0$ (M30)	$[H^+]_\mu$ (P30)	$\alpha_{6,0}$ (R49)	$\alpha_{6,\mu}$ (R53)
0.087 M	1.95×10^{-10} M	1.86×10^{-10} M	0.229	0.238

The most significant observation from these results is that fractions, α_6, of EDTA in form that react with metal ions are much larger than values without the buffer. In other words, a larger percentages of EDTA is in the fully deprotonated form, Y^{4-}, that reacts with metal ions.

Other observations are that the ionic strength is larger than that in any of the other solutions and that inclusion of 0.010 M H_2Y^{2-} causes a significant increase in the hydrogen ion concentration. The increased ionic strength results from larger percentages of the EDTA in more highly charged forms, HY^{3-} and Y^{4-}. As shown in Cells R49 and R53, 23 to 24% of the EDTA is in the form Y^{4-} form. Although not shown above, fractions of EDTA in the HY^{3-} form, Cells Q49 and Q53, are $\alpha_{5,0} = 0.77$ and $\alpha_{5,\mu} = 0.76$ corresponding to 76 to 77% of the EDTA in the triply charged form, HY^{3-}.

The increased hydrogen ion concentration results from stepwise deprotonation reactions of H_2Y^{2-}, i.e.,

$$H_2Y^{2-} \overset{K_{a5}}{\rightleftharpoons} H^+ + HY^{3-} \overset{K_{a6}}{\rightleftharpoons} H^+ + Y^{4-}$$

Each step in this reaction tends to increase the hydrogen ion concentration. Whereas some of the hydrogen ions produced by these reactions will react with ammonia to produce ammonium ion, some will remain unreacted producing the observed increase in the hydrogen ion concentration.

Now, let's talk about one additional feature of EDTA solutions used to titrate metal ions. The solutions are usually prepared in air-saturated solutions containing dissolved carbon dioxide which reacts with water to produce predominately carbonic acid and bicarbonate ion, i.e.,

$$CO_2 + H_2O \rightleftharpoons H_2CO_3 \overset{K_{a1}=4.45E-7}{\rightleftharpoons} H^+ + HCO_3^- \overset{K_{a2}=4.69E-11}{\rightleftharpoons} H^+ + CO_3^{2-}$$

We can use the weak acids program to determine if carbonic acid and bicarbonate ion resulting from dissolved carbon dioxide will have a significant effect on the hydrogen ion concentration.

It can be shown[10] that a saturated solution of carbon dioxide at 25°C will contain approximately 1.0×10^{-5} M carbonic acid and 2.1×10^{-6} M bicarbonate ion. Will these concentrations of carbonic acid and bicarbonate ion have a significant effect on the hydrogen ion concentration in solutions described above? To answer this question, set C29 to zero then set B34 and C34 to deprotonation constants for carbonic acid, $K_{a1} = 4.45E-7$ and $K_{a2} = 4.69E-11$, set B39 and C39 to concentrations of carbonic acid and bicarbonate, $C_0 = 1.0E-5$ and $C_1 = 2.1E-6$ and set C44 to n = 2. Then set C29 to 1 and wait for D46 to change to solved. Then read hydrogen ion concentrations from M30 and P30.

Results I obtained without and with carbon dioxide are summarized below.

Without Carbon Dioxide		**With Carbon Dioxide**	
$[H^+]$ (M)	$[H^+]_\mu$ (M)	$[H^+]$ (M)	$[H^+]_\mu$ (M)
1.95×10^{-10}	1.86×10^{-10}	1.95×10^{-10}	1.86×10^{-10}

Calculations predict that dissolved carbon dioxide has little or no effect on the hydrogen ion concentration or quantities that depend on the hydrogen ion concentration. On the one hand, this conclusion is predictable based on the small values of the carbonic acid and bicarbonate ion concentrations relative to concentrations of other acids and bases. On the other hand, it is helpful to be able to test predicted conclusions. Moreover, it is helpful to be able to do such calculations for situations for which it isn't so easy to predict the results.

Let's talk next about how the weak acids program calculates hydrogen ion concentrations and related quantities.

SOLVING THE CHARGE-BALANCE EQUATION FOR THE HYDROGEN ION CONCENTRATION

The program used to calculate hydrogen ion concentrations is based on a feature of charge-balance equations, CBE's, discovered during the writing of this text. Specifically, I observed that charge-balance equations written by equating differences between contributions of positive and negative charges to zero is negative for hydrogen ion concentrations smaller than the equilibrium concentration and positive for hydrogen ion concentrations larger than the equilibrium concentration.

A Two-Step Process

The weak acids program uses similar two-step processes to resolve hydrogen ion concentrations without or with correction for ionic strength. I will talk first about the two-step process in Cells H71–I72 for calculations without correction for ionic strength.

A *low-resolution step* in Cells H71 and I71 starts with very small trial value of the hydrogen ion concentration, e.g., 1×10^{-15} M, in I71 and increases it in 10-fold steps until the charge-balance equation in H71 changes from $-$ to $+$. The hydrogen ion concentration in I71 that causes the $-$ to $+$ change is the upper limit of a 10-fold range of concentrations that includes the equilibrium concentration.

Then, a *high-resolution step* in Cells H72 and I72 starts with a hydrogen ion concentration in I72 equal to one tenth of the concentration from the low-resolution step and increases it by small increments until the charge-balance equation in H72 changes from $-$ to $+$. The hydrogen ion concentration that causes the $-$ to $+$ change in the charge-balance equation is the equilibrium concentration resolved to the number of significant figures in I29.

With I29 set to 3, the hydrogen ion concentration in I72 is resolved to three significant figures. A hydrogen ion concentration of $[H^+] = 1.95 \times 10^{-10}$ M, is used below to explain how the program resolves the H^+ concentration to three significant figures.

Low-Resolution Step

In this step, values of the charge-balance equation are calculated in H71 for trial values of the H^+ concentration in I71. Hydrogen ion concentrations in I71 are controlled by the logical function

$$= IF\left(C29 = 0, F29, IF\left(H71 < 0, 10 * I71, I71\right)\right)$$

Using Excel syntax, this function reads as follows. If C29 is equal to zero, set I71 to the concentration in F29, e.g., 1.0×10^{-15}. Then, if the charge-balance equation in H71 is less than zero, multiply the concentration in I71 by 10. Then, if H71 is larger than zero, set I71 to the concentration in I71 that caused the $-$ to $+$ change in the CBE.

Given that the H^+ concentration is increased in 10-fold steps and that the concentration that produces the $-$ to $+$ change is the first concentration to exceed the equilibrium concentration, the concentration in I71 is the upper limit of a 10-fold range of concentrations that includes the equilibrium concentration.

For an equilibrium hydrogen ion concentration of 1.95×10^{-10}, the upper limit concentration is 1.0×10^{-9} M. This means that the equilibrium concentration is between 10^{-10} and 10^{-9} M. This is where the high-resolution step starts.

High-Resolution Step

With I29 set to 3, the high-resolution step in Cells H72 and I72 resolves H^+ concentrations in the 10-fold range found in the low-resolution step to three significant figures.

The formula in H72 calculates values of the charge-balance equation, CBE, for trial values of the H^+ concentration in I72. The logical function that controls the H^+ concentration in I72 is

$$= \Big(IF\big(H71 < 0, 0.1 * I71, IF\big(H72 < 0, I72 + 10 \wedge -I\$29 * I71, I72\big)\big)\Big)$$

With I29 set to 3, this logical function is translated into words as follows.

If the CBE in H71 is less than zero, set I72 to one-tenth the concentration in I71, i.e., to $0.1(1.0 \times 10^{-9}) = 1.0 \times 10^{-10}$. Otherwise, if the CBE in H72 is negative, set I72 to its current value $+ 10^{-3}(I71) =$ current value $+ 10^{-3}(1.0 \times 10^{-9}) =$ current value $+ 10^{-12}$ M. Otherwise, if H72 is positive, stop with the H^+ concentration that caused the $-$ to $+$ change in I72.

Based on this sequence, the hydrogen ion concentrations in I72 from start to finish of the high-resolution step are 1.00×10^{-10}, 1.01×10^{-10}, 1.02×10^{-10}, ..., 1.93×10^{-10}, 1.94×10^{-10}, and 1.95×10^{-10}. The concentration that causes the $-$ to $+$ change, 1.95×10^{-10} M, is the equilibrium concentration resolved to three significant figures. It follows that percentage errors resulting from the iterative process will vary from 1% at the lower limit of each 10-fold range to 0.1% at the upper limit of each range.

The same reasoning applies for each 10-fold range. For example, for an equilibrium concentration of $[H^+] = 8.35 \times 10^{-6}$ M with I29 = 3, trial concentrations from start to finish of the high-resolution step are 1.00×10^{-6}, 1.01×10^{-6}, 1.02×10^{-6}, ..., 8.33×10^{-6}, 8.34×10^{-6} and 8.35×10^{-6}.

In summary, with F29 set to 10^{-15} and I29 set to 3, the program will resolve any concentration larger than 10^{-15} M to 3 significant figures. More generally, with F29 set to 10^{-15}, the program will resolve any concentration larger than 10^{-15} M to the number of significant figures in I29.

Poorer resolution and shorter convergence times or better resolution and longer convergence times can be obtained by setting I29 to values smaller than or larger than 3, e.g., 2 or 4. However, given that maximum errors for three significant figures are well below errors imposed by uncertainties in quantities such as deprotonation constants, resolution to three significant figures is a good compromise between quantitative resolution and convergence time.

The same process is used in Cells H82–I83 for calculations with correction for ionic strength. The principal difference is that concentration-based (conditional) constants are calculated and used in place of activity-based (thermodynamic) constants.

CONCENTRATIONS OF DIFFERENT FORMS OF WEAK ACIDS

Hydrogen ion concentrations calculated in the foregoing processes are used to calculate concentration fractions which are summarized in Cells L49–R55. Concentration fractions are then used with total concentrations in Cells I64–I66 to calculate equilibrium concentrations of different forms of the weak acids which are summarized in Cells L35–R41. Concepts and equations associated with calculation and use of concentration fractions are discussed in Chapter 3.

SOME FEATURES OF THE ITERATIVE PROGRAM RELATIVE TO OTHER OPTIONS

The iterative program described above differs from procedures mentioned earlier[3–6] in that it doesn't require the use of graphical results nor does it involve any approximations. The program differs from Solver programs included with most spreadsheets and some programmable calculators in at least two distinct ways. First, it doesn't require and initial estimate of the hydrogen ion concentration. Second, it gives the user direct control of the number of significant figures to which the hydrogen ion concentration is resolved.

BASIS FOR CHARGE-BALANCE EQUATIONS

The spreadsheet program described above calculates the hydrogen ion concentration by solving a charge-balance equation and uses the hydrogen ion concentration to calculate concentration fractions and concentrations of different forms of the weak acids.

Charge-balance equations, CBE's, are based on the fact that the electrical charge on any solution at equilibrium must be zero. On the one hand, there is only one charge-balance equation for each solution. On the other hand, it is convenient to describe three equivalent forms of the CBE for each solution.

The *ionic form* of a CBE expresses the equation in terms of ionic concentrations of all reactants and products. This is the starting point for any situation. The *fractional form* of a charge-balance equation expresses the equation in terms of the H^+ concentration, the autoprotolysis constant, K_w, for water, diluted concentrations of strong electrolytes, fractions, α_i, of the different forms of each weak acid and the total concentration, C_T, of all forms of each weak acid. Fractional forms of charge-balance equations are used herein to illustrate some special features of the equations that facilitate adaptations to different situations.

The *computational form* of the charge-balance equation for each situation expresses the charge-balance equation in terms of the H^+ concentration and known quantities, including diluted concentrations of all reactants, cumulative constants, etc. As the name implies, the computational form of the charge-balance equation is the form solved for the hydrogen ion concentration.

SOME APPROXIMATE RESULTS

Let's use a simple example to associate some numbers with the concept of a charge-balance equation.

Given a solution containing 0.10 M each of acetic acid and sodium acetate, the hydrogen ion concentration can be approximated as

$$\left[H^+ \right] \cong K_a \frac{C_{Ac}}{C_{HAc}} \cong 1.8 \times 10^{-5} \frac{0.10}{0.10} \cong 1.8 \times 10^{-5} \text{ M}$$

It follows that equilibrium concentrations of all reactants and products can be approximated as $[H^+] \cong 1.8 \times 10^{-5}$ M, $[HAc] \cong 0.10 - 1.8 \times 10^{-5} \cong 0.10$ M, $[Ac^-] \cong 0.10 + 1.8 \times 10^{-5} \cong 0.10$ M, $[Na^+] = C_{NaAc} = 0.10$ M, and $[OH^-] \cong K_w/[H^+] \cong 1.0 \times 10^{-14}/1.8 \times 10^{-5}$ M $\cong 5.6 \times 10^{-10}$ M. The charge-balance equation can be approximated as the algebraic sum of concentrations of positive and negative ions, i.e.,

$$CBE = \left[Na^+ \right] + \left[H^+ \right] - \left[Ac^- \right] - \left[OH^- \right] \cong 0.10 \text{ M} + 1.8 \times 10^{-5} - 0.10 \text{ M} - 5.6 \times 10^{-10}$$
$$\cong 1.8 \times 10^{-5}$$

In other words, the numerical value of the charge-balance equation for this situation is approximately 1.8×10^{-5} M. Notice that the acetic acid concentration, $[HAc] \cong 0.10$ M, isn't included in the charge-balance equation because acetic acid is an uncharged acid.

I hope that these numbers will have given you a better understanding of a relatively simple charge-balance equation.

IONIC AND FRACTIONAL FORMS OF CHARGE-BALANCE EQUATIONS

It is assumed throughout this discussion that there are no side reactions such as volatilization or precipitation. It is also assumed that negatively charged ions such as acetate or phosphate are added as sodium salts, that positively charged ions such as the ammonium ion or protonated amines are added as chloride salts and that strong acids, strong bases, and salts such as sodium acetate and alanine hydrochloride dissociate completely.

The contribution of each ion in a solution to the total charge is proportional to the product of the signed charge on the ion times the equilibrium concentration of the ion. As examples, using signed charges on H_3PO_4, HB^+, H^+, Cl^-, CO_3^{2-}, HPO_4^{2-} and PO_4^{3-} it follows that contributions of these species to the charge on a solution containing them are proportional to $0 \times [H_3PO_4]$, $+1 \times [HB^+]$, $+1 \times [H^+]$, $-1 \times [Cl^-]$, $-2 \times [CO_3^{2-}]$, and $-3 \times [PO_3^{3-}]$.

The charge-balance equation for any solution can be written either by equating contributions of positive and negative charges to each other or by equating the algebraic sum of contributions of positive and negative ions to zero. Charge-balance equations are written herein by equating algebraic sums of contributions of positive and negative ions to zero.

I will derive ionic and fractional forms of the charge-balance equation for each of two situations and use patterns in those equations to illustrate some important features of charge-balance equations.

Solution Containing HCl, NaOH, and Different Forms of an Uncharged Triprotic Acid

Example 5.1A illustrates a process used to develop ionic and fractional forms of the charge-balance equation for a solution containing HCl, NaOH, and any or all forms of an uncharged triprotic acid such as phosphoric acid.

> **Example 5.1A Derive the ionic and fractional forms of a charge-balance equation for an aqueous solution containing concentrations equal to or larger than zero of HCl, NaOH, an uncharged triprotic acid, H_3X, and the sodium salts of the ionic forms of the acid, NaH_2X, Na_2X, and Na_3X.**

Step 1: Write dissociation reactions for water, HCl, NaOH, and the sodium salts of the three ionic forms of the acid, H_3X, and underline one **each** of the different ions.

$$H_2O \rightleftharpoons \underline{H^+} + \underline{OH^-}$$

$$HCl \rightarrow H^+ + \underline{Cl^-} \quad NaOH \rightarrow \underline{Na^+} + OH^-$$

$$NaH_2X \rightarrow Na^+ + \underline{H_2X^-} \quad Na_2HX \rightarrow 2Na^+ + \underline{HX^{2-}} \quad Na_3X \rightarrow 3\,Na^+ + \underline{X^{3-}}$$

Step 2: Write an ionic form of the charge-balance equation by equating the *algebraic sum* of products of signed charges times equilibrium concentrations of all underlined ions to zero (Equation 5.2a).

$$CBE = \left[H^+\right] - \left[OH^-\right] + \left[Na^+\right] - \left[Cl^-\right] - \left[H_2X^-\right] - 2\left[HX^{2-}\right] - 3\left[X^{3-}\right] = 0 \qquad (5.2a)$$

Step 3: Use a rearranged form of the autoprotolysis constant for water to express the OH^- concentration in terms of the H^+ concentration.

$$\left[OH^-\right] = \frac{K_w}{\left[H^+\right]}$$

Step 4: Use mass-balance equations to express the Cl^-, Na^+, and total concentration of all forms of the weak acid in terms of diluted concentrations of reactants.

$$\left[Na^+\right] = C_{NaOH} + C_{1x} + 2C_{2x} + 3C_{3x}, \qquad \left[Cl^-\right] = C_{HCl}, \qquad C_{TX} = C_{0x} + C_{1x} + C_{2x} + C_{3x}$$

Step 5: Write equations for the charged forms of the acid in terms of concentration fractions, α_i, and the total concentration, C_T.

$$\left[H_2X^-\right] = \alpha_{1x}C_T, \qquad \left[HX^{2-}\right] = \alpha_{2x}C_T, \qquad \left[X^{3-}\right] = \alpha_{3x}C_{Tx}$$

Step 6: Write a fractional form of the charge-balance equation by replacing ionic concentrations in Step 2 with corresponding equations in Steps 3–5 and rearranging the equation into the form below.

$$\overbrace{\left[H^+\right] - \frac{K_w}{\left[H^+\right]} + \left(C_{NaOH} - C_{HCl}\right)}^{\text{Term independent of the weak acid}} + \overbrace{\left(C_{1x} + 2C_{2x} + 3C_{3x}\right) - \left(\alpha_{1x} + 2\alpha_{2x} + 3\alpha_{3x}\right)C_{Tx}}^{\text{Weak–acid term for } H_3X} = 0 \qquad (5.2b)$$

Symbols, C_{1x} and α_{ix}, represent diluted concentrations and fractions of Acid X in forms that have lost i protons and C_{Tx} is the total diluted concentration of all forms of Acid X.

The first step is to write all reactions that produce different ions and to identify one each of the different ions using underlines or some other option. Only one concentration of each ion is included in the ionic form of the CBE because that concentration represents the total concentration of the ion from all sources. Also, assuming that the reaction of a salt of each charged form of the weak acid is included as is necessary to obtain a complete equation, it isn't necessary to include stepwise deprotonation reactions for the acid. As you can show for yourself, the stepwise deprotonation reactions for the triprotic acid don't add any ions different than those produced by dissociation of the salts.

Equations in Steps 3–6 are used to express ionic concentrations from different sources in terms of the H^+ concentration and known quantities, i.e., diluted concentrations and constants. I will trust you to review discussions in Chapter 3 of any equations in Steps 3–5 about which you may have questions.

Solution Containing HCl, NaOH, and a Mixture of Different Forms of Two Weak Acids

Example 5.1B extends the foregoing process to a solution containing HCl, NaOH, and any or all forms of a singly charged triprotic acid and an uncharged diprotic acid.

Example 5.1B Derive the charge-balance equation for an aqueous solution containing concentrations equal to or larger than zero of HCl, NaOH, H_3XCl, H_2X, NaHX, Na_2X, H_2Y, NaHY, and Na_2Y in which H_3X^+ is a singly charged triprotic weak acid such as fully protonated iminodiacetic acid and H_2Y is an uncharged diprotic acid such as oxalic acid.

Step 1: Write dissociation reactions for water, HCl, NaOH, the chloride and sodium salts of different forms of the singly charged triprotic acid, H_3X^+, and the sodium salts of the forms of an uncharged diprotic acid, H_2Y.

$$H_2O \underset{}{\overset{K_w}{\rightleftharpoons}} \underline{H^+} + \underline{OH^-}$$

$$HCl \rightarrow H^+ + \underline{Cl^-} \qquad NaOH \rightarrow \underline{Na^+} + OH^-$$

$$H_3XCl \rightarrow Cl^- + \underline{H_3X^+} \qquad NaHX \rightarrow Na^+ + \underline{HX^-} \qquad Na_2X \rightarrow 2Na^+ + \underline{X^{2-}}$$

$$NaHY \rightarrow Na^+ + \underline{HY^-}, \qquad Na_2Y \rightarrow 2Na^+ + \underline{Y^{2-}}$$

Note: The stepwise deprotonation reactions of H_3X^+ and H_2Y don't introduce any additional ions and therefore are not included.

Step 2: Write an ionic form of the charge-balance equation by equating the algebraic sum of products of signed charges times equilibrium concentrations of all ions to zero.

$$\mathrm{CBE} = \left[H^+\right] - \left[OH^-\right] + \left[Na^+\right] - \left[Cl^-\right] + \left[H_3X^+\right] - \left[HX^-\right] - 2\left[X^{2-}\right] - \left[HY^-\right] - 2\left[Y^{2-}\right] = 0 \quad (5.2c)$$

Step 3: Use procedures as in Steps 3–5 of Example 5.1A to express the ionic concentrations in terms of the H^+ concentration, diluted concentrations and concentration fractions.

$$\left[OH^-\right] = \frac{K_w}{\left[H^+\right]}, \qquad \left[Cl^-\right] = C_{HCl} + C_{0X},$$

$$\left[Na^+\right] = C_{NaOH} + C_{2X} + 2C_{3X} + C_{1Y} + 2C_{2Y}$$

$$\left[H_3X^+\right] = \alpha_{0X}C_{TX}, \quad \left[HX^-\right] = \alpha_{2X}C_{TX}, \quad \left[X^{2-}\right] = \alpha_{3X}C_{Tx}$$

$$[HY^-] = \alpha_{1Y}C_{TY}, \quad \left[Y^{2-}\right] = \alpha_{2Y}C_{Ty},$$

Step 4: Write a fractional form of the charge-balance equation by replacing ionic concentrations in Equation 5.2c with corresponding equations in Step 3 and collect terms into the form in Equation 5.2d.

$$\overbrace{\left[H^+\right] - \frac{K_w}{\left[H^+\right]} + \left(C_{NaOH} - C_{HCl}\right)}^{\text{Term independent of the weak acid}} \overbrace{- \left(C_{0X} - C_{2X} - 2C_{3X}\right) + \left(\alpha_{0X} - \alpha_{2X} - 2\alpha_{3X}\right)C_{TX}}^{\text{Weak−acid term for } H_3X^+}$$

$$\underbrace{+ \left(C_{1Y} + 2C_{2Y}\right) - \left(\alpha_{1Y} + 2\alpha_{2Y}\right)C_{TY}}_{\text{Weak−acid term for } H_2Y} = 0 \quad (5.2d)$$

Symbols, C_{ix} and α_{ix}, represent diluted concentrations and fractions of Acid X in forms that have lost i protons and C_{TX} is the total diluted concentration of all forms of Acid X. Symbols, C_{iY} and α_{iY}, represent diluted concentrations and fractions of Acid Y in forms that have lost i protons and C_{TY} is the total diluted concentration of all forms of Acid Y.

We will talk later about computational forms of charge-balance equations.

SOME USEFUL FEATURES OF CHARGE-BALANCE EQUATIONS

Equations 5.2b and 5.2d are used here to illustrate some useful features of charge-balance equations. The equations are reprinted below for your convenience.

$$\overbrace{\left[H^+\right] - \frac{K_w}{\left[H^+\right]} + \left(C_{NaOH} - C_{HCl}\right)}^{\text{Term independent of the weak acid}} + \overbrace{\left(C_{1x} + 2C_{2x} + 3C_{3x}\right) - \left(\alpha_{1x} + 2\alpha_{2x} + 3\alpha_{3x}\right)C_{Tx}}^{\text{Weak−acid term for } H_3X} = 0 \quad (5.2b)$$

$$\overbrace{\left[H^+\right] - \frac{K_w}{\left[H^+\right]} + \left(C_{NaOH} - C_{HCl}\right)}^{\text{Term independent of the weak acid}} \overbrace{- \left(C_{0X} - C_{2X} - 2C_{3X}\right) + \left(\alpha_{0X} - \alpha_{2X} - 2\alpha_{3X}\right)C_{TX}}^{\text{Weak−acid term for } H_3X^+}$$

$$\underbrace{+ \left(C_{1Y} + 2C_{2Y}\right) - \left(\alpha_{1Y} + 2\alpha_{2Y}\right)C_{TY}}_{\text{Weak−acid term for } H_2Y} = 0 \quad (5.2d)$$

Symbols are as defined in the last lines of the examples.

Additivity of Terms for Different Acids

Annotations in Equation 5.2d show that the equation consists of the algebraic sum of three terms, namely a term that is independent of the weak acids, a term for the triprotic acid, and a term for the diprotic acid. This observation can be generalized. The charge-balance equation for any solution containing two or more weak acids will consist of the sum of a set of terms that is independent of the weak acid plus a set of terms for each weak acid. This and other features of the equations discussed below are used later to write an equation for solutions containing HCl, NaOH, and any or all forms of three weak acids with up to six acidic hydrogens each as the sum of the independent term and a set of terms for each weak acid.

Weak-Acid Terms for Charged and Uncharged Acids with Equal Numbers of Acidic Hydrogens

Weak acid terms for uncharged and singly charged triprotic acids in Equations 5.2b and 5.2d appear to be different. However, it is relatively easy to show that these terms are, in fact, equivalent to each other. To do this, it will be helpful to express the fraction, α_0, in Equation 5.2d in terms of the other fractions, α_1, α_2, and α_3.

Starting with the fact that the sum of the fractions, α_0 to α_n, for any weak acid with n acidic hydrogens is equal to one, relationships among fractions for a triprotic weak acid can be written in the following forms

$$\alpha_0 + \alpha_1 + \alpha_2 + \alpha_2 = 1 \Rightarrow \alpha_0 = 1 - (\alpha_1 + \alpha_2 + \alpha_3)$$

In other words, the fraction, α_0, is one minus the sum of the fractions, α_1–α_3.

Example 5.2 shows how this relationship is used to compare the weak-acid term for the singly charged triprotic acid in Equation 5.2d to the analogous term for the uncharged acid in Equation 5.2b. I have dropped the X subscripts in Equation 5.2d to simplify and generalize the example.

> **Example 5.2 Show that the weak-acid term in Equation 5.2d for a singly charged triprotic acid, H_3X^+, is equivalent to the weak acid term in Equation 5.2b for an uncharged triprotic acid, H_3X.**
>
> *Step 1:* Write the weak-acid term for the charged triprotic weak acid.
>
> $$-(C_0 - C_2 - 2C_3) + (\alpha_0 - \alpha_2 - 2\alpha_3)C_T$$
>
> *Step 2:* Replace α_0 in the weak-acid term with $\alpha_0 = 1 - (\alpha_1 + \alpha_2 + \alpha_3)$ and expand the term into the form below.
>
> $$-(C_0 - C_2 - 2C_3) + (1 - (\alpha_1 + \alpha_2 + \alpha_3) - \alpha_2 - 2\alpha_3)C_T$$
>
> *Step 3:* Substitute $1C_T = C_0 + C_1 + C_2 + C_3$ and rewrite the equation in Step 2 as shown below.
>
> $$-(C_0 - C_2 - 2C_3) + (C_0 + C_1 + C_2 + C_3) - (\alpha_1 + \alpha_2 + \alpha_3)C_T - (\alpha_2 + 2\alpha_3)C_T$$
>
> *Step 4:* Collect terms into the form below and compare to Equation 5.2b.
>
> Singly charged triprotic acid (Eq. 5.2d)   Uncharged triprotic acid (Eq. 5.2b)
>
> $(C_1 + 2C_2 + 3C_3) - (\alpha_1 + 2\alpha_2 + 3\alpha_3)C_T$ $(C_1 + 2C_2 + 3C_3) - (\alpha_1 + 2\alpha_2 + 3\alpha_3)C_T$

TABLE 5.6A

Comparison of Weak-Acid Terms for Diprotic and Triprotic Acids

Weak Acid Terms for Diprotic and Triprotic Acids from Example 5.1B.	
$n = 2$	$C_1 + 2C_2 - (\alpha_1 + 2\alpha_2)(C_0 + C_1 + C_2)$
$n = 3$	$C_1 + 2C_2 + 3C_3 - (\alpha_1 + 2\alpha_2 + 3\alpha_3)(C_0 + C_1 + C_2 + C_3)$

Except for the subscripted X's in Equation 5.2b, the weak-acid terms for uncharged and singly charged triprotic acids are the same. This observation can be generalized. Weak-acid terms for uncharged and charged acids with equal numbers of acidic hydrogens are equivalent to each other. It also follows that computational forms of equations for charged and uncharged acids with equal numbers of acidic hydrogens are equivalent to each other. Therefore, until we begin to talk about effects of ionic strength, all that we learn about uncharged acids also applies for charged acids with equal numbers of acidic hydrogens.

Adapting the Weak-Acid Term for One Acid to an Acid with Fewer and More Acidic Hydrogens

Weak-acid terms for diprotic and triprotic acids from Examples 5.1A and 5.1B are summarized Table 5.6A.

Notice that the equation for the triprotic acid can be obtained from that for the diprotic acid by adding $3C_3$, $3\alpha_3$, and C_3 terms. Conversely, the weak-acid term for a diprotic acid can be obtained from that for a triprotic acid by dropping the same three terms. In general, *the weak acid term for any acid with any number of acidic hydrogens can be obtained from that for an acid with a different number of acidic hydrogens by dropping or adding appropriate iC_i, $i\alpha_i$, and C_i terms for i from 0 to the number, n, of acidic hydrogens, i.e., $0 \leq i \leq n$.*

As examples, weak-acid terms for monoprotic, tetraprotic, and hexaprotic acids obtained by dropping and adding appropriate terms from equations above for n = 2 are summarized in Table 5.6B.

I suggest that you work through processes used to obtain each set of terms for monoprotic, tetraprotic, and hexaprotic acids.

Putting It All Together

As mentioned at the outset, a goal of this chapter is an equation for solutions containing concentrations equal to or larger than zero of HCl, NaOH, and different forms of three uncharged or charged hexaprotic weak acids. We now have all the information needed to write the fractional form of the equation for such situations.

TABLE 5.6B

Weak-Acid Terms for Monoprotic, Tetraprotic, and Hexaprotic Acids

$n = 1$ $C_1 - \alpha_1(C_0 + C_1)$

$n = 4$ $C_1 + 2C_2 + 3C_3 + 4C_4 - (\alpha_1 + 2\alpha_2 + 3\alpha_3 + 4\alpha_4)(C_0 + C_1 + C_2 + C_3 + C_4)$

$n = 6$ $C_1 + 2C_2 + \cdots + 5C_5 + 6C_6 - (\alpha_1 + 2\alpha_2 + \cdots + 5\alpha_5 + 6\alpha_6)(C_0 + C_1 + \cdots + C_5 + C_6)$

Based on the foregoing discussion, an abridged form of the CBE for a solution containing HCl, NaOH, and all forms of an uncharged or charged hexaprotic acid, H_6X, H_6X^+, or H_6B^{2+}, is as follows:

$$\overbrace{\left[H^+\right] - \frac{K_w}{\left[H^+\right]} + \left(C_{NaOH} - C_{HCl}\right)}^{\text{Independent term}} + \overbrace{\left(C_{1x} + \cdots + 6C_{6x}\right) - \left(\alpha_{1x} + \cdots + 6\alpha_{6x}\right)\left(C_{0x} + \cdots + C_{6x}\right)}^{\text{Weak-acid term for } H_6X} \quad (5.3a)$$

Given that weak acid terms are additive, the equation for a solution containing these reactants plus different forms of two additional acids, H_6Y and H_6Z, can be written by adding weak acid terms like that for the acid, H_6X, with the x subscripts changed to ys and zs. I'll leave it to you to write the terms and limit the discussion from this point on to solutions containing HCl, NaOH, and different forms of one hexaprotic acid.

COMPUTATIONAL FORM OF A CHARGE-BALANCE EQUATION FOR HEXAPROTIC ACIDS

Concentration fractions in the weak-acid term for the hexaprotic acid in Equation 5.3a are the only parts of the term that aren't expressed in terms of known quantities. As discussed in Chapter 3, the denominator term for fractions, α_0–α_6, of a hexaprotic acid is

$$D_6 = \left[H^+\right]^6 + \kappa_1\left[H^+\right]^5 + \kappa_2\left[H^+\right]^4 + \kappa_3\left[H^+\right]^3 + \kappa_4\left[H^+\right]^2 + \kappa_5\left[H^+\right] + \kappa_6$$

in which κ_1–κ_6 are cumulative deprotonation constants.

As also discussed in Chapter 3, successive fractions, α_0–α_6, are successive denominator terms divided by the denominator. Therefore, the part of the weak-acid term involving concentration fractions can be written in the following abridged form

$$\left(\alpha_1 + 2\alpha_2 + \cdots + 6\alpha_6\right) = \frac{\kappa_1\left[H^+\right]^5 + 2\kappa_2\left[H^+\right]^4 + \cdots + 6\kappa_6}{\left[H^+\right]^6 + \kappa_1\left[H^+\right]^5 + \cdots + \kappa_6}$$

This term can be used to write an abridged form of the complete equation for a solution containing HCl, NaOH, and different forms of a hexaprotic acid, i.e.,

$$\overbrace{\left[H^+\right] - \frac{K_w}{\left[H^+\right]} + \left(C_{NaOH} - C_{HCl}\right)}^{\text{Independent term}}$$

$$+ \underbrace{\left(C_{1x} + \cdots + 6C_{6x}\right) - \left(\frac{\kappa_1\left[H^+\right]^5 + 2\kappa_2\left[H^+\right]^4 + \cdots + 6\kappa_6}{\left[H^+\right]^6 + \kappa_1\left[H^+\right]^5 + \cdots + \kappa_6}\right)\left(C_{0x} + \cdots + C_{6x}\right)}_{\text{Weak-acid term for } H_6X} \quad (5.3b)$$

To this point, I have been doing most of the work. Now, I have a question for you. Suppose we have a program to solve Equation 5.3a for a hexaprotic acid. How could we use that program to do calculations for a tetraprotic acid? To answer this question, consider what would happen to the equation if we were to set C_{5x}, C_{6x}, K_{a5x}, and K_{6x} to zero, i.e., $C_{5x} = C_{6x} = K_{a5x} = K_{6x} = 0$. I will trust you to show that the equation would reduce to that for a tetraprotic weak acid.

Whether you do or don't work through the process above, you should remember the following. The weak acids program can be adapted to situations ranging from pure water to situations containing concentrations equal to or larger than zero of HCl, NaOH, and any or all forms of up to three weak acids with up to six acidic hydrogens each by setting quantities not relevant to each situation to zero.

Generalized Form of the Complete Charge-Balance Equation

Even in the abridged form, these equations are beginning to get a bit bulky. The equation can be simplified and generalized by replacing the sums of terms using conventional summation symbols. Making use of the fact that the coefficients of the successive concentration and fractional terms, C_1 thru $6C_6$, α_1–$6\alpha_6$, correspond to the numbers, i, of protons lost from 0–6, Equation 5.3b can be written in the following generalized form,

$$\left[H^+\right] - \frac{K_w}{\left[H^+\right]} + \left(C_{NaOH} - C_{HCl}\right) + \sum_{i=0}^{n} iC_i - \overbrace{\left(\frac{\sum_{i=0}^{n} i\kappa_i\left[H^+\right]^{n-i}}{\sum_{i=0}^{n} \kappa_i\left[H^+\right]^{n-i}}\right)}^{\text{Weak-acid term}} \left(\sum_{i=0}^{n} C_i\right) = 0 \qquad (5.3c)$$

This is the origin of Equation 5.1a discussed earlier. This equation, like others before it, can be extended to two or more weak acids by adding two or more weak acid terms.

Some Hindsight

All of my weak acid programs were written initially using an expanded form of Equation 5.3c in which all individual terms such as the $\kappa_i[H^+]^{n-i}$ terms were included. More recently, I realized that programs written using the generalized equation are simpler, less prone to errors, and easier to debug than programs written using the expanded form of the equation. Charge-balance equations in some programs have been converted to the generalized form but CBEs in most programs are still based on the expanded equations. Programs written in the two forms give equivalent results.

In any event, the basis for the program to do calculations without correction for ionic strength was discussed earlier in connection with solutions containing ammonium ion, ammonia, carbonic acid, and bicarbonate and isn't discussed further here. Let's talk next about procedures used to correct for ionic strength.

CORRECTION FOR IONIC STRENGTH

A two-step process similar to that described earlier to calculate hydrogen ion concentrations in Cells H71–I72 without correction for ionic strength is used in Cells H82–I83 to calculate hydrogen ion concentrations with correction for ionic strength. The principal difference is that conditional constants, i.e., concentration-based constants rather than thermodynamic, i.e., activity-based, constants are used to correct for effects of ionic strength. Concepts and equations used to account for effects of ionic strength on deprotonation constants are discussed in this section.

Given that the discussion here builds on the more general discussion of ionic strength in Chapter 1, you may find it useful to refer to that chapter as you work through this discussion.

SOME CAVEATS

Some caveats regarding corrections for effects of ionic strength merit comment at the outset. First, given that hydrated diameters used to calculate activity coefficients are not available[8] for all ions with which we shall be concerned, I have used an average hydrated diameter, $D_{Ion} = 0.5$ nm, for all ions except H^+ and OH^-. Fortunately, errors introduced by this approximation are relatively small because activity coefficients are relatively insensitive to hydrated diameters. Hydrated diameters used for hydrogen ions and hydroxide ions are as reported by Kielland,[11] i.e., $D_H = 0.9$ nm and $D_{OH} = 0.35$ nm.

Second, given that the extended Debye-Hückle equation[8] is valid only for ionic strengths up to about 0.1 M, concentrations corrected for ionic strengths above 0.1 M are less reliable than those for ionic strengths less than 0.1 M. Even so, results corrected for ionic strengths up to about 0.2 M are likely more reliable than results calculated without correction for the same range of ionic strengths.

Third, a version of the weak acids program on Sheet 4 uses Davies equation to calculate ionic strengths. For reasons I don't understand, the program fails for some situations. Therefore, all results discussed in this chapter were obtained using the program on Sheet 1 based on the Debye-Hückle equation. However, anyone needing to do calculations for ionic strengths larger than about 0.10 may wish to determine if the program on Sheet 4 based on Davies' equation works satisfactorily for their problems.

Finally, although most tabulated deprotonation constants are for zero ionic strength, some such as those for EDTA are for ionic strengths larger than zero e.g., $\mu = 0.10$ M. The user can differentiate among these constants by setting Cells E42–E44 to Yes or No. Constants for acids for which the corresponding cell is set to "Yes" are adjusted for effects of ionic strength; constants for acids for which the corresponding cell is set to "No" are not adjusted for effects of ionic strength.

SOME OBSERVATIONS

As shown in Example 5.2, charge-balance equations for charged and uncharged acids with equal numbers of acidic hydrogens are equivalent to each other. This conclusion applies for equations written in terms of quantities that don't account for effects of ionic strength as well as quantities that do account for effects of ionic strength. Effects of ionic strength on hydrogen ion concentrations, pHs, and concentrations of different forms of charged and uncharged acids are accounted for by effects of ionic strength on several quantities, including activity coefficients, deprotonation constants, cumulative deprotonation constants, and concentration fractions. All these differences are taken into account by the parts of the weak acids program in Rows 80–143. Bases for calculations of quantities calculated in these rows are discussed below.

IONIC STRENGTH

The contribution of each ion to the ionic strength is one half the product of the squared charge on the ion times the concentration of the ion. The ionic strength of a solution is the sum of the contributions of the individual ions, i.e., one-half the sum of squared charges times concentrations of all ions in a solution.[12] Given that ionic strength depends on squared charges, contributions to the ionic strength of equal concentrations of ions with the same numerical values of positive and negative charges are the same. My goal here is to help you develop a generalized equation to account for contributions to the ionic strength of all ions in aqueous solutions containing hydrochloric acid, sodium hydroxide, sodium and chloride salts of ionic forms of weak acids with up to six acidic hydrogens and any charge on the fully protonated form of the acid.

Let's begin with an example.

Example 5.3   The weak acids program gave equilibrium concentrations summarized below for a solution containing $C_1 = 0.025$ M monosodium phosphate, NaH_2PO_4, and $C_2 = 0.010$ M disodium phosphate, Na_2HPO_4 and no other reactants except water. Use the concentrations to calculate the ionic strength of the solution.

Concentrations Corrected for Effects of Ionic Strength

$[H^+]$	$[OH^-]$	$[Na^+]$	$[H_3PO_4]$	$[H_2PO_4^-]$	$[HPO_4^{2-}]$	$[PO_4^{3-}]$
1.6×10^{-7}	6.3×10^{-8}	0.045	8.3×10^{-7}	0.0250	0.010	4.2×10^{-8}

Step 1: Write the equation for the ionic strength as products of charges squared equilibrium concentrations.

$$\mu = 1^2\,[H^+] + 1^2\,[OH^-] + 1^2\,[Na^+] + 0^2[H_3PO_4] + 1^2\,[H_2PO_4^-] + 2^2[HPO_4^{2-}] + 3^2[PO_4^{3-}]$$

Step 2: Substitute equilibrium concentrations and calculate the ionic strength.

$$\mu = 0.5[1.6 \times 10^{-7} + 6.3 \times 10^{-8} + 0.045 + 0.0250 + 2^2(0.010) + 3^2(4.2 \times 10^{-8}] = 0.055 \text{ M}$$

Given equilibrium concentrations corrected for effects of ionic strength as in Example 5.3, it is relatively easy to calculate the ionic strength. My purpose here is to help you understand how the weak acids program calculates ionic strengths for solutions containing known concentrations of hydrochloric acid, sodium hydroxide, and any or all forms of up to three uncharged, singly charged, or doubly charged weak acids with up to six acidic hydrogens each. Don't panic just yet. We will start with simpler situations and divide them into smaller parts. Having learned how the program calculates ionic strength for simpler situations, we will talk about how that information can be extended to more complex situations.

Given that we are working with aqueous solutions, all solutions will contain hydrogen ions and hydroxide ions. Let's talk first about contributions of these ions to the ionic strength.

CONTRIBUTIONS OF HYDROGEN IONS AND HYDROXIDE IONS

Hydrogen ion being a singly charged ion, its contribution to the ionic strength is equal to one half the equilibrium concentration, as expressed in Equation 5.4a:

$$\mu_H = 0.5\left(z_H^2\right)\left[H^+\right] = 0.5\left(1^2\right)\left[H^+\right] = 0.5\left[H^+\right] \tag{5.4a}$$

Similarly, hydroxide ion being a singly charged ion, its contribution to the ionic strength is equal to one half the equilibrium concentration, as expressed in Equation 5.4b:

$$\mu_{OH} = 0.5\left(z_{OH}^2\right)\left[OH^-\right] = 0.5\left(-1^2\right)\left[OH^-\right] = 0.5\frac{K_w}{\left[H^+\right]} \tag{5.4b}$$

All symbols have their usual significance.

The hydrogen ion concentration is calculated by the weak acids program.

CONTRIBUTIONS OF CHLORIDE IONS FROM HCL AND SODIUM IONS FROM NAOH

Assuming that hydrochloric acid dissociates completely, the contribution of chloride ion from HCl to the ionic strength is equal to one half the diluted HCl concentration, as expressed in Equation 5.4c:

$$\mu_{HCl} = 0.5\left(z_{Cl}^2\right)\left[Cl^-\right] = 0.5\left(-1^2\right)C_{HCl} = 0.5C_{HCl} \tag{5.4c}$$

in which C_{HCl} is the diluted HCl concentration.

Assuming that sodium hydroxide dissociates completely, the contribution of sodium ion from sodium hydroxide to the ionic strength is one half the diluted NaOH concentration, as shown in Equation 5.4d:

$$\mu_{NaOH} = 0.5\left(z_{Na}^2\right)\left[Na^+\right] = 0.5\left(1^2\right)C_{NaOH} = 0.5C_{NaOH} \tag{5.4d}$$

in which C_{NaOH} is the diluted NaOH concentration.

CONTRIBUTIONS OF SALTS OF WEAK ACIDS AND THEIR CONJUGATE BASES

We will talk about both uncharged and charged acids. As examples, phosphoric acid, glutamic acid, and protonated lysine are examples of uncharged, singly charged, and doubly charged triprotic acids. The fully protonated acids are represented herein by H_3B, H_3B^+, and H_3B^{2+}, respectively. We will assume that negative ions such as phosphate ion are added as the sodium salts, e.g., Na_3B, and positive ions such as fully protonated glutamic acid or lysine are added as chloride salts, e.g., H_3BCl, or H_3BCl_2. We will also assume that the salts dissociate completely.

Let's begin with a relatively simple way to calculate charges on different forms of weak acids.

Charges on Ions

Charges on different ions are used in two ways. First, the contribution of each ion to the ionic strength is proportional to the squared charge on the ion. Second, given that the charge on each form of an acid determines how many sodium ions or chloride ions will be associated with the ion, charges on different forms of an acid can be used to quantify numbers of sodium ions and chloride ions associated with the different forms of an acid.

Given that the contribution of each ion is proportional to its squared charge, z_i^2, we will use absolute values of charges rather than signed charges. Absolute values of charges on sodium ions and chloride ions are +1. Absolute values of charges on different forms of weak acids depend on the charge, z_0, on the fully protonated acid and the number, i, of protons lost by the ion.

Given that one hydrogen ion is lost in each step of a deprotonation reaction, charges on successive forms of acids will decrease by one for each step in the reaction. Therefore, representing the charge on the fully protonated acid as z_0 and numbers of protons lost by different forms as i, absolute values of charges on forms of acids that have lost i protons are given by

$$|z_i| = |z_0 - i| \tag{5.4e}$$

Let's look at Example 5.4.

> **Example 5.4 Fully protonated glutamic acid is a singly charged triprotic acid, i.e., $z_0 = 1$. Calculate absolute values of charges, $|z_i|$, on forms of glutamic acid that have lost i = 0–3 protons.**
>
> $$|z_i| = |z_0 - i|: \quad |z_0| = |1 - 0| = 1, \quad |z_1| = |1 - 1| = 0, \quad |z_2| = |1 - 2| = 1, \quad |z_3| = |1 - 3| = 2$$

The calculations are easy. Charges calculated in this way are correlated with charges on different forms of uncharged, singly charged, and doubly charged acids in Table 5.7.

I suggest that you work through enough of the situations to become familiar with the process.

Uncharged forms of the acids are included for sake of completeness. As you can see, by comparing charges on different ions with charges calculated using Equation 5.4e, charges calculated using the equation are consistent with charges on the different forms of the three acids.

TABLE 5.7
Correlations of Charges Calculated Using Equation 5.4E with Ionic Forms of Uncharged, Singly Charged, and Triply Charged Triprotic Weak Acids Produced by Dissociations of the Sodium and Chloride Salts of the Charged Forms of the Acids

i = 0	i = 1	i = 2	i = 3
H_3B	$NaH_2B \to Na^+ + H_2B^-$	$Na_2HB \to 2Na^+ + HB^{2-}$	$Na_3B \to 3Na^+ + B^{3-}$
$|z_0| = |0 - 0| = 0$	$|z_1| = |0 - 1| = 1$	$|z_2| = |0 - 2| = 2$	$|z_3| = |0 - 3| = 3$
$H_3BCl \to Cl^- + H_3B^+$	H_2B	$NaHB \to Na^+ + HB^-$	$Na_2B \to 2Na^+ + B^{2-}$
$|z_0| = |1 - 0| = 1$	$|z_1| = |1 - 1| = 0$	$|z_2| = |1 - 2| = 1$	$|z_3| = |1 - 3| = 2$
$H_3BCl_2 \to 2Cl^- + H_3B^{2+}$	$H_2BCl \to Cl^- + H_2B^+$	**HB**	$NaB \to Na^+ + B^-$
$|z_0| = |2 - 0| = 2$	$|z_1| = |2 - 1| = 1$	$|z_2| = |2 - 2| = 0$	$|z_3| = |2 - 3| = 1$

It is easy to adapt Equation 5.4e to any acid with any number, n, of acidic hydrogens and any charge, z_0, on the fully protonated form of the acid. For example, protonated ethylenediaminetetraacetic acid, EDTA, is a doubly charged hexaprotic acid. Charges calculated using Equation 5.4e are compared below with charges on forms of the acid that have lost 0–6 protons.

i	0	1	2	3	4	5	6
Ions	H_6Y^{2+}	H_5Y^+	H_4Y	H_3Y^-	H_2Y^{2-}	HY^{3-}	Y^{4-}
$|z_0 - i| = |2 - i|$	2	1	0	1	2	3	4

Absolute values of charges calculated using Equation 5.4e are consistent with unsigned values of charges on forms of the acid that have lost 0–6 protons.

The weak acids program uses charges calculated as described above in two ways. First, the program uses charges on different forms of an acid to calculate numbers of sodium ions or chloride ions associated with salts of the acids. Second, the program uses the charges on different forms to calculate contributions of the charged forms of the acids to the ionic strength.

Sodium Ions and Chloride Ions

Given that the absolute charge on each sodium ion and each chloride ion is one, the number of sodium ions or chloride ions associated with each form of an acid is the same as the charge on that form. As examples, two chloride ions will be associated with the fully protonated form of EDTA, H_6Y^{2+}, and three sodium ions will be associated with the fully deprotonated form of phosphoric acid, PO_4^{3-}.

The total concentration of sodium ions and chloride ions associated with different forms of an acid is the sum of the concentrations associated with the individual ions. For example, dissociation reactions of the salts of a singly charged triprotic weak acid are annotated below to correlate sodium ion and chloride ion concentrations with diluted concentrations, C_i, of the different forms of the acid.

$$\overset{C_0}{\overbrace{H_3BCl}} \to \overset{C_0}{\overbrace{Cl^-}} + H_3B^+ \qquad \overset{C_1}{\overbrace{H_2B}} \qquad \overset{C_2}{\overbrace{NaHB}} \to \overset{C_2}{\overbrace{Na^+}} + \mathbf{HB^-} \qquad \overset{C_3}{\overbrace{Na_2B}} \to \overset{2C_3}{\overbrace{2Na^+}} + B^{2-}$$

It follows that the sum of sodium ion and chloride ion concentrations for all forms of the singly charged acid is

$$\left[Cl^-\right] + \left[Na^+\right] = 1C_0 + 0C_1 + 1C_2 + 2C_3 = \sum_{i=0}^{n} |z_0 - i| C_i$$

It also follows that contributions of sodium ions and chloride ions to the ionic strength for the different forms of any singly charged triprotic acid can be expressed as

$$\mu_{Na,Cl} = 0.5\left[1\left(1^2\right)C_0 + 0\left(1^2\right)C_1 + 1\left(1^2\right)C_2 + 2\left(1^2\right)C_3\right] = 0.5\left[C_0 + 0C_1 + C_2 + 2C_3\right]$$

This equation can be written in the general form

$$\mu_{Na,Cl} = 0.5\sum_{i=0}^{n} |z_0 - i| C_i \tag{5.4f}$$

in which $\mu_{Na,Cl}$ is the part of the ionic strength resulting from the sodium ions and chloride ions, i is the number of protons lost, n is the number of acidic hydrogens, and z_0 is the charge on the fully protonated acid. Although developed here using a singly charged triprotic weak acid, Equation 5.4f applies for any acid with any charge, z_0, and number, n, of acidic hydrogens.

As an example, tabulated results for EDTA, a doubly charged hexaprotic acid, are repeated here.

i	0	1	2	3	4	5	6
Ions	H_6Y^{2+}	H_5Y^+	H_4Y	H_3Y^-	H_2Y^{2-}	HY^{3-}	Y^{4-}
$|z_0 - i| = |2 - i|$	$2Cl^-$	$1Cl^-$	$0Cl^-$	$1Na^+$	$2Na^+$	$3Na^+$	$4Na^+$

It follows that the total contribution to the ionic strength of sodium ions and chloride ions from concentrations, C_0–C_6, of forms of EDTA that have lost 0–6 protons is

$$\mu_{Na,Cl} = 0.5\sum_{i=0}^{n}|z_0 - i|C_i = 0.5\left(2C_0 + C_1 + 0C_2 + C_3 + 2C_4 + 3C_5 + 4C_6\right)$$

It is common practice to prepare EDTA solutions using the sodium salt of the form that has lost four protons, i.e., $Na_2Y \cdot 2H_2O$. It follows that the contribution of sodium ions to the ionic strength of a solution prepared to contain $C_4 = 0.010$ M EDTA using the sodium salt is $\mu_{Na} = 0.5|z_4|C_4 = 0.5(2)$ $(0.010$ M$) = 0.010$ M.

Contributions of Different Forms of Weak Acids

The contribution of each form of an acid can be calculated as the product of the charge squared, z_i^2, times the fraction, α_i, in that form times the total concentration, C_T, of all forms of the acid. The contribution of all forms of an acid is the sum of contributions of the individual forms.

The contribution to the ionic strength of the conjugate bases of a weak acid, μ_{CB}, is the sum of charges squared times fractions in individual forms times the total concentration. This is expressed as follows in Equation 5.4g:

$$\mu_{CB} = 0.5\left[\left(\sum_{i=0}^{n}|z_0 - i|^2\alpha_i\right)C_T\right] \tag{5.4g}$$

in which i is the number of protons lost by each form, n is the number of acidic hydrogens on the fully protonated acid, z_0 is the charge on the fully protonated acid, α_i is the fraction of the total concentration in a form that has lost i protons and C_T is the total concentration of all forms of the acid (Exercise 5.3).

EXERCISE 5.3

Given that fully protonated EDTA is a doubly charged hexaprotic acid, H_6Y^{2+}, use stepwise deprotonation reactions to show that the equation for the contribution of all forms of EDTA to the ionic strength is:

$$\mu_{EDTA} = 0.5\left[\sum_{i=0}^{6}(z_0 - i)^2\alpha_i\right]C_T = 0.5\left(2^2\alpha_0 + 1^2\alpha_1 + 0^2\alpha_2 + 1^2\alpha_3 + 2^2\alpha_4 + 3^2\alpha_5 + 4^2\alpha_6\right)C_T$$

Hint: Write the stepwise deprotonation reaction for protonated EDTA and use charges on the different forms as a guide.

Complete Equation

The complete equation for the ionic strength is the sum of contributions of all ions in a solution. As noted earlier, the weak acids program does calculations for up to three weak acids with up six acidic hydrogens each. Therefore, the complete equation includes terms for three weak acids. An abridged form of the complete equation is

$$\mu = 0.5\left\{\left[H^+\right] + \frac{\overbrace{\left[OH^-\right]}{K_w}}{\left[H^+\right]} + \underbrace{\left(C_{HCl} + C_{NaOH}\right)}_{\text{Strong AB part}} + \left[\overbrace{\sum_{i=0}^{n}\left(\left|z_0 - i\right|C_i\right)}^{\text{Na}^+\text{ and/or Cl}^-} + \overbrace{\sum_{i=0}^{n}\left(z_0 - i\right)^2\alpha_i}^{\text{Conjugate bases}}\right]C_T\right\}\qquad(5.5)$$

$$\underbrace{}_{\text{Three terms, one for each of three weak acids.}}$$

in which n is the number of acidic hydrogens for each fully protonated acid, z_0 is the charge on each fully protonated acid, i is the number of protons lost by each form of each acid, α_i is the fraction of the concentration in forms of each acid that have lost i protons and C_T is the total concentration of all forms of each acid.

CALCULATING IONIC STRENGTH

Ionic strengths are calculated in Cell C112 using Equation 5.5. This equation is more easily implemented and checked for errors in any spreadsheet program than the fully expanded equation. For example, with unsigned values, $|z_0 - i|$, in Cells C88–I88 and diluted concentrations, C_i, in Cells B64–H64, the term for weighted concentrations is expressed in Excel syntax as SUM(C88:I88*B64:H64). Similarly simple expressions are used for other terms in Equation 5.5.

As with charge-balance equations, the ionic-strength equation for hexaprotic acids can be adapted to acids with fewer acidic protons by setting to zero all concentrations and constants beyond those corresponding to the number, n, of acidic hydrogens for each acid.

Ionic strengths for the low-resolution and high-resolution steps are calculated in Cells C85 and C112, respectively, of the program on Sheet 1 of the weak acids program. The user, you, must (a) set C29 to 0, (b) enter concentrations and constants larger than zero into appropriate cells, (c) set to zero all concentrations and constants not relevant to each problem, (d) enter values of n and z_0 in Cells C42–D44, (e) set Cells E42 through E44 to "Yes or No" to identify sets of deprotonation constants that do or don't correspond to zero ionic strength and f) set C29 to 1 and wait for D46 to change to "Solved." The program does the rest.

Most quantities used to correct for ionic strength are in Rows 84–108 for the low-resolution step and in Rows 111–131 for the high-resolution step. Fractions used for both steps are in Rows 135–143. The somewhat fragmented layout of the program results from my decision late in the process to use generalized charge-balance and ionic-strength equations in place of term-by-term equations.

ACTIVITY COEFFICIENTS

Two options commonly used to calculate activity coefficients are a theoretically based equation by Debye-Hückel[8] and an empirical equation by Davies.[7] de Levie[4] has discussed Davies' equation as well as applications of activity coefficients calculated using it. Activity coefficients are calculated herein using the extended Debye-Hückel equation,[8] EDHE.

The exponential form of the EDHE for any reactant or product is expressed in Equation 5.6a:

$$f = 10 \wedge \left(\frac{-0.51z_i^2\sqrt{\mu}}{1 + 3.3D_i\sqrt{\mu}}\right)\qquad(5.6a)$$

in which f is the activity coefficient, z_i is the charge on the ion, D is the hydrated diameter[11] *in nanometers* and μ is the ionic strength. Activity coefficients for the H$^+$ and OH$^-$ ions are represented by f_H and f_{OH}; and activity coefficients for forms of acids that have lost i acidic protons are represented by f_i.

Hydrated diameters for the H^+ and OH^- ions are $D_H = 0.9$ nm and $D_{OH} = 0.35$ nm. Given that hydrated diameters are not known for all the ions of interest in this discussion and that activity coefficients are relatively insensitive to changes in hydrated diameters, an average value of $D_i = 0.5$ nm is used for all forms of all acids. In other words, activity coefficients for all ions except the hydrogen ion and hydroxide ion are calculated as expressed in Equation 5.6b:

$$f = 10 \wedge \left(\frac{-0.51z_i^2 \sqrt{\mu}}{1+3.3D_i \sqrt{\mu}} \right) \cong 10 \wedge \left(\frac{-0.51z_i^2 \sqrt{\mu}}{1+3.3(0.5)\sqrt{\mu}} \right) \tag{5.6b}$$

in which z_i is the charge on an ion, D_i is the hydrated diameter, and μ is the ionic strength.

RELATIONSHIPS AMONG ACTIVITY-BASED AND CONCENTRATION-BASED CONSTANTS

Activity coefficients are used to convert activity-based constants, more commonly called thermodynamic constants, to concentration-based constants, more commonly called conditional constants. The equilibration reaction and relationship between activity-based and concentration-based constants for any monoprotic acid-base pair can be represented as follows:

$$HB \overset{K_{a,0}}{\rightleftharpoons} H^+ + B^- \qquad K_{a,0} = \frac{f_H \left[H^+ \right] f_B \left[B^- \right]}{f_{HB} \left[HB \right]} = \frac{f_H f_B}{f_{HB}} \frac{\left[H^+ \right] \left[B^- \right]}{\left[HB \right]} = \frac{f_H f_B}{f_{HB}} K_{a,\mu}$$

in which K_{a0} and $K_{a\mu}$ are the activity-based and concentration-based deprotonation constants, respectively, and f_H, f_B, and f_{HB} are activity coefficients of the reactants. A simple rearrangement shows that the concentration-based constant, $K_{a,\mu}$, is related to the activity-based constant, $K_{a,0}$, as follows:

$$K_{a,\mu} \cong \frac{f_{HB}}{f_H f_B} K_{a,0} \tag{5.6c}$$

in which symbols are as described above.

Although described here in the context of a monoprotic acid, the same form of the equation applies for any reaction between a conjugate acid-base pair with any number of acidic hydrogens. For example, assuming that sodium salts of negatively charged ions and chloride salts of positively charged ions are used to prepare solutions, information needed to apply all these concepts to solutions containing HCl, NaOH, and the sodium and/or chloride salts of all forms of weak acids with six or fewer acidic protons is summarized in Table 5.8.

Numbers of protons lost are in the first row and charges on fully protonated acids are in the first column. Forms of acids that could, if available, be used to prepare solutions are in the first group of three rows. For sake of completeness, it is assumed that all forms of all acids are available. Initial concentrations of any forms not used are set to zero in calculation steps.

Forms of acids produced by dissociations of the salts are included in the second set of three rows. When assigning numerical values to initial concentrations, C_i, it is assumed that salts dissociate completely before equilibration reactions begin. I will trust you to write dissociation reactions for the salts.

Each concentration-based constant is a conversion factor analogous to that in Equation 5.6c times the activity-based constant. Relationships for acids with fewer than six acidic protons are obtained by dropping terms from right to left in each row for each decrease by one in the number, n, of acidic hydrogens. Concentration-based constants are calculated on Sheet 1 in Cells C100–H102 for the low-resolution step and Cells C123–H125 for the high-resolution step. These constants are

TABLE 5.8

Summary of Information Used to Convert Activity-Based or Thermodynamic Constants, $K_{ai,0}$, to Concentration-Based or Conditional Constants, $K_{ai,\mu}$, for Hexaprotic Acids

i →	i = 0	i = 1	i = 2	i = 3	i = 4	i = 4	i = 6
z_0 ↓	Forms of Hexaprotic Acids[a]						
0	(H_6B)	NaH_5B	Na_2H_4B	Na_3H_4B	Na_4H_4B	Na_5HB	Na_6B
+1	H_6BCl	(H_5B)	NaH_4B	Na_2H_3B	Na_3H_2B	Na_4HB	Na_5B
+2	H_6BCl_2	H_5BCl	(H_4B)	NaH_3B	Na_2H_2B	Na_3HB	Na_4B
	Forms of the Acids Remaining after Dissociations of the Salts Are Complete						
0	(H_6B)	H_5B^-	H_4B^{2-}	H_3B^{3-}	H_2B^{4-}	HB^{5-}	B^{6-}
+1	H_6B^+	(H_5B)	H_4B^-	H_3B^{2-}	H_2B^{3-}	HB^{4-}	B^{5-}
+2	H_6B^{2+}	H_5B^+	(H_4B)	H_3B^-	H_2B^{2-}	HB^{3-}	B^{4-}
	Products of Conversion Factors Times Activity-Based Constants						
	$K_{a1,\mu}$	$K_{a2,\mu}$	$K_{a3,\mu}$	$K_{a4,\mu}$	$K_{a5,\mu}$	$K_{a6,\mu}$	
All[b]	$=\dfrac{f_0}{f_H f_1}K_{a1,0}$	$=\dfrac{f_1}{f_H f_2}K_{a1,0}$	$=\dfrac{f_2}{f_H f_3}K_{a1,0}$	$=\dfrac{f_3}{f_H f_4}K_{a4,0}$	$=\dfrac{f_4}{f_H f_5}K_{a5,0}$	$=\dfrac{f_5}{f_H f_6}K_{a6,0}$	

[a] Diluted concentrations of any forms not used are set to zero.

[b] f_i is the activity coefficient for the form of an acid that has lost i protons.

used in Cells H82 and H83 to calculate charge-balance equations for the low-resolution and high-resolution steps, respectively.

As discussed earlier, algorithms in Cells I82 and I83 use a two-step process to resolve hydrogen ion concentrations to the number of significant figures in I29, usually 3. Templates used to adapt the program to different situations as well as to help you learn to use the program were also described earlier.

SUMMARY

Charge-balance equations are combined with mass-balance, concentration-fraction, and autoprotolysis constant equations to obtain exact equations for the hydrogen ion concentration in terms of initial concentrations and deprotonation constants. Systematic patterns in these equations are used to adapt them to a variety of situations ranging from pure water to different forms of up to three hexaprotic acids with HCl and NaOH. An iterative spreadsheet program based on these equations is adapted to different situations by setting appropriate initial concentrations and deprotonation constants to zero. Hydrogen ion concentrations are calculated and used with deprotonation constants to calculate concentration fractions that are then used with total concentrations to calculate equilibrium concentrations of the different forms of each acid.

Activity-based deprotonations constants are used for calculations without correction for ionic strength and concentration-based constants are used for calculations with correction for ionic strength. A generalized equation is developed for ionic strength. The generalized equation accounts for numbers of protons in the fully protonated acid, numbers of protons lost, and charges on fully protonated acids. Ionic strengths calculated in this way are used with the extended Debye-Hückle equation to calculate activity coefficients that are then used to calculate concentration-based deprotonation constants. Concentration-based deprotonation constants are used to calculate cumulative constants, concentration fractions, and equilibrium concentrations of different forms of acids corrected for effects of ionic strength.

The observation that charge-balance equations written as algebraic sums of positive and negative charges are negative for H^+ concentrations less than the equilibrium concentration and positive for

H^+ concentrations larger than the equilibrium concentration is used as a convergence criterion in an iterative program to solve charge-balance equations. Digital logic functions are used to vary the H^+ concentration in small increments and to detect the point at which the sign of the charge-balance equation changes from negative to positive values. The equilibrium concentration is taken as the concentration that produces the − to + transition. With I29 set to 3, the program resolves the H^+ concentration to three significant figures with calculation times of several seconds to one or two minutes.

REFERENCES

1. G. Frison, A. Calatroni, www.academia/518142/ Some Notes on pH Computation and Theories, 1–29.
2. J. E. Ricci, *Hydrogen Ion Concentration, New Concepts in a Systematic Treatment*, Princeton U.P.: Princeton, NJ, 1952.
3. H. Freiser, *Concepts and Calculations in Analytical Chemistry, A Spreadsheet Approach*, CRC Press: Boca Raton, FL, 1992.
4. R. de Levie, *Principles of Quantitative Chemical Analysis*, McGraw-Hill: New York, NY, 1997.
5. H. A. Laitinen, W. E. Harris, *Chemical Analysis*, Second Edition, McGraw-Hill: New York, NY, 1975.
6. C. G. Enke, *The Art & Science of Chemical Analysis*, Preliminary Edition, John Wiley and Sons: New York, NY, 2000.
7. C. W. Davies, *Ion Association*, Butterworths: Washington DC, 1962.
8. P. Debye and E. Hückel, *Phys. Z.*, **1923,** *24*, 185–206.
9. H. S. Harned and F. C. Hickey, *J. Am. Chem. Soc.*, **1937,** *59*, 1284–1288.
10. www.ion.chem.usu.edu/~sbialkow/Classes/3650/Carbonate/Carbonic%20Acid.html (2014).
11. J. Kielland, *J. Am. Chem. Soc.*, **1937,** *59*, 1675–1678.
12. G. N. Lewis, *J. Am. Chem. Soc.*, **1912,** *34*, 1631–1644.

6 Some Applications of Acid-Base Equilibria

This chapter describes adaptations of the more general weak acids program to acid-base titration curves as well as a discussion of how buffer compositions influence their abilities to resist changes in hydrogen ion concentrations and pH.

Regarding titration curves, a program is described to plot titration curves for up to six sets of conditions on each plot. This facilitates comparisons of effects of factors such as concentration and magnitudes of deprotonation constants on shapes of titration curves. The program also calculates and plots first and second derivatives of titration curves using five-point smoothed first- and second-derivative algorithms.[1]

Regarding buffers, three measures of buffer performance are discussed. In addition to the usual measure, buffer capacity,[2,3,4] two new measures called *pH sensitivity* and *H+ sensitivity* are introduced. These measures of buffer performance show that features of buffers that favor control of hydrogen ion concentration and pH are very different.

Spreadsheet programs discussed in this chapter are available on the website, http://crcpress.com/978113836722

Let's begin with titration curves.

TITRATION CURVES

The program on Sheet 2 of the weak acids program uses concepts and procedures discussed in earlier chapters to calculate values of pH vs. titrant volume for titrations of acids and bases. The program is designed to plot three sets of plots for each set of conditions, namely pH vs. volume, the first derivative, dpH/dV, vs. volume, and the second derivative, d^2pH/dV^2, vs. volume. The program includes a feature that allows the user to plot curves for up to six sets of conditions on each figure. It's a moderately complex program that is relatively easy to use after a little practice.

It is assumed that all acids are titrated with a monoprotic strong base such as NaOH and that all bases are titrated with a monoprotic strong acid such as HCl. Types of situations for which the program can plot titration curves are as follows:

a. A monoprotic strong acid alone.
b. A mixture of different forms of a weak acid with one to six acidic hydrogens.
c. A mixture of a monoprotic strong acid with a mixture of different forms of a weak acid with one to six acidic hydrogens.
d. A monoprotic strong base alone.
e. A mixture of different forms of a weak base that can accept up to six protons per molecule.
f. A mixture of a monoprotic strong base with a mixture of different forms of a weak base that can accept one to six protons per molecule.

The program can be used to identify situations that are and aren't likely to represent practical titrations as well as effects of different variables such as concentrations, deprotonation constants, and numbers of acidic hydrogens on shapes of titration curves.

Effects of Concentration on Titration Curves

We will talk about two groups of titrations used to illustrate effects of concentrations on shapes of titration curves. We will talk first about a group of solutions containing a wide range of hydrochloric acid concentrations, 10^{-5} M to 0.10 M, each titrated with a solution containing a sodium hydroxide ion concentration equal to each HCl concentration. We will then talk about titrations of solutions containing a smaller range of acetic acid concentrations, 0.0070 M–0.012 M, each titrated with 0.10 M sodium hydroxide.

Wide Range of HCl Concentrations

Example 6.1 describes the problem to be solved.

> **Example 6.1 Plot titration curves for titrations of 50.0 mL volumes of solutions containing (A) 1.0×10^{-5} M, (B) 1.0×10^{-4} M, (C) 1.0×10^{-3} M, (D) 0.010 M, and (E) 0.10 M hydrochloric acid with solutions containing sodium hydroxide concentrations equal to the HCl concentrations.**

I will describe the titration curves for this example and then describe a process used to obtain them.

Full Scale Titration Curves

Figure 6.1A includes titration curves for NaOH volumes from 0 to 100 mL. The dashed horizontal lines on the figure represent pH's at the start and end of the yellow to blue color change, pH = 6.0–7.6, for bromothymol blue. You can superimpose dashed lines representing the transition range for any indicator on any set of titration curves by setting D45 to "Yes" and Cells G45 and I45 to pH's at the beginning and end of the transition range for the indicator.

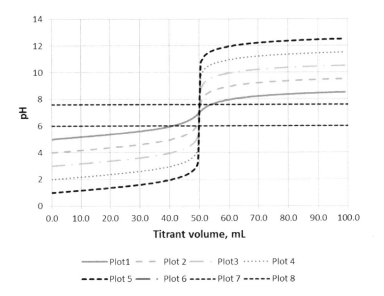

FIGURE 6.1A Titration curves for titrations of hydrochloric acid with sodium hydroxide.

Plots 1–5: Titrations of 50-mL volumes of solutions containing (1) 1.0×10^{-5} M, (2) 1.0×10^{-4} M, (3) 1.0×10^{-3} M, (4) 0.010 M, and (5) 0.10 M hydrochloric acid with sodium hydroxide concentrations equal to the HCl concentrations. Plots 7 and 8: Transition range for bromothymol blue.

TABLE 6.1

Sequence of Cell Settings to Plot Titration Curves for the Five HCl Concentrations in Example 6.1

Set D87 to Yes, G45 to 6.0, and I45 to 7.6 to superimpose the transition range for bromothymol blue on the titration curves. Then set H28 to 0 and then to 1 and follow sequences below for Parts A–F of Example 6.1.						
Cells	Part A	Part B	Part C	Part D	Part E	Part F
C29	0	0	0	0	0	0
F28 and F42	1.0E-05	1.0E-4	1.0E-3	1.0E-2	1.0E-1	0
C29	1	1	1	1	1	0
Wait for F45 to change from "Solving" to "Solved" and then change H28 to values below.						
H28	2	3	4	5	5.5	0

You can find transition ranges for a variety of indicators by searching for "acid base indicators" on the internet and selecting one of several sites, including "Acid-Base Indicators-Boundless." We will talk later about how slopes of titration curves near the equivalence point can be used to estimate changes in titrant volumes corresponding to transition ranges of indicators. First, however, let's talk about a set of cell settings used to plot the titration curves in Figure 6.1A.

Scenario for Titration Curves in Figure 6.1A

To begin, open the weak acids program and select Sheet 2. A template for Part A of the example is included in Cells A68–I87 of the program on Sheet 2. To use the template, copy it and paste it into Cells A26–I45 using the Paste (P) option. Then follow the scenario in Table 6.1 to plot the five titration curves as well as first and second derivatives of the curves. Plotted figures are in Columns AF–BT Rows 3–22 on Sheet 2.

The sequence for Part A is explained briefly. Setting C29 to zero presets the iterative program used for the calculations to its initial conditions. Setting F28 and F42 to 1E-5 each sets sample and titrant concentrations to the values for Part A of the example. Setting H28 to 0 clears all the cells where volumes and pH's are stored for plotting; setting H28 to 1 prepares cells in Columns AE and AF to accept results for Part A. Setting C29 to 1 initiates the iterative calculations. Setting H28 to 2 after iterations are complete tells cells in Columns AE and AF to retain results for Part A until H28 is set to some number smaller than 2.

Analogous sequences apply for other parts of this and other situations.

One additional point merits comment. Notice that H28 for Part E is set to 5.5 rather than 6. This prevents volume vs. pH values for Part E from being plotted twice on the figure. If the example had involved six titrations, H28 would have been set to 6 and 6.5, respectively, for Parts E and F. If the example had involved fewer than five titrations, H28 for the last titration would have been set to a number between that of the last and next titration, e.g., 2.5 for two titration curves, 3.5 for three titration curves, etc.

For the record, first- and second-derivative curves are plotted at the same time the titration curves are plotted and are controlled by settings in H28.

Expanded-Scale Titration Curves

Figure 6.1B is an expanded-scale plot of the portions of the titration curves near the equivalence point.

The plot for 10^{-5} M HCl is virtually flat on this plot and doesn't intersect the dashed lines corresponding to the transition range for bromothymol blue at any point. I used points at which the other plots cross the dashed lines at the beginning and end of the transition range of the indicator to estimate volume changes corresponding to the transition range of the indicator.

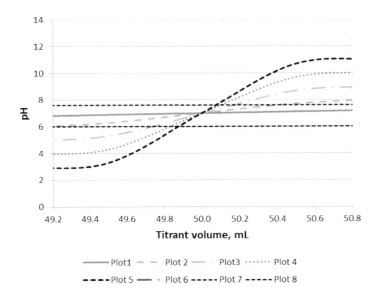

FIGURE 6.1B Expanded-scale titration curves for titrations of hydrochloric acid with sodium hydroxide.

Results for Parts B–E are ΔV_{NaOH} = 1.2 mL, 0.47 mL, 0.30 mL, and 0.20 mL. These volume differences correspond to about 2.4%, 0.9%, 0.6%, and 0.4%, respectively, of the equivalence-point volumes of 50 mL. In other words, the larger the slope of a titration curve at the equivalence point, the smaller the change in titrant volume corresponding to the transition range of an indicator.

Differences in the shapes of the plots in Figure 6.1B are reflected in *first derivatives*, dpH/dV, of the titration curves. Figure 6.1C includes first derivative plots for the titration curves.

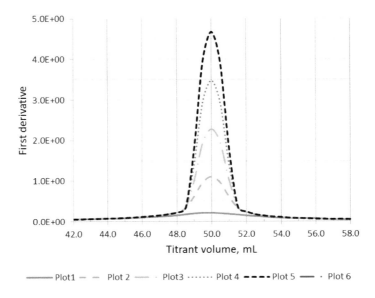

FIGURE 6.1C First derivatives of titration curves for titrations of hydrochloric acid with sodium hydroxide.

Plots 1–5: Titrations of 50-mL volumes of solutions containing (1) 1.0×10^{-5} M, (2) 1.0×10^{-4} M, (3) 1.0×10^{-3} M, (4) 0.010 M, and (5) 0.10 M hydrochloric acid with sodium hydroxide concentrations equal to the HCl concentrations.

These plots show that slopes of the titration curves increase with increasing HCl concentrations. This observation can be generalized. All other factors being the same, slopes of titration curves near equivalence points will increase with increasing concentrations of the reactant being titrated, hydrochloric acid in this case.

Example 6.1 represents a special case in the sense that all samples were diluted by the same amounts throughout each titration. We will talk next about a more practical situation for which sample dilution increases with increasing acid concentrations.

TYPICAL SET OF TITRATIONS

The usual approach to a set of titrations is to select a titrant concentration close to the acid or base concentrations of interest and to titrate all samples with the same titrant concentration. Example 6.2 illustrates such a set of titrations.

Example 6.2 Plot titration curves for 50.0-mL volumes of solutions containing (A) 0.07 M, (B) 0.080 M, (C) 0.090 M, (D) 0.10 M, (E) 0.11 M, and (F) 0.12 M acetic acid with 0.10 M sodium hydroxide.

A template for Part A of the example is included in Cells A93–I112. A scenario similar to that described in Table 6.1 for Example 6.1 can be used to obtain the complete set of titration curves, including first and second derivative plots by doing calculations with Cell B37 set to the desired acetic acid concentrations.

Figure 6.2A includes titration curves for the six acetic acid concentrations. The dashed horizontal lines represent the transition range, pH = 8.3 to 10.0, for the colorless to pink color change for phenolphthalein. All six plots exhibit sharp breaks near equivalence points meaning that relatively sharp end points can be obtained.

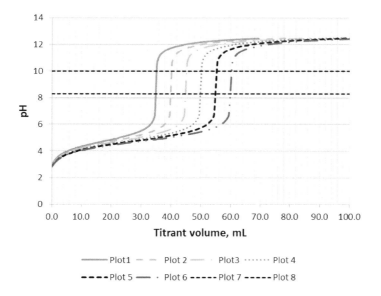

FIGURE 6.2A   Titration curves for titrations of acetic acid with sodium hydroxide.

Plots 1–6: Titrations of 50-mL volumes of solutions containing (1) 0.070 M, (2) 0.080 M, (3) 0.090 M, (4) 0.10 M, (5) 0.11 M, and (6) 0.12 M hydrochloric acid with 0.10 M sodium hydroxide.
Plots 7 and 8: Transition range for phenolphthalein.

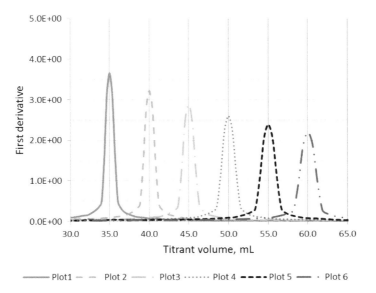

FIGURE 6.2B   First derivatives of titration curves for titrations of acetic acid with sodium hydroxide.

Plots 1–6: Titrations of 50-mL volumes of solutions containing (1) 0.070 M, (2) 0.080 M, (3) 0.090 M, (4) 0.10 M, (5) 0.11 M, and (6) 0.12 M hydrochloric acid with 0.10 M sodium hydroxide.

First Derivatives of Titration Curves

As noted earlier, we can use first derivatives of the titration curves to visualize how slopes of the titration curves near equivalence points depend on acetic acid concentrations.

Figure 6.2B is an expanded-scale plot of first derivatives of the titration curves in Figure 6.2A. Whereas I had expected slopes to increase with increasing acetic acid concentrations, slopes decrease with increasing concentrations. Can you explain this? I'll try.

The most plausible explanation for the decrease slopes of titration curves with increasing acetic acid concentrations is related to different dilution factors for the titrant, sodium hydroxide, as the acetic acid concentration is increased. For example, the change in the sodium hydroxide concentration at the equivalence point before reaction occurs can be calculated as

$$\Delta C_{NaOH} = \frac{\Delta V_{NaOH} C^0_{NaOH}}{V_{HB} + V_{NaOH}} = \frac{0.050 \text{ mL} \times 0.10 \text{ M}}{50.0 \text{ mL} + V_{NaOH,Eq}}$$

From Figure 6.2B, equivalence-point volumes for solutions containing 0.070 M and 0.12 M acetic acid are $V_{ep} = 35$ mL and 60 mL, respectively, corresponding to total volumes at equivalence points of $V_T = 85$ and 110 mL, respectively.

It follows that changes in NaOH concentrations produced by adding 0.050 mL of 0.10 M NaOH to the solutions at equivalence points for solutions containing 0.070 M and 0.12 M acetic acid are

$$\overbrace{\Delta C_{NaOH} = \frac{0.050 \text{ mL} \times 0.10 \text{ M}}{50.0 \text{ mL} + 35 \text{ mL}} = 5.9 \times 10^{-5} \text{M}}^{0.070 \text{ M acetic acid}} \text{ and } \overbrace{\Delta C_{NaOH} = \frac{0.050 \text{ mL} \times 0.10 \text{ M}}{50.0 \text{ mL} + 60 \text{ mL}} = 4.6 \times 10^{-5} \text{M},}^{0.12 \text{ M acetic acid}}$$

respectively. Analogous differences for other volumes before and after equivalence points will result in different changes in pH and differences in slopes of titration curves.

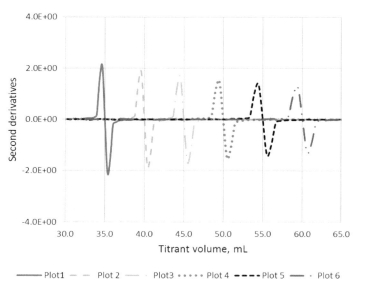

FIGURE 6.2C   Second derivatives of titration curves for titrations of acetic acid with sodium hydroxide.

Plots 1–6: Titrations of 50-mL volumes of solutions containing (1) 0.070 M, (2) 0.080 M, (3) 0.090 M, (4) 0.10 M, (5) 0.11 M, and (6) 0.12 M hydrochloric acid with 0.10 M sodium hydroxide.

Second Derivatives of Titration Curves

Figure 6.2C includes second-derivative plots for the titration curves in Figure 6.2A.

The most significant point about these plots is that the second derivatives all pass through zero at equivalence points. As a side note, my thesis advisor in graduate school, Prof. H. V. Malmstadt, having served as a radar officer on a destroyer during the Second World War, used his knowledge of electronics to develop an automatic end-point detection method based on the use of electronic circuitry to detect the point at which second derivatives passed through zero. The resulting titrator was commercialized by E. H. Sargent Co., and, to the best of my knowledge, is still available from the Sargent-Welch company.

EFFECTS OF DEPROTONATION CONSTANTS ON TITRATION CURVES

Example 6.3 is used to illustrate effects of deprotonation constants on shapes of titration curves for weak acids and weak bases.

> **Example 6.3 Plot on one figure titration curves for 50.0 mL volumes of 0.10 M solutions of (A) ammonium ion, $K_a = 5.7 \times 10^{-10}$, (B) hypochlorous acid, $K_a = 3.0 \times 10^{-8}$, (C) acetic acid, $K_a = 1.8 \times 10^{-5}$, (D) chloroacetic acid, $K_a = 1.3 \times 10^{-3}$, and (E) trichloroacetic acid, $K_a = 0.22$.**

A template for Part A of this example is included in Cells A118–I137 on Sheet 2 of the acid-base program. Calculations for the other parts of the example are done by entering the desired deproton-ation constants into Cell B32.

Plots for the five situations are illustrated in Figure 6.3.

As you can see, pH changes at equivalence points increase with increasing values of deprot-onation constants. pH changes for ammonium ion and hypochlorous acid are too small to permit

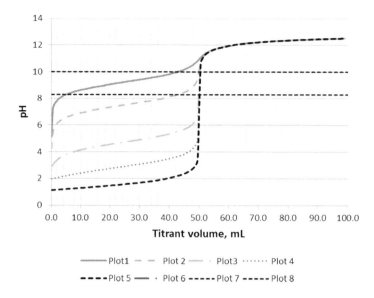

FIGURE 6.3   Effects of deprotonation constants on titration curves.

Plots 1–5: Titration curves for titrations with 0.10 M sodium hydroxide of 50.0 mL volumes of solutions containing 0.10 M each of (1) ammonium ion, $K_a = 5.7 \times 10^{-10}$, (2) hypochlorous acid, $K_a = 3.0 \times 10^{-8}$, (3) acetic acid, $K_a = 1.8 \times 10^{-5}$, (4) chloroacetic acid, $K_a = 1.3 \times 10^{-3}$, and (5) trichloroacetic acid, $K_a = 0.22$.
Plots 7 and 8: Transition range for phenolphthalein.

reliable detection of end points. Given that conjugate bases of weak acids are strong bases, an alternative procedure for these acids would be to add a small excess of sodium hydroxide to convert the ammonium ion and hypochlorous acid to ammonia and hypochlorite and to back titrate the excess sodium hydroxide and ammonia or hypochlorite with a strong acid such as hydrochloric acid.

As shown by the dashed lines, phenolphthalein would be a satisfactory indicator for titrations of acetic acid, chloroacetic acid, and trichloroacetic acid. An indicator with a transition range at lower pH's, e.g., bromothymol blue, transition range 6.0–7.6, would be a better indicator for the two stronger acids.

TITRATION CURVES FOR PHOSPHORIC ACID AND PHOSPHATE ION

Example 6.4 is used to illustrate titration curves for titrations of triprotic acids with a strong base and titrations of triprotic weak bases with a strong acid.

Example 6.4   Plot on one figure titration curves for titrations of solutions containing 0.10 M, 0.010 M, and 0.0010 M phosphoric acid with equal concentrations of sodium hydroxide and solutions containing 0.10 M, 0.010 M, and 0.0010 M phosphate ion with equal concentrations of hydrochloric acid.

A template for Part A of this example is in Cells A143–I162 on Sheet 2 of the weak acids program.
Titration curves for the six situations in Example 6.4 are plotted in Figure 6.4.
Plots 1–3 are for titrations of 25.0 mL each of solutions containing 0.10 M, 0.010 M, and 0.0010 M phosphoric acid with titrants containing 0.10 M, 0.010 M, and 0.0010 M NaOH, respectively. Plots 4–6 are for titrations of 25.0 mL each of solutions containing 0.10 M, 0.010 M, and 0.0010 M phosphate ion with titrants containing 0.10 M, 0.010 M, and 0.0010 M HCl, respectively.

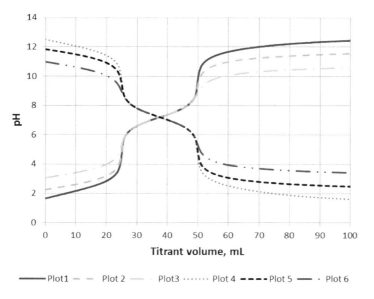

FIGURE 6.4 Titration curves for titrations of phosphoric acid with sodium hydroxide and phosphate ions with hydrochloric acid.

Plots 1–3: Titration curves for 25.0 mL volumes of solutions containing (1) 0.10 M, (2) 0.010 M, and (3) 0.0010 M phosphoric acid with (1) 0.10 M, (2) 0.010 M, and (3) 0.0010 M sodium hydroxide.

Plots 4–6: Titration curves for 25.0 mL volumes of solutions containing (1) 0.10 M, (2) 0.010 M, and (3) 0.0010 M phosphate ion with (1) 0.10 M, (2) 0.010 M, and (3) 0.0010 M hydrochloric acid.

Although phosphoric acid has three acidic hydrogens and phosphate ion can accept up to three hydrogen ions, each set of plots includes only two breaks in the titration curves. Reasons for this can be explained using the stepwise reactions and associated constants. Stepwise reactions and deprotonation constants for phosphoric acid are as follows:

$$H_3PO_4 \overset{K_{a1}=7.1E-3}{\rightleftharpoons} H^+ + H_2PO_4^- \overset{K_{a2}=6.2E-8}{\rightleftharpoons} H^+ + HPO_4^{2-} \overset{K_{a3}=4.5E-13}{\rightleftharpoons} H^+ + PO_4^{3-}$$

Whereas constants K_{a1} and K_{a2} are large enough that sodium hydroxide will react completely with H_3PO_4 and $H_2PO_4^-$, the third constant, $K_{a3} = 4.5 \times 10^{-13}$, is so small that hydroxide ion doesn't react completely with HPO_4^{2-}.

Stepwise protonation reactions and associated constants for phosphate ion are as follows:

$$PO_4^{3-} + H^+ \overset{K_{b1}=1.4E-2}{\rightarrow} HPO_4^{2-} + H^+ \overset{K_{b2}=1.6E-7}{\rightarrow} H_2PO_4^- + H^+ \overset{K_{b3}=1.4E-12}{\rightleftharpoons} H_3PO_4$$

The first two protonation constants are large enough that hydrogen ion reacts completely with phosphate ion and monohydrogen phosphate. However, the third protonation constant is too small to force complete reaction between dihydrogen phosphate and the hydrogen ion.

Differences between the first two constants in each case are large enough that breaks in the titration curves are well separated from each other. The situation would have been quite different if I had chosen citric acid and citrate ion for the example. Stepwise deprotonation constants for citric acid, $K_{a1} = 7.55 \times 10^{-4}$, $K_{a2} = 1.73 \times 10^{-5}$, and $K_{a3} = 4.02 \times 10^{-7}$, are all so large and differences among the constants are so small that the titration curve for citric acid with sodium hydroxide gives a plot with only one break corresponding to reactions of all three forms of the acid. Protonation constants are so small and differ by such small amounts that no break is observed in the calculated titration curve for a titration of citrate ion with HCl.

DETECTING END POINTS IN ACID-BASE TITRATIONS

Two methods commonly used to detect end points in acid-base titrations include color indicators and potentiometry. Each option is discussed briefly below.

Color Indicators

Color indicators are weak acids that have different colors in their protonated and deprotonated forms. For example, the reaction of methyl orange can be depicted as follows:

$$\underset{\text{pH}\leq3.1}{\overset{\text{Red}}{\underbrace{HIn}}} \rightleftharpoons H^+ + \underset{\text{pH}\geq4.4}{\overset{\text{Yellow}}{\underbrace{In^-}}}$$

in which HIn is the protonated form of the indicator and In$^-$ is the deprotonated form. In the titration of a strong acid such as hydrochloric acid with a strong base such as sodium hydroxide, the indicator will change from red to yellow at the endpoint.

As with other weak acids, acid-base indicators are characterized by deprotonation constants. It is desirable that the pK_a of the indicator be close to the equivalence point pH of the titration.

There is far more detailed information on acid-base indicators than I can describe in a few pages. You can access this information by searching for "acid base indicators" and selecting one or more of the several options. One option, Acid-Base Indicators-Boundless, includes a chart of transition ranges of several indicators with colors of the protonated and deprotonated forms of the indicators.

Potentiometric Detection

A variety of reference-electrode/indicator-electrode combinations has been used with a variety of voltage measuring devices for potentiometric acid-base titrations. However, the most common option is to use a pH meter with a saturated calomel reference electrode and a glass indicator electrode. pH vs. volume can be recorded manually for fixed increments of titrants or constant flow burets can be used with continuous measurement and recording of pH vs. volume. pH vs. volume can be plotted in the usual way or it can be used to calculate first and second derivatives. As mentioned earlier, zero-crossing points of second derivatives can be used to identify end-point volumes manually or automatically.

You can access a wealth of information relating to potentiometric detection of end points in titrations by searching for "potentiometric detection of acid-base end points" or other similar search items.

Standardizing Acid-Base Titrants

Hydrochloric acid and sodium hydroxide are common titrants for acid-base titrations. There are several reasons why it isn't possible to prepare hydrochloric acid or sodium hydroxide solutions with accurately known concentrations. Therefore, it is necessary to standardize titrant concentrations by titrating known amounts of standard solutions. Two compounds used as *primary standards* for the calibration of acid-base titrants are potassium hydrogen phthalate for basic titrants, e.g., NaOH, and sodium carbonate for acidic titrants, e.g., HCl. You can access information relating to standardization procedures by searching for "standardizing acid-base titrants" or other similar targets.

OVERVIEW OF ACID-BASE BUFFERS

As you probably know, acid-base buffers are designed to resist changes in the hydrogen ion concentration and pH. As you probably also know, acid-base buffers contain mixtures of conjugate acid-base pairs. Some conjugate acid-base pairs commonly used as buffers are acetic acid and acetate ion, bicarbonate ion and carbonate ion, and dihydrogen phosphate and monohydrogen phosphate.

Buffers are important in a wide variety of situations, including natural processes and laboratory studies. Regarding natural processes, bicarbonate/carbonate buffers control hydrogen ion concentrations and pH in natural waters, including lakes and oceans as well as in fluids in our bodies. Many reactions in our bodies, including enzyme catalyzed reactions, require careful control of hydrogen ion concentrations and pH.

POTENTIAL SOURCES OF HYDROGEN IONS AND HYDROXIDE IONS

Some reactions produce reaction products that can alter the hydrogen ion or hydroxide ion concentrations. For purposes of this discussion, it will be convenient to have independent control of added hydrogen ion and hydroxide ion concentrations. To do this, we will assume that concentrations, C_H of hydrogen ions and C_{OH} of hydroxide ions, are added to buffer solutions as hydrochloric acid and sodium hydroxide, both of which dissociate completely. We will also assume that the concentrations, C_H and C_{OH}, of hydrogen ions and hydroxide ions added to solutions are diluted concentrations, i.e., concentrations after all dilutions are taken into account and before any reactions occur. In other words, added concentrations, $C_H = 0.010$ M H^+ or $C_{OH} = 0.020$ M, mean that 0.010 M hydrogen ion or 0.020 M hydroxide ion is available to react with the buffer.

COMPARISONS OF UNBUFFERED AND BUFFERED SOLUTIONS

The weak acids program was used to calculate hydrogen ion concentrations in (a) pure water, (b) 0.010 M hydrochloric acid in otherwise pure water, (c) a buffer solution containing 0.050 M each of acetic acid, $K_a = 1.8 \times 10^{-5}$, and 0.050 M acetate ion, and (d) the same buffer with 0.010 M hydrochloric acid. Results are summarized below.

Pure Water	0.010 M HCl	Buffer Alone	Buffer + 0.010 M HCl
$[H^+] = 1.0 \times 10^{-7}$ M	$[H^+] \cong 0.010$ M	$[H^+] \cong 1.8 \times 10^{-5}$	$[H^+] \cong 2.7 \times 10^{-5}$

Whereas hydrogen ion concentrations in pure water and 0.010 M HCl differ by five orders of magnitude, hydrogen ion concentrations in the buffer solution without and with 0.010 M hydrochloric acid differ by a factor of about two. On the one hand, the buffer doesn't totally nullify the effect of added hydrochloric acid. On the other hand, it reduces its effect on the hydrogen ion concentration relative to that in pure water by about five orders of magnitude. Why the difference between pure water and the buffer?

When hydrochloric acid is added to pure water, it dissociates completely giving a hydrogen ion concentration equal to the HCl concentration, i.e.,

$$HCl \rightarrow H^+ + Cl^-$$

However, when hydrochloric acid is added to the acetic acid/acetate buffer, it reacts with acetate ion to convert it to acetic acid, i.e.,

$$HCl \rightarrow H^+ + Cl^- \qquad H^+ + Ac^- \rightarrow HAc$$

Most of the added hydrogen ion concentration is consumed by reacting with acetate ion converting it to acetic acid and therefore doesn't change the equilibrium hydrogen concentration by as much as when HCl is added to pure water. You might think of the acetate ion as a "chemical sponge" that "absorbs" the hydrogen ions. However you want to think about it, the effect is to reduce the change in the equilibrium hydrogen ion concentration by several orders of magnitude relative to the change when the same HCl concentration is added to pure water.

What happens if sodium hydroxide is added to the buffer? As shown below, the sodium hydroxide will dissociate completely and the hydroxide ion will react with acetic acid to convert it to acetate ion, i.e.,

$$NaOH \rightarrow Na^+ + OH^- \qquad HAc + OH^- \rightarrow H_2O + Ac^-$$

Most of the added hydroxide ion concentration is consumed by reacting with acetic acid converting it to acetate ion and therefore doesn't change the equilibrium hydrogen ion concentration by as much as when sodium hydroxide is added to pure water. Acetic acid might be viewed as a "chemical sponge" to "absorb" the hydroxide ion to reduce its effects on the equilibrium hydrogen concentration and pH.

My purpose in the remainder of this chapter is to help you understand some features of acid-base buffers that influence their abilities to control hydrogen ion concentrations and pH. A program on Sheet 6 of the acid-base program is designed to help us do this. The program includes three templates in Cells A80–I142. The template in Cells A80–I94 is for a buffer consisting of 0.060 M acetic acid and 0.040 M sodium acetate. The template in Cells A105–I119 is for a buffer consisting of 0.060 M bicarbonate and 0.040 M carbonate ion. The template in Cells A128–I142 is for a buffer consisting of 0.060 M dihydrogen phosphate and 0.040 M monohydrogen phosphate.

After pasting each template into Cells A28–I42, it can be adapted to other situations by changing deprotonation constants in Cells B32–D32 and/or concentrations in B37–E37. The associated program is designed to permit the user to plot measures of buffer performance for up to six sets of conditions on each figure. Procedures for using this feature are discussed later.

SOME SYMBOLS USED IN SUBSEQUENT DISCUSSIONS

I will attempt to identify symbols used in the discussions as we progress through this part of the chapter. However, some symbols used in the discussions are summarized in Table 6.2.

Symbols in the last two rows merit some additional comments.

TABLE 6.2

Selected Symbols Used in Our Discussion of Buffers

C_{HB}	Diluted concentration of a weak acid, HB.
C_B	Diluted concentration of a weak base, B.
C_T	Sum of the diluted concentrations of the acid-base pair, i.e., $C_T = C_{HB} + C_B$.
C_H	Diluted concentration of hydrogen ion added to a solution.
C_{OH}	Diluted concentration of hydroxide ion added to a solution.
K_a	Deprotonation constant for a monoprotic weak acid.
K_{a1}, K_{a2}	Stepwise deprotonation constants for a diprotic weak acid.
$K_{a1}–K_{a3}$	Stepwise deprotonation constants for a triprotic weak acid.
Special	$\kappa_1' = 1/K_{a2}$, $\kappa_2' = 1/(K_{a2}K_{a1})$ for a diprotic weak acid.
Special	$\kappa_1' = 1/K_{a3}$, $\kappa_2' = 1/(K_{a2}K_{a1})$, $\kappa_3' = 1/(K_{a3}K_{a2}K_{a1})$ for a triprotic weak acid.
β	Buffer capacity.
$S_{pH,H}, S_{pH,OH}$	pH sensitivities to added hydrogen ion, $S_{pH,H}$, and added hydroxide ion, $S_{pH,OH}$, concentrations.
$S_{H,H}, S_{H,OH}$	Hydrogen ion sensitivities to added hydrogen ion, $S_{H,H}$, and added hydroxide ion, $S_{H,OH}$, concentrations.

MEASURES OF BUFFER PERFORMANCE

The traditional measure of the ability of a buffer to resist changes in pH, introduced almost a century ago, is a quantity originally called *buffer value*[2] and now called *buffer capacity*.[3,4] For reasons to be discussed later, I will introduce two complementary measures of buffer performance called *pH sensitivity* and *H^+ sensitivity* herein. Symbols in the last two rows are the symbols used herein for these quantities. The symbol, $S_{pH,H}$, translates to "sensitivity of pH to changes in hydrogen ion concentrations." The symbol, $S_{H,OH}$, translates to "sensitivity of the hydrogen ion concentration to changes in hydroxide ion concentrations."

To facilitate comparisons of the three measures of performance, I will describe a program to calculate and plot the three measures of performance for the same six sets of conditions. Plots for the six sets of conditions will be controlled by numbers, 0–6.5 in Cells E28, G28, and I28. To save time, you can plot the three measures of performance simultaneously by using the same numbers in the three cells for each set of conditions. For example, what I do is to set G28 = E28 and I28 = G28 and then control all three sets of the plots by changing numbers in E28.

Let's begin with buffer capacity.

BUFFER CAPACITY

Exact equations have been described for buffer capacities prepared using acid-base pairs of monoprotic, diprotic, and triprotic acids.[3,4] The equations are described in terms of the total concentrations of each acid-base pair, the equilibrium hydrogen ion concentration and reciprocals of products of deprotonation constants.

Concentrations of added hydrogen ions and hydroxide ions are represented by the symbols, C_H and C_{OH}, respectively. As also discussed earlier, it is assumed that the concentrations, C_H and C_{OH}, are diluted concentrations, i.e., the added H^+ and OH^- concentrations after all dilutions have been taken into account. Finally, it is assumed that H^+ and OH^- concentrations, C_H and C_{OH}, are added without changing the solution volume.

Equations for Buffer Capacity

Buffer capacity, commonly represented by, β, is usually defined as the change in equivalents per liter of a strong acid or strong base per unit of change in pH. Given that strong acids and strong bases dissociate completely and that one equivalent per liter of a strong base or strong acid produces one mole per liter of added OH^- or H^+, buffer capacity can be described as the change in added OH^- or H^+ concentration per unit of change in pH, as expressed in Equation 6.1:

$$\beta = \frac{dC_{OH}}{dpH} = -\frac{dC_H}{dpH} \tag{6.1}$$

in which β is the buffer capacity, C_{OH} is the added hydroxide ion concentration and C_H is the added hydrogen ion concentration. In other words, absolute values of buffer capacities for the same concentrations of added hydroxide ions and hydrogen ions in buffers with the same compositions are identical to each other. As noted earlier, hydroxide ions and hydrogen ions may be added externally or they may be produced by the reaction of interest.

In any event, the larger the buffer capacity, the larger the added H^+ or OH^- concentration must be to change the pH by a given amount, e.g., 1 pH unit.

King and Kester[3] and Urbansky and Schock[4] have derived equations for buffer capacity for a variety of situations. Modified forms of Urbansky and Schock's equations[4] for monoprotic, diprotic, and triprotic buffers are summarized in Table 6.3.

TABLE 6.3

Buffer-Capacity Equations for Monoprotic, Diprotic, and Triprotic Acid-Base Pairs[4]

Monoprotic Weak Acid-Base Pairs	Diprotic Weak Acid-Base Pairs
$\beta = 2.3\left[\dfrac{K_w}{[H^+]} + [H^+] + C_T \dfrac{K_a[H^+]}{\left([H^+] + K_a\right)^2}\right]$ (6.2a)	$\beta = 2.3[H^+]C_T\left[\dfrac{\kappa_1' + 4\kappa_2'[H^+] + \kappa_1'\kappa_2'[H^+]^2}{\left(1 + \kappa_1'[H^+] + \kappa_2'[H^+]^2\right)^2}\right]$ (6.2b)

Triprotic Weak Acid-Base Pairs	
$\beta = 2.3[H^+]C_T\left[\dfrac{\kappa_1' + 4\kappa_2'[H^+] + \left(\kappa_1'\kappa_2' + 9\kappa_3'\right)[H^+]^2 + 4\kappa_1'\kappa_3'[H^+]^3 + \kappa_2'\kappa_3'[H^+]^4}{\left(1 + \kappa_1'[H^+] + \kappa_2'[H^+]^2 + \kappa_3'[H^+]^3\right)^2}\right]$	(6.2c)

Selected symbols

C_T–Total concentration of the acid-base pair.

For diprotic acids: $\kappa_1' = 1/K_{a2}$, $\kappa_2' = 1/(K_{a2}K_{a1})$.

For triprotic acids: $\kappa_1' = 1/K_{a3}$, $\kappa_2' = 1/(K_{a2}K_{a1})$, $\kappa_3' = 1/(K_{a3}K_{a2}K_{a1})$.

Note: κ_1'–κ_3' are the same as Urbansky and Schock's β_1–β_3.

The quantities κ_1'–κ_3' in Equations 6.2a–6.2c, called *inverse cumulative constants* herein, are substituted for Urbansky and Schock's β_1–β_3 to avoid confusion with the usual symbol, β, for buffer capacity. We will talk later about a spreadsheet program to calculate and plot buffer capacities vs. pH.

Example 6.5 illustrates an application of Equation 6.2b.

Example 6.5 Calculate the hydrogen ion concentration, the pH, and the buffer capacity for a solution containing 0.060 M sodium bicarbonate and 0.040 M sodium carbonate. For carbonic acid: $K_{a1} = 4.5 \times 10^{-7}$, $K_{a2} = 4.7 \times 10^{-11}$.

Step 1: Calculate an approximate value of the hydrogen ion concentration.

$$\left[H^+\right]_{Buf} \cong K_{a2}\frac{C_{HB}}{C_B} \cong 4.7 \times 10^{-11}\frac{0.06}{0.04} \cong 7.05 \times 10^{-11}$$

Step 2: Calculate the approximate pH

$$pH = -\log\left[H^+\right] \cong 10.15$$

Step 3: Calculate the reverse cumulative deprotonation constants.

$$\kappa_1' = \frac{1}{K_{a2}} = 2.13 \times 10^{10}; \quad \kappa_2' = \frac{1}{K_{a1}K_{a2}} = 4.79 \times 10^{16}$$

Step 4: Substitute $\kappa_1' = 1/k_{a2} = 1/4.69 \times 10^{-11} = 2.13 \times 10^{10}$, $\kappa_2' = 1/((4.5 \times 10^{-7})(4.7 \times 10^{-11})) = 4.7 \times 10^{16}$, $[H^+] = 7.0 \times 10^{-11}$, and $C_T = 0.060\ M + 0.040\ M = 0.10\ M$ into Equation 6.2b and solve.

$$\beta = 2.303[H^+]C_T\left[\frac{\kappa_1' + 4\kappa_2'[H^+] + \kappa_1'\kappa_2'[H^+]^2}{\left(1 + \kappa_1'[H^+] + \kappa_2'[H^+]^2\right)^2}\right] = 0.055$$

I will trust you to do the substitutions and calculations. I used a spreadsheet program to do the calculations; you may wish to do the same.

What does the value of the buffer capacity, $\beta = 0.055$, mean in terms of pH changes? Based on the definition of buffer capacity, $\beta = dC_B/dpH = -dC_H/dpH$, it means that 0.055 M added hydrogen ion concentration will cause the pH to decrease by one pH unit, i.e., from 10.15 to 9.15. It also means that 0.055 M added hydroxide ion concentration will cause the pH to increase by one pH unit, i.e., from 10.15 to 11.15.

It is equally interesting to ask what the buffer capacity means in terms of changes in hydrogen ion concentrations. As shown below, equal changes in pH correspond to very different changes in hydrogen ion concentrations, i.e.,

$$\Delta\left[H^+\right]_H = 10^{-9.15} - 10^{-10.15} = 7.05 \times 10^{-10} - 7.05 \times 10^{-11} = 6.34 \times 10^{-10}$$

$$\Delta\left[H^+\right]_{OH} = 10^{-11.15} - 10^{-10.15} = 7.05 \times 10^{-12} - 7.05 \times 10^{-11} = -6.34 \times 10^{-11}$$

In other words, addition of 0.055 M hydrogen ion produces a 10-fold larger change in the equilibrium hydrogen ion concentration than addition of the same hydroxide ion concentration. Why is this important? It is important because many properties of solutions such as concentration fractions and rates of reactions depend more directly on the hydrogen ion concentration.

Some Graphical Results

A program on Sheet 6 of the weak acids program is designed to calculate and plot buffer capacities vs. pH for ranges of added hydrogen ion and added hydroxide ion concentrations from near zero to near 100% of the conjugate base concentration for added hydrogen ions and 100% of the conjugate acid concentration for added hydroxide ion.

Example 6.6 describes a set of situations for which buffer capacities are compared below.

Example 6.6 Plot on one figure buffer capacities vs. pH for added hydrogen ion concentrations from 0.0 M to 0.038 M for Buffers 1–3 below and added OH⁻ concentrations from 0.0 M to 0.038 M for Buffers 4–6 below.

Concentration ratios, C_1/C_2, for dihydrogen phosphate and monohydrogen phosphate:

Buffer (1) $C_1/C_2 = 0.060/0.040$ M, Buffer (2) $C_1/C_2 = 0.050/0.050$ M, Buffer (3) $C_1/C_2 = 0.040/0.060$ M,

Buffer (4) $C_1/C_2 = 0.060/0.040$ M, (Buffer 5) $C_1/C_2 = 0.050/0.050$ M, Buffer (6) $C_1/C_2 = 0.040/0.060$ M. For phosphoric acid: $K_{a1} = 7.1 \times 10^{-3}$, $K_{a2} = 6.3 \times 10^{-8}$, $K_{a3} = 4.5 \times 10^{-13}$.

A template for Buffer 1 is included in Cells A128–I142 of the program on Sheet 6 of the acid-base programs. To plot curves for Example 6.6, begin by copying the template in Cells A128–I142 and pasting it into Cells A28–I42 using the Paste (P) option. Then follow the sequence in Table 6.4.

TABLE 6.4
Sequence Used to Plot Buffer Capacities vs. pH for Phosphate Buffers

Parts 1–3 of Example 6.6				Parts 4–6 of Example 6.6			
I33	C37	D37	D28	I33	C37	D37	D28
A	0.06	0.04	0→1→2	B	0.06	0.04	4→5
A	0.05	0.05	2→3	B	0.05	0.05	5→6
A	0.04	0.06	3→4	B	0.04	0.06	6→6.5

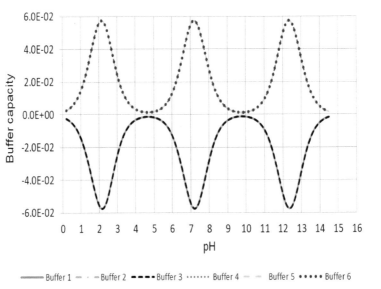

FIGURE 6.5   Buffer capacities vs. pH for phosphate buffers with hydrogen ions added to buffers 1–3 and hydroxide ions added to buffers 4–6.

Buffers 1 and 4: 0.060 M $H_2PO_4^-$ plus 0.040 M HPO_4^{2-}.
Buffers 2 and 5: 0.050 M $H_2PO_4^-$ plus 0.050 M HPO_4^{2-}.
Buffers 3 and 6: 0.040 M $H_2PO_4^-$ plus 0.060 M HPO_4^{2-}.

Plots of buffer capacity vs. pH should be plotted in the figure in Cells AG5–AN22 on Sheet 6 of the weak acids program. Figure 6.5 includes plots you should see when you complete the foregoing scenario. Peaks and valleys from left to right occur at pK_a values of the stepwise deprotonation constants, i.e., $pK_{a1} = 2.15$, $pK_{a2} = 7.20$, and $pK_{a3} = 12.35$.

Negative going peaks correspond to additions of hydrogen ions to Buffers 1–3 and positive going peaks correspond to additions of hydroxide ions to buffers 4–6.

The three peaks for the phosphate buffers are well resolved because there are large differences among deprotonation constants. This will not be the case for acids with smaller differences among deprotonation constants. For example, the analogous plot for citric acid, $K_{a1} = 7.4 \times 10^{-4}$, $K_{a2} = 1.7 \times 10^{-5}$, and $K_{a3} = 4.0 \times 10^{-7}$, has one broad peak for the first two acid-base pairs and a second peak for the last acid-base pair, HB^{2-} and B^{3-}.

Some Perspectives

On the one hand, buffer capacities have been the accepted measure of buffer performance since their introduction almost a century ago.[2] On the other hand, buffer capacities aren't without some limitations. For example, we know that abilities of buffers to resist changes in pH and hydrogen ion concentrations depend on ratios of conjugate acid-base pairs. Yet, all plots in each of the two sets in Figure 6.5 are completely overlapped, thus masking effects of relative concentrations of the acid-base pair on abilities of the buffers to control pH. Moreover, these plots don't give any direct information relevant to abilities of the buffers to resist changes in hydrogen ion concentrations.

Why is this latter point important? It is important because many properties of chemical reactions in which chemists and others are interested depend more directly on hydrogen ion concentrations than on pH. As examples, fractions of monoprotic weak acids in fully protonated and fully deprotonated forms are expressed more directly in terms of the hydrogen ion concentration than pH, i.e.,

$$\alpha_{HB} = \frac{\left[H^+\right]}{\left[H^+\right] + K_a} \qquad\qquad \alpha_B = \frac{K_a}{\left[H^+\right] + K_a}$$

in which α_{HB} and α_B represent fractions of the total concentration of the acid in the fully protonated and deprotonated forms, respectively. Similar equations apply for fractions of polyprotic acids and bases in different forms. Other quantities such as concentrations of different forms of acids and bases, enzyme activities and rate constants for pH-dependent reactions also depend more directly on hydrogen ion concentrations than on pH.

These observations led me to consider two complementary measures of buffer performance called *pH sensitivity* and *H^+ sensitivity* herein. Each is discussed below starting with pH sensitivity. As a reminder, procedures were described earlier for plotting buffer capacity, pH sensitivity, and hydrogen ion sensitivity simultaneously for the same sets of conditions.

pH SENSITIVITY

pH sensitivity is defined herein as the change in pH per unit of change in added hydrogen ion or added hydroxide ion concentration. For small changes in added hydrogen ion or hydroxide ion concentrations, pH sensitivities can be approximated as changes in pH, ΔpH, per unit of change in added hydrogen ion concentration, ΔC_H, or added hydroxide ion concentration, ΔC_H. Equations 6.3a and 6.3b are as follows:

$$S_{pH,H} = \frac{dpH}{dC_H} \cong \frac{\Delta pH}{\Delta C_H} \qquad (6.3a) \qquad\qquad S_{pH,OH} = \frac{dpH}{dC_{OH}} \cong \frac{\Delta pH}{\Delta C_{OH}} \qquad (6.3b)$$

in which $S_{pH,H}$ and $S_{pH,OH}$ are sensitivities of pH to changes in added hydrogen ion or hydroxide ion concentrations, ΔpH is the change in pH, and ΔC_H and ΔC_{OH} are changes in added hydrogen ion and added hydroxide ion concentrations.

To simplify the discussion, I will assume that hydrogen ions are added as hydrochloric acid and that hydroxide ions are added as sodium hydroxide.

Calculation, Storage, and Plotting pH Sensitivities

The program in Columns J–T on Sheet 6 of the weak acids program uses a three-step iterative process to calculate equilibrium hydrogen ion concentrations for HCl and NaOH concentrations from near zero to about 120% of the concentrations of the conjugate acid-base pair used to prepare the buffer. The first step in Columns L–N finds upper limits of 10-fold ranges of hydrogen ion concentrations that include equilibrium concentrations. The second step in Columns O–Q resolves each hydrogen ion concentration to two significant figures. Then, with I29 set to 3, the third step in Columns R–T resolves equilibrium hydrogen ion concentrations to three significant figures. Hydrogen ion concentrations in Column T are used to calculate pH's in Column U.

pH's calculated in this way are used in Column W to calculate pH sensitivities. pH sensitivities calculated in this way are stored for plotting in Columns CB–CM.

Effects of Ratios of Acid-Base Pairs on pH Sensitivities

Example 6.7 describes a set of situations for which pH sensitivities are compared below.

> ### Example 6.7 Plot on one figure pH sensitivities vs. HCl concentrations from 0.0 M to 0.038 M for Buffers 1–3 below and NaOH concentrations from 0.0 M to 0.038 M for Buffers 4–6 below.
>
> Concentration ratios, C_1/C_2, for dihydrogen phosphate and monohydrogen phosphate:
>
> (1) $C_1/C_2 = 0.060/0.040$ M, (2) $C_1/C_2 = 0.050/0.050$ M, (3) $C_1/C_2 = 0.040/0.060$ M,
>
> (4) $C_1/C_2 = 0.060/0.040$ M, (5) $C_1/C_2 = 0.050/0.050$ M, (6) $C_1/C_2 = 0.040/0.060$ M.
>
> For phosphoric acid: $K_{a1} = 7.1 \times 10^{-3}$, $K_{a2} = 6.3 \times 10^{-8}$, $K_{a3} = 4.5 \times 10^{-13}$.

TABLE 6.5
Scenario Used to Plot pH Sensitivities for Example 6.7

	C29	I33	C37	D37	C29	G28[a]
			Parts 1–3			
Part 1	0	A	6.00E-2	4.00E-2	1	2
Part 2	0	A	5.00E-2	5.00E-2	1	3
Part 3	0	A	4.00E-2	6.00E-2	1	4
			Parts 4–6			
	C29	I33	C37	D37	C29	G28[a]
Part 4	0	B	6.00E-2	4.00E-2	1	5
Part 5	0	B	5.00E-2	5.00E-2	1	6
Part 6	0	B	4.00E-2	6.00E-2	1	6.5

[a] Change G28 only after D46 changes to from Solving to Solved.

A template for Part 1 of this example is included in Cells A128–I142 of the program on Sheet 6 of the acid-base programs.

To solve Example 6.7, copy the template in Cells A128–I142 on Sheet 6, paste it into Cells A28–I42 on the same sheet using the Paste (P) option, change G28 from 0 to 1 and follow the scenario in Table 6.5 starting with Part 1.

Plots you should obtain for pH sensitivities for conditions in Example 6.7 are plotted in interactive format in Cells BZ5–CG22 and in picture format in CI5–CP22 on Sheet 6 as well as in Figure 6.6.

Unlike buffer capacities, pH sensitivities are different for different ratios of concentrations of acid-base pairs used to prepare buffers.

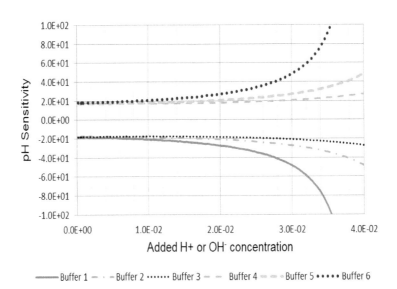

FIGURE 6.6   Effects of added hydrogen ion and hydroxide ion concentrations on pH sensitivities.

Plots 1–3: Hydrogen ion added, Plots 4–6: hydroxide ion added.
Buffers 1 and 4: 0.060 M $H_2PO_4^-$ plus 0.040 M HPO_4^{2-}.
Buffers 2 and 5: 0.050 M $H_2PO_4^-$ plus 0.050 M HPO_4^{2-}.
Buffers 3 and 6: 0.040 M $H_2PO_4^-$ plus 0.060 M HPO_4^{2-}.

Looking first at the shape of the lowermost plot, we observe that the pH becomes increasingly sensitive to added hydrogen ion concentrations as the added hydrogen ion concentration increases. Reasons for this can be explained using the reaction below.

$$HB^{2-} + H^+ \rightarrow H_2B^- + H_2O$$

As added hydrogen ion concentrations are increased, the monohydrogen phosphate concentration decreases. Remember, it is the reaction of added hydrogen ion concentrations with monohydrogen phosphate that allows the buffer to resist changes in pH. As added hydrogen ion concentrations increase, the monohydrogen phosphate concentration decreases and the buffer becomes less effective in controlling the pH.

Looking next at relative positions of the three plots below zero, you will observe that pH sensitivities increase with decreasing concentrations of monohydrogen phosphate. Given that it is the monohydrogen phosphate that permits the buffer to resist changes in pH, it follows that the ability of the buffer to control pH will decrease as the monohydrogen phosphate concentration decreases.

Analogous reasoning applies for effects of hydroxide ion concentrations on pH sensitivities.

Additional Points regarding Figure 6.6

Ranges of the X and Y axes in Figure 6.6 were selected to illustrate general trends. As such, they mask some important features of the results. Figures illustrating plots for narrower and wider ranges of X and Y axes are included in Cells CR5–CY22 and DA5–DH22 on Sheet 6 of the weak acids program along with discussions of the plots. The most relevant observations regarding plots in Cells CR5–CY22 are that (a) the solution containing 0.050 M each of the two forms of phosphate has the smallest pH sensitivity before any strong acid or strong base is added and (b) the plot for the solution containing 0.060 M dihydrogen phosphate and 0.040 M monohydrogen phosphate passes through a minimum when 0.010 M hydroxide has been added. By reacting with dihydrogen phosphate, this hydroxide ion concentration is sufficient to produce 0.050 M each of dihydrogen phosphate and monohydrogen phosphate. These observations are consistent with the fact that abilities of buffers containing equal concentrations of acid-base pairs are most resistant to changes in pH.

Regarding plots in Cells DA5–DH22 for larger ranges of added hydrogen ion and hydroxide ion concentrations, the most relevant observations are that negative peaks occur when sufficient hydrogen ion has been added to convert all the monohydrogen phosphate in each solution to dihydrogen phosphate and positive peaks occur when sufficient hydroxide ion has been added to convert all the dihydrogen phosphate in each solution to monohydrogen phosphate. Changes in hydrogen ion and hydroxide ion concentrations beyond these points are controlled exclusively by diluted concentrations of the strong acid or strong base and the autoprotolysis of water.

Hydrogen Ion Sensitivity

Hydrogen ion sensitivity is defined herein as the change in hydrogen ion concentration per unit of change in added hydrogen ion or hydroxide ion concentration. For small changes in added hydrogen ion or hydroxide ion concentrations, hydrogen ion sensitivities can be approximated as expressed in Equations 6.4a and 6.4b:

$$S_{H,H} = \frac{d[H^+]}{dC_H} \cong \frac{\Delta[H^+]}{\Delta C_H} \qquad (6.4a) \qquad\qquad S_{H,OH} = \frac{d[H^+]}{dC_{OH}} \cong \frac{\Delta[H^+]}{\Delta C_{OH}} \qquad (6.4b)$$

in which $S_{H,H}$ and $S_{H,OH}$ are sensitivities of equilibrium hydrogen ion concentrations to changes in added hydrogen ion or added hydroxide ion concentrations, $\Delta[H^+]$ is the change in the equilibrium hydrogen ion concentration, and ΔC_H and ΔC_{OH} are changes in added hydrogen ion and added hydroxide ion concentrations.

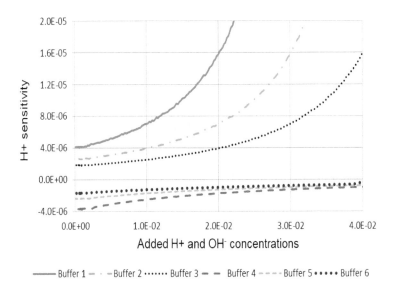

FIGURE 6.7   Effects of added hydrogen ion and hydroxide ion concentrations on hydrogen ion sensitivities.

Plots 1–3: Hydrogen ion added, Plots 4–6: hydroxide ion added.
Buffers 1 and 4: 0.060 M $H_2PO_4^-$ plus 0.040 M HPO_4^{2-}.
Buffers 2 and 5: 0.050 M $H_2PO_4^-$ plus 0.050 M HPO_4^{2-}.
Buffers 3 and 6: 0.040 M $H_2PO_4^-$ plus 0.060 M HPO_4^{2-}.

Hydrogen ion concentrations calculated in Column T on Sheet 6 of the weak acids program are used to calculate hydrogen ion sensitivities in Column V on the same sheet.

Hydrogen ion sensitivities are stored in Columns BF–BQ for plotting; plotted in interactive format in the figure in Cells BC5–BJ22 and in picture format in BL5–BS22.

The scenario in Table 6.5 can be adapted to plots of hydrogen ion sensitivities for the same buffers by using the same setting in I28 as shown for G28 in the table.

Plots you should obtain for hydrogen ion sensitivities for conditions in Example 6.7 are plotted in interactive format in the figure in Cells BC5–BJ22 and in picture format in BL5–BS22 on Sheet 6 as well as in Figure 6.7.

Unlike buffer capacities and pH sensitivities, Plots 1–6 in Figure 6.7 show that hydrogen ion sensitivities are different for additions of the same hydrogen ion and hydroxide ion concentrations. Also, unlike buffer capacities but like pH sensitivities, hydrogen ion sensitivities are different for different ratios of concentrations of acidic and basic forms of the buffer acid. Hydrogen ion sensitivities become increasingly positive with increasing hydrogen ion concentrations and less negative with increasing hydroxide ion concentrations.

Reasons for this behavior are related to the absolute concentrations of the hydrogen ion concentration. As examples, consider effects of 0.050 M each of added hydrochloric acid and sodium hydroxide to aliquots on a buffer containing 0.10 M each of acetic acid and sodium acetate. The program on Sheet 1 of the weak acids program can be used to show that 0.050 M added HCl will produce a change in the hydrogen ion concentration of $\Delta[H^+] = 5.4 \times 10^{-5} - 1.8 \times 10^{-5}$ M $= 3.6 \times 10^{-5}$ M. The program can also be used to show that 0.050 M added NaOH will produce a change of $\Delta[H^+] = 6.0 \times 10^{-6} - 1.8 \times 10^{-5}$ M $= 1.2 \times 10^{-5}$ M. In other words, the added hydrogen ion concentration produces a three-fold larger change in the equilibrium hydrogen ion concentration than the same added hydroxide ion concentration.

Given that most properties of chemical reactions, including concentration fractions, equilibrium concentrations, and reaction rates depend more directly on hydrogen ion concentrations than on pH, I believe that H^+ sensitivity is a more useful measure of buffer performance than either buffer capacity or pH sensitivity.

SUMMARY

After a brief introduction, we begin with a description of a scenario to plot up to six titration curves on a single plot and use the scenario to plot titration curves for six acetic acid concentrations titrated with a fixed concentration of sodium hydroxide. We then talk about effects of concentration on titration curves as well as a way that second derivatives of titration curves have been used to detect end points in acid-base and other titrations. We then talk about effects of deprotonation constants on shapes of titration curves for monoprotic weak acids titrated with a strong base and monoprotic weak bases titrated with a strong acid. We also discuss shapes of titration curves for different concentrations of phosphoric acid titrated with a strong monoprotic base and different concentrations of phosphate ion titrated with a monoprotic strong acid.

We then discuss three measures of buffer performance, namely buffer capacity, pH sensitivity, and H^+ sensitivity. Buffer capacity is the change in added hydrogen ion concentration or added hydroxide ion concentration per unit of change in pH. For any fixed total concentration of an acid-base pair, it is shown that buffer capacity depends directly on the total concentration of the acid-base pair and depends indirectly on the ratio of the concentrations of the weak acid and conjugate base.

We define pH sensitivity as the change in pH, per unit of change in added hydrogen ion or added hydroxide ion concentrations. We define H^+ sensitivity as the change in the equilibrium hydrogen ion concentration per unit of change in added hydrogen ion or added hydroxide ion concentration. We show that, unlike buffer capacity, both pH sensitivity and H^+ sensitivity give information about how buffer performance depends on ratios and total concentrations of conjugate acid-base pairs as well as concentrations of added hydrogen ion and added hydroxide ion concentrations. We also show that H^+ sensitivity gives more complete information about buffer performance than either buffer capacity or pH sensitivity.

REFERENCES

1. A. Savitzky and M. J. E. Golay, *Anal. Chem.*, **1964**, *36*, 1627–1639.
2. D. D. Van Slyke, *J. Biol. Chem.*, **1922**, *52*, 525–570.
3. D. W. King and D. R. Kester, *J. Chem. Ed.*, **1990**, *67*, 932–933.
4. E. T. Urbansky and M. R. Schock, *J Chem. Ed.*, **2000**, *77*, 1640–1643.

7 Solubility Equilibria

This chapter is concerned with equilibria involving sparingly soluble compounds in water. As an example of solubility equilibria that are important to us, hydroxyapatite, $Ca_5(PO_4)_2OH$, is the primary component of bones and teeth. Prevention and treatment of tooth decay and osteoporosis in the elderly are among the most important reasons to understand equilibria involving slightly soluble compounds. Other reasons more relevant to this text are discussed below.

Some salts such as nitrates, halides except those of Ag^+, Pb^{2+}, and $Hg(I)$, and numerous other salts are highly soluble. Examples of highly soluble salts include sodium chloride, zinc chloride, potassium nitrate, and lead nitrate. Solubilities of these and many other highly soluble salts exceed one mole per liter.

Other solids including some salts of strong acids, metal hydroxides, and salts of weak acids are only slightly soluble in water. Examples of slightly soluble salts include silver chloride, zinc iodate, calcium hydroxide, Mn(II) hydroxide, and lead phosphate, to mention a few. Solubilities of these and other salts can range from very low, e.g., $< 10^{-20}$ mol/L, to relatively high, e.g., > 0.01 mol/L. Although these solids are sometimes called insoluble solids, they are more properly called *slightly soluble* or *sparingly soluble solids* because they are not totally insoluble but are only slightly soluble.

Some ways that solubility equilibria influence other types of equilibria are discussed in other chapters. For the moment, suffice it to say that *solubility equilibria are important in many aspects of analytical chemistry, including gravimetry, titrimetry, colorimetry, spectrophotometry, separations, and electroanalytical chemistry*. This chapter is concerned primarily with understanding solubility equilibria and procedures used to calculate solubilities and concentrations of ions associated with the solubility reactions.

The discussion includes descriptions of iterative programs designed to do calculations for both simple and complex situations. It also includes procedures that can be used to identify situations for which simpler calculations can be done using handheld calculators as well as procedures for doing those calculations.

SOME BACKGROUND INFORMATION

This section discusses some aspects of solubility reactions and equilibria that are relevant to all aspects of this chapter.

SIMPLIFYING ASSUMPTIONS

Some simplifying assumptions that apply only to selected situations are discussed at more relevant points in this chapter. Let's talk here about some assumptions that apply throughout the chapter.

Dissolution and Dissociation

It is assumed that all of each salt that dissolve also dissociates into the ions. For example, it is assumed that magnesium phosphate will dissolve and dissociate as follows:

$$Mg_3PO_4 \rightleftharpoons 3Mg^{2+} + 2PO_4^{3-}$$

Hydrolysis of Metal Ions

As you probably know, some metal ions react with water to form metal-hydroxide complexes. As an example, at sufficiently high pH, Cr^{3+} forms the complexes, $CrOH^{2+}$ and $Cr(OH)_2^+$ with hydroxide ions. Other metal ions form analogous complexes. Fortunately, it is possible to reduce effects of these reactions by controlling the hydrogen ion concentration at sufficiently high values that relatively small percentages of the metal ion concentrations form hydroxide complexes. Therefore, *it is assumed throughout this chapter that conditions are such that effects of complex ions can be ignored.*

Sparingly Soluble Solids to Be Discussed

Three groups of slightly soluble solids are discussed. *One group* involves a metal ion, M^{z+}, and the conjugate bases, B^{z-}, of strong acids such as hydrochloric or hydrobromic acid. Silver chloride, AgCl, and lead bromate, $Pb(BrO_3)_2$, are typical examples of salts in this group. A *second group* involves metal ions, M^{z+}, and hydroxide ions, OH^-. Calcium hydroxide, $Ca(OH)_2$ and ferric hydroxide, $Fe(OH)_3$, are typical examples of bases in this group. A *third group* involves metal ions, M^{z+}, and conjugate bases, B^{z-}, of weak acids. Typical examples are lithium fluoride, LiF, lead iodide, PbI_2, and magnesium phosphate, $Mg_3(PO_4)_2$.

Generic Reactions and Solubility Product Constants

Each of these solids can be represented by the general symbol, M_mB_b, in which M^{z+} and B^{z-} represent the metal ion and basic forms of the solids and m and b represent the numbers of metal ions and bases in the solid. Using magnesium phosphate as an example, the solubility reaction and solubility product constant can be written as follows:

$$Mg_3PO_4 \rightleftharpoons 3Mg^{2+} + 2PO_4^{3-} \qquad K_{sp} = \left[Mg^{2+}\right]^3\left[PO_4^{3-}\right]^2$$

in which K_{sp} is the solubility product constant, $[Mg^{2+}]$ and $[PO_4^{3-}]$ are equilibrium concentrations of the ions, and 3 and 2 are coefficients of the reactants in the balanced reaction.

On the one hand, we could develop concepts and write different programs for solids with different numbers of metal ions and bases. However, in order to generalize the discussion and reduce the number of programs we must develop, we will write reactions and solubility product constants for all solids in each of the three groups mentioned earlier in a generic form. Examples include silver chloride, calcium oxalate, and magnesium phosphate. The anionic part of each solid is a base, e.g., Cl^-, $C_2O_4^{2-}$, and PO_4^{3-}. Recognizing this, we will use symbols M and B, to represent the metallic and basic parts, respectively, of each solid. With this in mind, generic reactions and solubility product constants are written as follows:

$$M_mB_b \rightleftharpoons mM^{z_m} + bB^{z_b} \qquad K_{sp} = \left[M^{z_m}\right]^m\left[B^{z_b}\right]^b$$

In this way, we will be able to limit the number of programs to three, one each for salts of strong acids, hydroxides, and salts of weak acids.

Symbols Used in Our Discussion

We will use a symbolic approach in order to adapt the treatment to as many different situations as possible. I will try to identify symbols representing different quantities as we encounter them in text. However, some of the more relevant symbols are summarized in Table 7.1.

Diluted Concentrations of Reactants and Products

Diluted metal ion, weak base, and hydroxide ion concentrations are represented herein by symbols C_M, C_B, and C_{OH}, respectively. As examples, diluted silver ion and chloride ion concentrations are represented as C_{Ag} and C_{Cl}.

Equilibrium Concentrations and Solubilities

As is common practice, equilibrium concentrations of reactants and products are represented by square bracketed symbols. As examples, equilibrium Pb^{2+} and phosphate concentrations are

TABLE 7.1

Symbols Used in This Chapter Involving Solubility Equilibria

$Pb_3(PO_4)_2 \rightleftharpoons 3Pb^{2+} + 2PO_4^{3-}$	$M_mB_b \rightleftharpoons mM^{z+} + bB^{z-}$
m, b	Coefficients of ions in a balanced solubility reaction.
z_m, z_b	Charges on metal ions, M^{z_m}, and basic ions, B^{z_b}.
C_M^0, C_B^0	Initial or starting metal ion and basic ion concentrations.
C_M, C_B	Diluted metal ion and basic ion concentrations.
C_M', C_B'	Net metal ion and basic ion concentrations after reactions of excesses of each.
$[M^{z_m}]$, $[B^{z_b}]$, $[OH^-]$	Equilibrium metal ion, basic ion, and hydroxide ion concentrations.
$\langle M^{z_m} \rangle$, $\langle B^{z_b} \rangle$, $\langle OH^- \rangle$	Pre-equilibrium metal ion and hydroxide ion concentrations varied by iterative programs to calculate equilibrium concentrations.
$[H^+]$, $[OH^-]$	Equilibrium hydrogen ion and hydroxide ion concentrations.
K_{sp}, Q_{sp}	Solubility product constant and solubility reaction quotient.
S and S'	Equilibrium, S, and non-equilibrium, S', solubilities.
n(Acid)	Number, 1–3, of acidic protons in an acid.
K_{a1}, K_{a2}, K_{a3}	Stepwise deprotonation constants for a triprotic weak acid.
C_{0B}, C_{1B}, C_{2B}, and C_{3B}	Initial concentrations of forms of a triprotic weak acid that have 0, 1, 2, and 3 protons, respectively.
α_0, α_1, α_2, α_3	Fractions of a weak acid with up to three acidic protons in forms that have lost i = 0–3 protons.
$[H_nB]...[H_{n-3}B]$	Equilibrium concentrations of forms of a weak acid with up to three acidic protons in forms that have lost i = 0–3 protons.
μ	Ionic strength
f_1, f_2, f_3	Activity coefficients of forms of an uncharged acid that have lost i = 1, 2, and 3 protons.
$K_{w,\mu}$, $K_{sp,\mu}$, $K_{a1,\mu}$, $K_{a2,\mu}$, $K_{a3,\mu}$	Equilibrium constants adjusted for effects of ionic strength.

represented by $[Pb^{2+}]$ and $[PO_4^{3-}]$. More generally, equilibrium metal ion and base concentrations are represented by symbols, [M] and [B].

Equilibrium solubilities are represented by S. We will talk later about relationships between equilibrium solubilities and concentrations of the ions.

Non-Equilibrium Concentrations and Solubilities

Each of the three programs to be discussed starts with a set of non-equilibrium concentrations and solubilities and varies the concentrations and/or solubility until equilibrium concentrations are resolved. For the magnesium phosphate reaction, non-equilibrium Mg^{2+} and PO_4^{3-} are represented by $\langle Mg^{2+} \rangle$ and $\langle PO_4^{3-} \rangle$. More generally, non-equilibrium concentrations for metal ions and weak bases and hydroxides are represented by $\langle M^{z_m} \rangle$, $\langle B^{z_b} \rangle$, and $\langle OH^- \rangle$. Equilibrium and non-equilibrium solubilities are represented by S and S', respectively.

Reaction Quotients Used as Convergence Criteria

Some programs to be discussed use a quantity called a *reaction quotient* as a convergence criterion for iterative calculations.

As you may recall, the reaction quotient, Q, for a reaction has the same form as the equilibrium constant expressed in terms of non-equilibrium concentrations. Although reaction quotients are more commonly used in connection with gas-phase reactions, the concept is extended here to include solubility reactions.

Using the magnesium phosphate reaction as an example, equations for the solubility product constant, K_{sp}, and reaction quotient, Q_{sp}, are written as follows:

$$Mg_3PO_4 \rightleftharpoons 3Mg^{2+} + 2PO_4^{3-} \qquad K_{sp} = \left[Mg^{2+} \right]^3 \left[PO_4^{3-} \right]^2 \qquad Q_{sp} = \langle Mg^{2+} \rangle^3 \langle PO_4^{3-} \rangle^2$$

in which Q_{sp} is the reaction quotient, $\langle Mg^{2+} \rangle$ and $\langle PO_4^{3-} \rangle$ are non-equilibrium magnesium ion and phosphate ion concentrations, and other symbols have their usual significance. Some programs to be discussed start with very small trial values of the solubility and increase it by small amounts until the ratio of the reaction quotient to the solubility product constant is equal to one, i.e., $Q_{sp}/K_{sp} = 1$. The solubility and reactant concentrations corresponding to the condition, $Q_{sp}/K_{sp} = 1$, represent equilibrium values. A program for salts of weak acids on Sheet 3 finds equilibrium conditions by varying hydrogen ion concentrations until the sign of a charge-balance equation changes from − to +. A program for salts of weak acids on Sheet 4 finds equilibrium conditions by varying the solubility until the ratio of the reaction quotient to the solubility product constant changes from less than one to slightly larger than one.

By analogy with other programs discussed earlier, users control degrees of resolution by setting a cell in each program to the number of significant figures to which solubilities or hydrogen ion concentrations are to be resolved. Three significant figures are a good compromise between convergence time and uncertainties associated with equilibrium constants used in the calculations.

COMMON IONS AND NET CONCENTRATIONS

Calculations for each group of salts with or without correction for ionic strength can be divided into two subgroups, namely situations without a common ion or situations with a common ion. Regardless of initial conditions, any situation can be simplified to one without a common ion or with just one common ion.

If a solution is prepared to contain concentrations of both common ions initially as is done in precipitation titrations, the ions will react until the concentration of at least one ion is very close to zero. Although neither concentration will decrease to exactly zero, calculations are simplified by assuming that the initial concentration of at least one ion decreases to zero and then calculating concentrations after equilibrium conditions are reestablished.

For example, consider a solution prepared to contain diluted concentrations, C_{Pb} of Pb^{2+} and C_{PO_4}, of PO_4^{3-} ions. The ions will react as follows until the concentration of one or both ions approaches zero.

$$Pb_3PO_4 \rightleftharpoons 3Pb^{2+} + 2PO_4^{3-}$$

Diluted concentrations, C_{Pb} or C_{PO_4}, and balancing coefficients, 3 and 2, are used in Equations 7.1a and 7.1b to calculate net diluted concentrations, C_{Pb}' or C_{PO_4}', to be used in solubility calculations.

$$C_{Pb}' = C_{Pb} - \left(\frac{3}{2} \right) C_{PO_4} \tag{7.1a}$$

$$C_{PO_4}' = C_{PO_4} - \left(\frac{2}{3} \right) C_{Pb} \tag{7.1b}$$

Although negative values can be avoided by limiting the calculation to the ion that is in excess, it is just as easy to do the calculations for both ions and set any negative concentrations to zero.

You may find it useful to do Exercise 7.1A.

EXERCISE 7.1A

Given a solution containing $C_{Pb} = 0.050$ M Pb^{2+} and $C_{PO_4} = 0.010$ M PO_4^{3-} ions initially, show that the *net* Pb^{2+} and PO_4^{3-} concentrations are $C_{Pb}' = 0.035$ M and $C_{PO_4}' = -0.023 \Rightarrow 0.000$ M, respectively.

Now, let's generalize these thoughts. To do this, let's begin by representing initial metal ion concentrations by C_M and net concentrations by C_M'. Given that most anions such as phosphate, PO_4^{3-}, and iodate, IO_3^-, are bases, B, let's represent diluted anion concentrations by C_B and net concentrations as C_B'. Let's also represent coefficients of metal ions and bases in balanced reactions by the symbols, m and b, respectively, e.g.,

$$mM + bB \rightleftharpoons M_m B_b$$

Using these symbols for a salt of the form, $M_m B_b$, the corresponding Equations 7.2a and 7.2b for net concentrations are

$$C_M' = C_M - \left(\frac{m}{b}\right) C_M \tag{7.2a}$$

$$C_B' = C_B - \left(\frac{b}{m}\right) C_M \tag{7.2b}$$

in which C_M' and C_B' are net metal ion and base concentrations, C_M and C_B are diluted metal ion and base concentrations and m and b are coefficients of the metal ion and base in the solid.

For the Pb^{2+}/PO_4^{3-} reaction, the ratios, $m/b = 3/2$ and $b/m = 2/3$ reflect the facts that three Pb^{2+} ions react for each two phosphate ions and vice versa.

I suggest that you do Exercise 7.1B to test your understanding of the foregoing discussion.

EXERCISE 7.1B

Given a solution containing initial concentrations, $C_M = C_{Cr} = 0.050$ M Cr^{3+} and $C_B = C_{IO_3} = 0.010$ M IO_3^- ions initially, show that the *net* Cr^{3+} and IO_3^- concentrations are $C_{Cr}' = 0.047$ M and $C_{IO_3}' = -0.103 \Rightarrow 0.000$ M, respectively.

Spreadsheet programs to be discussed later calculate net concentrations as described above and use them in subsequent calculations.

CRAZY NUMBERS

If applied indiscriminately, procedures to be described herein and elsewhere can give impossible results for some situations. For example, using seemingly valid procedures to calculate the solubility of calcium hydroxide, $K_{sp} = 5.5 \times 10^{-6}$, in a solution "buffered" at pH = 2 gives a solubility, $S = 6.5 \times 10^6$ M. Numbers such as this mean that hidden assumptions used in the calculations are not valid. In this case, the supply of calcium hydroxide would be depleted long before such excessive solubilities occurred and/or hydroxide ion from the solubility reaction would exceed the ability of the buffer to control the pH.

Some situations involving very small solubility product constants can yield impossibly small numbers. Either way, I trust that you will interpret impossibly large solubilities as "very soluble" and impossibly small solubilities "very small solubilities."

BUFFERED SOLUTIONS

Programs on Sheets 2 and 3 of the solubilities program include calculations in buffered solutions. Calculations for buffered solutions don't account for effects of ionic strength because it isn't possible to account for effects of ionic strength without knowing the compositions of the buffers.

Let's begin the remainder of this discussion with the simpler of the three groups, namely salts of strong acids.

SALTS OF STRONG ACIDS

Examples of salts in this group include cuprous chloride, $CuCl$, lead bromide, $PbBr_2$, lead chloride, $PbCl_2$, mercury(I) chloride, Hg_2Cl_2, and thallium(I) chloride, $TlCl$. Both chloride and bromide are anions of strong acids, HCl and HBr.

I suggest that you open the program, *Solubility equilibria*, and you select Sheet 1 after reading the discussion on Sheet 5.

Let's begin with a conventional approach to a typical situation.

CONVENTIONAL APPROACH TO SOLUBILITY CALCULATIONS

Most tabulated values of solubility product constants are activity-based constants, i.e., they are constants at zero ionic strength. Conventional procedures don't differentiate between activity-based constants and concentration-based constants. In other words, activity-based constants are equated to concentration products rather than activity products. We will follow this convention initially and talk later about conversions of activity-based constants to concentration-based constants.

Situation without a Common Ion

Example 7.1A illustrates calculations for a situation involving the salt of a strong acid without a common ion.

> **Example 7.1A Given a saturated solution of lead bromide in otherwise pure water, calculate the solubility of lead bromide and the Pb^{2+} and Br^- concentrations. For $PbBr_2$: $K_{sp} = 4.0 \times 10^{-5}$.**
>
> *Step 1:* Write the solubility reaction and solubility product constant.
>
> $$PbBr_2 \rightleftharpoons Pb^{2+} + 2Br^- \qquad K_{sp} = \left[Pb^{2+}\right]\left[Br^-\right]^2 = 4.0 \times 10^{-5}$$
>
> *Step 2:* Express Pb^{2+} and Br^- concentrations in terms of the solubility, S.
>
> $$\left[Pb^{2+}\right] = S; \left[Br^-\right] = 2S$$
>
> *Step 3:* Write the solubility product constant in terms of the concentrations and solubilities.
>
> $$K_{sp} = \left[Pb^{2+}\right]\left[Br^-\right]^2 = S(2S)^2 = 4S^3 = 4.0 \times 10^{-5}$$
>
> *Step 4:* Rearrange the foregoing equation and calculate the solubility.
>
> $$S = \left(\frac{K_{sp}}{4}\right)^{1/3} = \left(\frac{4.0 \times 10^{-5}}{4}\right)^{1/3} = 0.0215 \text{ M}$$
>
> *Step 5:* Calculate the Pb^{2+} and Br^- concentrations.
>
> $$\left[Pb^{2+}\right] = S = 0.0215 \text{ M}; \left[Br^-\right] = 2S = 2(0.0215) = 0.0430 \text{ M}$$

Individual steps in the calculation are judged to be obvious and are not discussed further. However, there is a "hidden assumption" in this example that lead hydroxide will not precipitate from the solution. The product of the lead ion concentration times the hydroxide ion concentration squared in otherwise pure water is $Q_{sp} = [Pb^{2+}][OH^-]^2 = (0.022)(1 \times 10^{-7})^2 = 2.2 \times 10^{-16}$. Given that this ion product is less than the solubility product constant for lead hydroxide, $K_{sp} = 1.2 \times 10^{-15}$, the assumption is valid.

Situation with a Common Ion

Example 7.1B illustrates a conventional approach to calculations for a solution containing bromide ion as a common ion. For the record, this example is intended to illustrate a situation for which conventional approximations fail.

Example 7.1B Calculate the solubility of lead bromide and the concentrations of Pb^{2+} and Br^- in a saturated solution of lead bromide containing a diluted concentration, $C_{Br} = 0.010$ M, of bromide ion initially. For $PbBr_2$: $K_{sp} = 4.0 \times 10^{-5}$.

Step 1: Write the solubility reaction.

$$PbBr_2 \rightleftharpoons Pb^{2+} + 2Br^- \qquad K_{sp} = 4.0 \times 10^{-5}$$

Step 2: Express Pb^{2+} and Br^- concentrations in terms of the solubility, S, and the initial bromide ion concentration.

$$\left[Pb^{2+}\right] = S; \left[Br^-\right] = C_{Br} + 2S = 0.010 + 2S$$

Step 3: Write the solubility product constant in terms of the concentrations and solubilities.

$$K_{sp} = \left[Pb^{2+}\right]\left[Br^-\right]^2 = S(0.010 + 2S)^2 = 4.0 \times 10^{-5}$$

Step 4: Calculate an approximate value of the solubility by assuming that the solubility is much less than the initial bromide ion concentration, i.e., that $2S \ll 0.010$.

$$K_{sp} = S(0.010 + 2S)^2 \cong S(0.010)^2 = 4.0 \times 10^{-5} \Rightarrow S \cong \frac{K_{sp}}{(0.010)^2} \cong \frac{4.0 \times 10^{-5}}{(0.010)^2} \cong 0.40$$

Step 5: Test the assumption that the approximate solubility is much less than the initial bromide ion concentration.

Given that the initial bromide ion concentration is $C_{Br} = 0.010$ M and the approximate solubility is 0.040 M, it follows that the assumption in Step 4 that the solubility is much less than the added bromide ion concentration, i.e., $S \ll C_{Br}$, fails badly.

On the one hand, procedures used in this example are consistent with conventional approximation procedures. On the other hand, conditions in this example were selected to illustrate types of errors that can occur when assumptions on which conventional approximation procedures are based fail.

Having illustrated a situation for which conventional approximations fail, my goal here is to describe an approach that will give accurate results for situations for which simplifying assumptions associated with conventional approximations do and don't give satisfactory results. We will use iterative calculations to do this.

I will begin by introducing a program designed to do the calculations and then describe the basis for the program after you have used it to solve some problems.

Generic Reactions and Symbols

Although described in Examples 7.1A and 7.1B for a specific solid, lead bromide, the programs to be described are designed to do calculations for a variety of solids with different numbers of metal ions and basic anions per molecule of a solid. To do this, I have used generic symbols for different quantities in the reactions and corresponding solubility product constant. As examples, the lead bromide reaction and solubility product constant are written below in their conventional and generic forms

$$PbBr_2 \rightleftharpoons Pb^{2+} + 2Br^- \qquad K_{sp} = \left[Pb^{2+} \right]\left[Br^- \right]^2$$

$$M_m B_b \rightleftharpoons mM^{z_m} + bB^{z_b} \qquad K_{sp} = \left[M^{z_m} \right]^m \left[B^{z_b} \right]^b$$

For lead bromide, $z_m = 2$, $z_b = 1$, $m = 1$, and $b = 2$.

As another example, consider mercury(I) bromide, Hg_2Br_2.

$$Hg_2Br_2 \rightleftharpoons 2Hg^+ + 2Br^- \qquad K_{sp} = \left[Hg^+ \right]^2 \left[Br^- \right]^2$$

$$M_m B_b \rightleftharpoons mM^{z+} + bB^{z-} \qquad K_{sp} = \left[M^{z_m} \right]^m \left[B^{z_b} \right]^b \quad (z_+ = 1, z_- = 1, m = 2, b = 2)$$

The use of generic symbols may be a bit confusing at times. However, it will make it possible for us to adapt each program to different solids with different numbers of metal ions and basic anions rather than having to write a different program for each group of compounds with different numbers of metal ions and basic anions.

Iterative Program

Let's begin with calculations for Example 7.1A, i.e., lead bromide without a common ion.

Calculations for Lead Bromide without a Common Ion

To begin, open the program, *Solubility Equilibria*, from the programs website http://crcpress.com/9781138367227 into an unused Excel file; select Sheet 1 and *correlate symbols in Cells B28–B36 with descriptions of the symbols in Table 7.1*.

Then select File/Options/Formulas and confirm that Workbook Calculation is set to Automatic, there is a check mark in the *Enable iterative calculations* box, *Maximum Iterations* is set to 2,000 and that *Maximum Change* is set to 1E–35.

Then copy the template for Example 7.1A in Cells A76–C87 and paste it into Cells A25–C36 using the Paste (P) option. Then set C25 to 1 and Press F9 repeatedly if necessary until H34 changes to "Solved." Then read the solubility, Pb^{2+} concentration and Br^- concentrations without, WO, and with, W, correction for ionic strength from Cells F30–G32. Results you should see are summarized in Table 7.2A.

TABLE 7.2A

Comparison of Results Calculated without and with Correction for Ionic Strength

Results without Correction for μ			Results with Correction for μ		
S	$[Pb^{2+}]$	$[Br^-]$	S_μ	$[Pb^{2+}]_\mu$	$[Br^-]_\mu$
2.16×10^{-2}	2.16×10^{-2}	4.32×10^{-2}	3.92×10^{-2}	3.92×10^{-2}	7.84×10^{-2}

Notice that the solubility and ionic concentrations corrected for effects of ionic strength are larger than those without correction for ionic strength. Why the difference?

We will show later that the relationship between constants with and without correction for ionic strength is as follows:

$$K_{sp,\mu} = \frac{K_{sp}}{f_M^m f_B^b}$$

in which $K_{sp,\mu}$ is the constant adjusted for effects of ionic strength and K_{sp} is the constant without correction for ionic strength and f_{Pb} and f_{Br} are activity coefficients of Pb^{2+} and Cl^-. Activity coefficients for bromide and Pb^{2+} ions calculated in Cells E60 and F60 are $f_1 = f_{Br} = 0.742$ and $f_2 = f_{Pb} = 0.302$. Using these activity coefficients with the coefficients, $m = 1$ and $b = 2$, it is easily shown that the solubility product constant corrected for effects of ionic strength is $K_{sp,\mu} = 4.0 \times 10^{-5}/[(0.302)(0.742)^2 = 2.4 \times 10^{-4}$. Therefore, the reason the solubility and equilibrium concentrations corrected for ionic strength are larger than those calculated without correction for ionic strength is that the solubility product constant corrected for ionic strength is about 6 times larger than the activity-based constant. i.e., $2.4 \times 10^{-4}/4.0 \times 10^{-5} = 6.0$.

Calculations for Lead Bromide with 0.010 M Bromide as a Common Ion as in Example 7.1B

The simplest option is to set C25 to 0, set C30 to 0.010, reset C25 to 1 and if necessary, press F9 one or more times until H34 changes to "Solved" and read the solubility and concentrations from Cells F30–F32. Alternatively, copy the template in Cells A92–C103 and paste it into Cells A25–C36 using the Paste (P) option. Then set C25 to 1 and, if necessary, press F9 repeatedly until H34 changes to "Solved." For calculations without correction for ionic strength, read the solubility, $S = 1.84 \times 10^{-2}$, lead ion concentration, $[M^{z_m}] = 1.84 \times 10^{-2}\,M$, and bromide ion concentration, $[B^{z_b}] = 4.68 \times 10^{-2}\,M$, from Cells F30–F32.

I will trust you to compare the solubility and concentrations calculated without and with correction for ionic strength.

Now, let's check the results for consistency with expected results. Based on the stoichiometry of the balanced reaction, the initial bromide ion concentrations, $C_{Br} = 0.010$ M, the equilibrium bromide ion concentration should be

$$\left[Br^- \right] = C_{Br} + 2S = 1.0 \times 10^{-2}\,M + 2\left(1.84 \times 10^{-2}\,M\right) = 4.7 \times 10^{-2}\,M$$

That is consistent with the calculated concentration, 4.68×10^{-2} M, in F32.

COMPARISON OF RESULTS

Results without and with a common ion are summarized in Table 7.2B to facilitate comparisons.

TABLE 7.2B
Summary of Results for Examples 7.1A and 7.1B

	S	[Pb^{2+}]	[Br$^-$]
Part A	2.15×10^{-2} M	2.15×10^{-2} M	4.31×10^{-2} M
Part B	1.84×10^{-2} M	1.84×10^{-2} M	4.68×10^{-2} M

Notice that the inclusion of a common ion suppresses the solubility of lead bromide. This observation can be generalized. The inclusion of a common ion will always reduce the solubility of a sparingly soluble solid relative the solubility without a common ion. The larger the concentration of the common ion, the more the solubility will be decreased.

AN EXERCISE

Now, I suggest that you do Exercise 7.2 to test your understanding of what we have discussed to this point.

EXERCISE 7.2

Given solutions containing (A) solid Hg_2Br_2 and (B) solid Hg_2Br_2 with 0.010 M Hg^+, show that the solubilities and equilibrium concentrations are as summarized below. For Hg_2Br_2: $K_{sp} = 5.6 \times 10^{-23}$.

	S	[Hg⁺]	[Br⁻]
Part A	1.37×10^{-6} M	2.74×10^{-6} M	2.74×10^{-6} M
Part B	3.75×10^{-10} M	1.00×10^{-2} M	7.50×10^{-10} M

On the one hand, templates for Parts A and B of this exercise are included in Cells A109–C137 on Sheet 1 of the solubilities program. On the other hand, I suggest that you test your understanding of the input part of the program by trying to do the exercise without using the template.

BASIS FOR THE PROGRAM

Net concentrations are calculated in Cells C45 and D45 using procedures described earlier. Iterative calculations are done in Cells E48–I49 using a two-step process based on the net concentrations in Cells C45 and D45. The convergence criterion in each case is the ratio of a quantity called a reaction quotient, Q_{sp}, to the solubility product constant, K_{sp}.

Reaction Quotients Used as Convergence Criteria

As you may recall, the *reaction quotient, Q*, for a reaction has the same form as the equilibrium constant expressed in terms of non-equilibrium concentrations. Although reaction quotients are more commonly used in connection with gas-phase reactions, the concept is extended here to include solubility reactions.

Using the lead bromide reaction as an example, the reaction, solubility product constant and reaction quotient are as shown in Equations 7.3a and 7.3b:

$$PbBr_2 \rightleftharpoons Pb^{2+} + 2Br^-$$

$$K_{sp} = \left[Pb^{2+} \right]\left[Br^- \right]^2 \tag{7.3a}$$

$$Q_{sp} = \langle Pb^{2+} \rangle \langle Br^{-2} \rangle \tag{7.3b}$$

in which K_{sp} is the solubility product constant, $[Pb^{2+}]$ and $[Br^-]$ are equilibrium Pb^{2+} and Br^- concentrations, Q_{sp} is the reaction quotient, and $\langle Pb^{2+} \rangle$ and $\langle Br^- \rangle$ are pre-equilibrium concentrations which increase as trial solubilities are increased by the iterative program. The process continues until the reaction quotient exceeds the solubility product constant by a small amount controlled by the significant figures in C26. The reaction quotient should be equal to the solubility product constant at equilibrium. Therefore, the ratio of the reaction quotient to the solubility product constant should be equal to one at equilibrium, i.e., $Q_{sp}/K_{sp} = 1$ at equilibrium.

The reaction quotient is used in a two-step process to find equilibrium concentrations. A *low resolution step* is used to find the upper limit of a 10-fold range of solubilities that includes the equilibrium solubility. Then a *high resolution step* is used to resolve the equilibrium concentration to the number of significant figures in C26 on Sheet 1 of the program.

Low Resolution Step

The low resolution step is in Cells E48 and F48 on Sheet 1 of the solubility program. The program begins with a very small solubility from Cell C30, e.g., $S' = 1 \times 10^{-30}$ M, and increases it in 10-fold steps until the ratio, Q_{sp}/K_{sp}, in E48 is larger than one, i.e., $Q_{sp}/K_{sp} > 1$. Given that trial values of the solubility are increased by factors of 10 in each iteration, the first solubility for which $Q_{sp}/K_{sp} > 1$ is the upper limit of a 10-fold range that includes the equilibrium solubility.

High Resolution Step

The high resolution step is implemented in Cells E49 and F49 on Sheet 1 of the program. The high resolution step starts with a trial solubility equal to 10% of the solubility from the low resolution step and increases it by amounts determined by the number of significant figures in C26 until the ratio, Q_{sp}/K_{sp}, first exceeds one. The solubility that causes the change, $Q_{sp}/K_{sp} < 1$ to $Q_{sp}/K_{sp} > 1$, is the equilibrium solubility resolved to the number of significant figures in C26.

With C26 set to 2, 3 or 4, the equilibrium solubility is resolved to 2, 3 or 4 significant figures. As shown in Step 5 of Example 7.1A, coefficients in balanced reactions entered by the user as m and b in Cells C33 and C34 are used to convert solubilities to ionic concentrations.

CORRECTING FOR EFFECTS OF IONIC STRENGTH

The primary difference between calculations without and with correction for ionic strength is the use of activity-based and concentration-based constants.

The following is a stepwise procedure used to correct for ionic strength.

a. Use activity-based constants to calculate a first estimate of the solubility and equilibrium concentrations of all ions.
b. Use the first estimates of concentrations to calculate the ionic strength, μ.
c. Calculate activity coefficients, f_i, of the ions involved in the solubility reaction.
d. Use the activity coefficients to calculate a concentration-based solubility product constant.
e. Repeat the solubility calculations using the concentration-based constant.

Calculation procedures using concentration-based constants are the same as illustrated above using activity-based constants.

Lead chloride is used here as an illustrative example. The reaction and solubility product equation for lead chloride are as follows:

$$PbCl_2 \rightleftharpoons Pb^{2+} + 2Cl^- \qquad K_{sp} = \left[Pb^{2+}\right]\left[Cl^-\right]^2$$

in which all symbols have their usual significance. A generic equivalent of this reaction and solubility product equation are as follows:

$$M_m B_b \rightleftharpoons mM^{z_m} + bB^{z_b} \qquad K_{sp} = \left[M^{z_m}\right]^m \left[B^{z_b}\right]^b$$

in which m and b are numbers of metal ions and basic ions in the solid, z_m and z_b are charges on the ions and square-bracketed quantities are equilibrium concentrations of the ions.

Ionic Strength

The program is designed to allow the user to include an "inert" ionic strength in C35 equal to or larger than zero in addition to that produced by the water and the ions from common ions and the solubility reaction. The ionic strength will be the sum of the inert ionic strength plus one half the sum of products of charges squared times equilibrium concentrations of all ions. Assuming that common ions are added as nitrate salts of metal ions and sodium salts of basic anions, some typical and generic reactions are as summarized below.

$$H_2O \rightleftharpoons H^+ + OH^- \quad Pb(NO_3)_2 \rightarrow Pb^{2+} + 2NO_3^- \qquad NaCl \rightarrow Na^+ + Cl^-$$

$$H_2O \rightleftharpoons H^+ + OH^- \quad M(NO_3)_{z_m} \rightarrow M^{z+} + z_m NO_3^- \quad Na_{z_b}B \rightarrow z_b Na^+ + B^{z_b}$$

The equation below for the ionic strength in C35 is annotated to identify terms used to calculate the ionic strength in Equation 7.4.

$$\mu = \overbrace{C\$35}^{\mu_{Inert}} + 0.5*\left(2*SQRT(C27) + \overbrace{C31*C29}^{z_m[NO_3^-]} + \overbrace{C32*C30}^{z_b[Na^+]} + \overbrace{C31 \wedge 2*F31}^{z_m^2[M^{z+}]} + \overbrace{C32^2*F32}^{z_b^2[B^{z-}]}\right) \quad (7.4)$$

The term, $2*SQRT(C27) = 2*SQRT(K_w)$ accounts for hydrogen ion and hydroxide ion concentrations from water in a solution presumed to contain anions of strong acids. The C31*C29 term accounts for nitrate ions from a nitrate salt of metal ion added as a common ion. The C32*C30 term accounts for sodium ions from a sodium salt of a basic ion added as a common ion. The C31^2*F31 and C32^2*F32 terms account for contributions of equilibrium concentrations of the metal ion and basic anion, respectively.

Activity Coefficients of the Ions

The extended Debye-Hückle equation,[1] EDHE, is used to calculate *activity coefficients* for ionic strengths up to 0.1 M and the Davies modification[2,3] of the EDHE is used for larger ionic strengths. Exponential forms of the EDHE and Davies equations for activity coefficients are expressed in Equations 7.5a and 7.5b:

$$f_i = 10 \wedge \left(-\frac{0.51z_i^2\sqrt{\mu}}{\left(1 + 0.3D\sqrt{\mu}\right)}\right) \quad \left(EDHE^1\right) \tag{7.5a}$$

$$f_i = 10 \wedge \left(-\frac{0.51z_i^2\sqrt{\mu}}{1 + \sqrt{\mu}} - 0.3\mu\right) \quad \left(Davies^{2,3}\right) \tag{7.5b}$$

in which z_i is the charge on each ion, D is the mean hydrated diameter[4] of the ion and other symbols are as defined above. Given that hydrated diameters are not known for all ions of interest, an average value of $D_{Av} = 0.5$ nm is used for all ions.

Concentration-Based Constants, $K_{sp,\mu}$

Activity coefficients calculated as described above are used to convert activity-based constants, K_{sp}, to concentration-based constants, $K_{sp,\mu}$. For lead chloride, relationships between activity-based and concentration-based constants are as expressed in Equations 7.6a–7.6c:

$$K_{sp} = \{Pb^{2+}\}\{Cl^-\}^2 = f_{Pb}[Pb^{2+}]f_{Cl}^2[Cl^-]^2 = f_{Pb}f_{Cl}^2\overbrace{[Pb^{2+}][Cl^-]^2}^{K_{sp,\mu}} \tag{7.6a}$$

in which K_{sp} is the activity-based constant, $\{Pb^{2+}\}$ and $\{Cl^-\}$ are activities of the ions, f_{Pb} and f_{Cl} are activity coefficients of the ions, and $[Pb^{2+}]$ and $[Cl^-]$ are equilibrium concentrations. This equation can be rearranged into the form

$$K_{sp,\mu} = \frac{K_{sp}}{f_{Pb} f_{Cl}^2} = \left[Pb^{2+} \right]\left[Cl^- \right]^2 \tag{7.6b}$$

in which all symbols are as defined above.

The reaction and relationship between concentration-based and activity-based constants can be written more general forms as follows:

$$M_m B_b \rightleftharpoons mM^{z+} + bB^{z-}, \qquad K_{sp,\mu} = \frac{K_{sp}}{f_M^m f_B^b} = \left[M^{z_m} \right]^m \left[B^{z_b} \right]^b \tag{7.6c}$$

in which m and b are the numbers of metal ions and bases in the solid and f_M and f_B are activity coefficients of the metal ion and base and z_m and z_b are charges on the ions.

Although developed here using lead chloride as an example, this latter equation applies for any solubility reaction.

SPARINGLY SOLUBLE HYDROXIDES

We will use two metal hydroxides, calcium hydroxide, $Ca(OH)_2$, $K_{sp} = 5.5 \times 10^{-6}$, and tin(II) hydroxide, $Sn(OH)_2$, $K_{sp} = 1.4 \times 10^{-28}$, as examples. Let's begin with a brief discussion of the acid-base properties of water and how those properties influence solubilities of hydroxides in otherwise pure water.

EFFECTS OF THE ACID-BASE BEHAVIOR OF WATER ON SOLUBILITIES OF HYDROXIDES

As you probably recall, the autoprotolysis constant for water is $K_w = 1.0 \times 10^{-14}$. As you probably also recall, the hydrogen ion and hydroxide ion concentrations in pure water are equal to the square root of the autoprotolysis constant, i.e., $[H^+] = [OH^-] = K_w^{1/2} = 1.0 \times 10^{-7}$ M.

Now, consider what happens if we add either calcium hydroxide or stannous hydroxide to water. Some of each metal hydroxide will dissolve and dissociate as follows:

$$Ca(OH)_2 \rightleftharpoons Ca^{2+} + 2OH^-; \qquad Sn(OH)_2 \rightleftharpoons Sn^{2+} + 2OH^-$$

The hydroxide ion concentration will increase in each case. These observations lead to a simple but important conclusion. The addition of a metal hydroxide to otherwise pure water or to water containing either the metal ion or hydroxide ion as a common ion will always result in a hydroxide ion concentration equal to or larger than 1.0×10^{-7} M, i.e., $[OH^-] \geq 1.0 \times 10^{-7}$ M. The only situations to be discussed herein for which hydroxide ion concentrations will be less than 1.0×10^{-7} M will involve solutions buffered at pH's less than 7.0.

EXAMPLE CALCULATIONS FOR METAL HYDROXIDES IN OTHERWISE PURE WATER

Let's use an example to help us understand how the acid-base properties of water influence the ways we solve solubility problems involving sparingly soluble hydroxides. As a starting point, let's do a first approximation by ignoring effects of the hydroxide ion concentration from water and then see how we interpret the results.

Calculations for calcium hydroxide and tin(II) hydroxide are done in parallel below to facilitate comparisons. I suggest that you work through the two parts of the problem independently.

Example 7.2 Calculate the solubility, S, and hydroxide ion concentration, [OH⁻], in saturated solutions of (A) calcium hydroxide and (B) stannous hydroxide in otherwise pure water by ignoring effects of the dissociation of water on the solubilities and concentrations. For $Ca(OH)_2$, $K_{sp} = 5.5 \times 10^{-6}$ M; for $Sn(OH)_2$, $K_{sp} = 1.4 \times 10^{-28}$

Part A	Part B

Step 1: Write the solubility reactions and solubility product constants.

$$Ca(OH)_2 \rightleftharpoons \overbrace{Ca^{2+}}^{S} + \overbrace{2OH^-}^{2S} \qquad\qquad Sn(OH)_2 \rightleftharpoons \overbrace{Sn^{2+}}^{S} + \overbrace{2OH^-}^{2S}$$

$$K_{sp} = \left[Ca^{2+}\right]\left[OH^-\right]^2 \qquad\qquad K_{sp} = \left[Sn^{2+}\right]\left[OH^-\right]^2$$

Step 2: Write equations for the metal ion and hydroxide ion concentrations in terms of solubilities.

$$\left[Ca^{2+}\right] = S \qquad \left[OH^-\right] \cong 2S \qquad\qquad \left[Sn^{2+}\right] = S \qquad \left[OH^-\right] \cong 2S$$

Step 3: Substitute equations for the ions in terms of solubilities into the solubility product equation.

$$K_{sp} \cong S(2S)^2 \cong 4S^3 \cong 5.5 \times 10^{-6} \qquad\qquad K_{sp} \cong S(2S)^2 \cong 4S^3 \cong 1.4 \times 10^{-28}$$

Step 4: Rearrange the equations and calculate the solubilities.

$$S \cong \left(\frac{K_{sp}}{4}\right)^{1/3} \cong \left(\frac{5.5 \times 10^{-6}}{4}\right)^{1/3} \cong 1.1 \times 10^{-2} \text{ M} \qquad S \cong \left(\frac{K_{sp}}{4}\right)^{1/3} \cong \left(\frac{1.4 \times 10^{-28}}{4}\right)^{1/3} \cong 3.3 \times 10^{-10} \text{ M}$$

Step 5: Calculate the approximate hydroxide ion concentrations as $[OH^-] \cong 2S$

$$\left[OH^-\right] \cong 2S \cong 2\left(1.1 \times 10^{-2} \text{ M}\right) \cong 2.2 \times 10^{-2} \text{ M} \qquad \left[OH^-\right] \cong 2S \cong 2\left(3.3 \times 10^{-10} \text{ M}\right) \cong 6.6 \times 10^{-10} \text{ M}$$

Now, let's talk about the results starting with results for calcium hydroxide.

Given that the approximate hydroxide ion concentration from calcium hydroxide is much larger than that from otherwise pure water, 10^{-7} M, we accept the results, i.e., $S = 1.1 \times 10^{-2}$, $[OH^-] = 2S = 2.2 \times 10^{-2}$ M, and $[Ca^{2+}] = S = .1 \times 10^{-2}$ M.

The situation is very different for stannous hydroxide. The first approximation of the hydroxide concentration is much less than that from the dissociation of water, i.e., $[OH^-] \cong 6.6 \times 10^{-10}$ M $<< [OH^-]_{H_2O} \cong K_w^{1/2} \cong 1.0 \times 10^{-7}$ M. So what do we do? We recalculate the solubility by assuming that the hydroxide concentration is that from otherwise pure water, i.e., $[OH^-] \cong K_w^{1/2} \cong 1.0 \times 10^{-7}$. Recalculated results for the hydroxide ion concentration, solubility, and Sn^{2+} concentration are summarized below.

$$[OH^-] \cong \sqrt{K_w} \cong 1.0 \times 10^{-7} \text{ M}; \qquad S = \frac{K_{sp}}{\left[OH^-\right]^2} \cong \frac{1.4 \times 10^{-28}}{\left(1.0 \times 10^{-7}\right)^2} \cong 1.4 \times 10^{-14} \text{ M};$$

$$\left[Sn^{2+}\right] = S \cong 1.4 \times 10^{-14} \text{ M}$$

A spreadsheet program to do the calculations for metal hydroxides with different solubility product constants is introduced below.

A SPREADSHEET PROGRAM FOR SPARINGLY SOLUBLE HYDROXIDES

A program on Sheet 2 of the solubilities program is designed to do calculations for the following types of situations.

- Metal hydroxides in otherwise pure water.
- Metal hydroxides in unbuffered solutions with an excess concentration, C_M', of a metal ion as a common ion.

- Metal hydroxides in unbuffered solutions with an excess concentration, C_B', of hydroxide ion a common ion.
- Metal hydroxides in buffered solutions without common ions.

I will describe the basis for the program later. For now, let's use the program to solve problems in each of these four situations.

Metal Hydroxides in Otherwise Pure Water

Let's use the program to solve the two parts of Example 7.2. To solve part A of the example for calcium hydroxide, open the solubility equilibria program, select Sheet 2, copy the template in Cells A85–C99 and paste it into Cells A25–A39 using the Paste (P) option. Then set C25 to 1 and wait for Cells H33 and H34 to change to "Solved." Then compare the solubility, Ca^{2+} concentration, and OH^- concentration in Cells F27–F29 to those calculated for Part A of Example 7.2. You should see virtually identical results for the two sets of results.

To solve Part B of the example for stannous hydroxide, set A25 to zero set the solubility product constant in C28 to 1.4E-28, reset A25 to 1 and wait for H33 and H34 to change to "Solved." Then compare the solubility and concentrations in Cells F27–F29 to the recalculated values above. You should see results virtually identical to the recalculated results for this example that account for hydroxide ion concentration from water.

Solutions Containing an Excess of the Metal Ion as a Common Ion in Unbuffered Solutions

Let's begin with calcium hydroxide with $C_M = 0.010$ M Ca^{2+} as the common ion. To do this, set C25 to 0, set the solubility product constant in C28 to 5.5E-6 and set C_M in C29 to 0.010. Then set C25 to 1 and wait for H33 to change to solved and read the solubility and concentrations from F27–F29. Values you should obtain are S = 8.6E-3, [M] = $[Ca^{2+}]$ = 1.86E-2 and $[OH^-]$ = 1.72E-2.

As a measure of internal consistency, notice that the sum of the solubility plus the diluted Ca^{2+} concentration is equal to the equilibrium calcium concentration, i.e., $C_{Ca} + S = 0.010$ M + 8.6E-3 = 0.0186. The product, $[Ca^{2+}][OH^-]^2 = (1.86 \times 10^{-3})(1.72 \times 10^{-2})^2 = 5.5 \times 10^{-6}$ is the same as the solubility product constant.

For the record, neither of these measures of internal consistency applies for situations involving very small solubility product constants for which the hydroxide ion concentration is controlled by the autoprotolysis of water.

Solutions Containing an Excess of the Hydroxide Ion as a Common Ion in Unbuffered Solutions

To illustrate the effect $C_B = C_{OH} = 0.010$ M hydroxide ion as a common ion, set C25 to 0, C29 to 0 and C30 to 0.010. Then set C25 to 1 and wait for H33 to change to "Solved." Numbers in F27–F29 should be S = 8.07E-3, $[M^{z_m}] = [Ca^{2+}]$ = 8.07E-3 and [B] = $[OH^-]$ = 2.61E-2. I will trust you to show that the product, $[M^{z_m}][OH^-]^2$ is equal to the solubility product constant.

You can observe effects of 0.010 M hydroxide ion concentration on the solubility of stannous hydroxide by setting C28 to the solubility product constant, $K_{sp} = 1.4$E-28, and repeating the foregoing calculations. If you do this, don't forget to reset C28 to the solubility product constant for calcium hydroxide, i.e., 5.5×10^{-6}.

Buffered Solutions without Common Ions

To observe effects of a solution buffered at pH = 12.0 on the solubility of calcium hydroxide without a common ion, set C25 to 0, C29 and C30 to zero, C36 to "Yes" and C37 to 12.0. Then set C25 to 1 and wait for H33 to change to "Solved." Results in Cells F27–F29 should be S = 5.50E-2, M^{z_m} = 5.50E-2 and $[OH^-] = K_w/10^{-pH}$ = 1.00E-2, respectively. I will trust you to confirm that the ion product, $[Ca^{2+}][OH^-]^2$, is equal to the solubility product constant.

Extension to Other Hydroxides

The program described above can be extended to any metal hydroxide of the form, $M(OH)_b$ by setting C28 to the value of the solubility product constant, C32 to the charge on the metal ion, and C35 to the number of hydroxides per metal ion. Templates for selected situations are included in Cell A82–K99 on Sheet 2 of the solubility equilibria program. For example, Cells C28, C32, and C35 are set to 6.3E-46, 3, and 3, respectively, for $Tl(OH)_3$. The charge on the metal ion in C32 is relevant only for calculations with correction for ionic strength.

Basis for the Hydroxides Program without Correction for Ionic Strength

The solubility equilibria program uses the ratio of the reaction quotient to the solubility product constant, Q_{sp}/K_{sp}, as the convergence criterion. The program starts with a very small value of the solubility and increases it until the ratio of the reaction quotient to the solubility product constant is equal to one, i.e., $Q_{sp}/K_{sp} = 1$. The solubility for which $Q_{sp}/K_{sp} = 1$ is accepted as the equilibrium solubility, S.

As with other programs, the program is implemented in two steps. A low resolution step is used to find the upper limit of a 10-fold range of solubilities that includes the equilibrium solubility. A high resolution step is used to resolve the equilibrium solubility to the number of significant figures in C26 on Sheet 2 of the solubilities program.

I will use the solubility of calcium hydroxide to describe the basis for the program and then show how the resulting equations can be extended to other hydroxides. The reaction, solubility product constant, and reaction quotient are

$$Ca(OH)_2 \rightleftharpoons Ca^{2+} + 2OH^-; \qquad K_{sp} = \left[Ca^{2+} \right]\left[OH^- \right]^2; \qquad Q_{sp} = \langle Ca^{2+} \rangle\langle OH^{-2} \rangle$$

in which Q_{sp} is the reaction quotient, $\langle Ca^{2+} \rangle$ and $\langle OH^- \rangle$ are non-equilibrium Ca^{2+} and OH^- concentrations varied by the iterative program, and other symbols have their usual significance.

Calcium Ion Concentration

Given the one to one relationship between the calcium ion concentration and the solubility, the Ca^{2+} concentration is expressed as the sum of the net concentration added as a common ion, C_{Ca}', plus each trial value of the solubility. Equation 7.7a is:

$$\langle Ca^{2+} \rangle = C_{Ca}' + S' \tag{7.7a}$$

in which $\langle Ca^{2+} \rangle$ is a trial value of the Ca^{2+} concentration, C_{Ca}' is the net calcium ion concentration added as a common ion, and S' is the trial value of the solubility at each step in the iterative process.

Hydroxide Ion Concentration

It will be helpful to divide our discussion of trial values of hydroxide ion concentrations into three groups. One group involves buffered solutions for which the hydroxide ion concentration is controlled by the buffer. Another group involves unbuffered solutions with larger solubility product constants for which the hydroxide ion concentration is controlled by the solubility reaction. The third group involves unbuffered solutions with smaller solubility product constants for which the hydroxide ion concentration is controlled by the dissociation of water.

For buffered solutions, the hydroxide ion concentration is equal to the autoprotolysis constant for water divided by the hydrogen ion concentration, i.e.,

$$\langle OH^- \rangle = \frac{K_w}{\left[H^+ \right]} = \frac{K_w}{10^{-pH}} \tag{7.7b}$$

in which $\langle OH^- \rangle$ is the trial hydroxide ion concentration and other symbols have their usual significance.

For unbuffered solutions with larger solubility product constants for which the hydroxide ion concentration is controlled by the solubility reaction, the hydroxide ion concentration is equal to the net concentration plus that from the solubility reaction. The equation is

$$\langle OH^- \rangle = C_{OH}' + 2_{S'} \tag{7.7c}$$

in which $\langle OH^- \rangle$ is the trial hydroxide ion concentration varied by the iterative program, C_{OH}' is the net hydroxide ion concentration and S' is a trial value of the solubility.

For unbuffered solutions with smaller solubility product constants for which the hydroxide ion concentration is controlled by the dissociation of water, the hydroxide ion concentration is equal to the square root of the autoprotolysis constant, i.e.,

$$\langle OH^- \rangle = \sqrt{K_W} \tag{7.7d}$$

in which $\langle OH^- \rangle$ is the trial hydroxide ion concentration, and K_w is the autoprotolysis constant for water.

These equations are implemented in Cells E50 and E51 for trial solubilities in F50 and F51. I will spare you details. However, IF/THEN functions are used in Cells E50, E51, and F51 to choose among Equations 7.7b–7.7d for each trial solubility in F50 and F51.

Adapting the Foregoing Equations to Hydroxides of the Form, M(OH)$_B$

Concepts and equations described above for calcium hydroxide apply for any metal hydroxide with two hydroxides per metal ion, i.e., $M(OH)_2$. These equations can be extended to metal hydroxides, $M(OH)_b$, with b hydroxides per metal ion by replacing the number, 2, in the equations with b. More specifically, consider the reaction

$$M(OH)_b \rightleftharpoons M^{z_m} + bOH^-$$

The only equations that are different from those given above for calcium hydroxide are as expressed in Equations 7.8a–7.8d:

$$K_{sp} = \left[M^{z_m} \right] \left[OH^- \right]^b \tag{7.8a}$$

$$Q_{sp} = \langle M^{z_m} \rangle \langle OH^- \rangle^b \tag{7.8b}$$

$$\langle M^{z_m} \rangle = C_M' + S' \tag{7.8c}$$

$$\langle OH^- \rangle = C_{OH}' + bS' \tag{7.8d}$$

in which square-bracketed symbols are equilibrium concentrations, $\langle M^{z_m} \rangle$ and $\langle OH^- \rangle$ are non-equilibrium concentrations varied by the program, C_M' and C_{OH}' are net metal ion and hydroxide ion concentrations and S' is the non-equilibrium solubility varied by the iterative program to adjust the ratio, Q_{sp}/K_{sp}, to one. As discussed in the background section, net metal ion and hydroxide ion concentrations are the concentrations that remain after any initial metal ion and hydroxide ion concentrations react to completion.

The user must enter proper values of z_m and b in Cells C32 and C35 on Sheet 2 of the solubilities program. The number, m, of metal ions per molecule of the solid and the charge, z_b, on the hydroxide ion are both equal to one, i.e., m = 1 and $z_b = 1$.

BASIS FOR THE HYDROXIDES PROGRAM WITH CORRECTION FOR IONIC STRENGTH

The only difference between calculations without and with correction for ionic strength is that activity-based constants are used without correction and concentration-based constants are used with correction. Ionic strength is calculated in F31 using procedures described earlier. Activity coefficients calculated in Cells E63–G63 as described earlier are used to calculate concentration-based constants, $K_{w,\mu}$ and $K_{sp,\mu}$, in Cells C68 and D68. These constants are used in Cells E71 and E72 in the same way as described earlier for activity-based constants. Solubilities and ionic concentrations corrected for ionic strength are calculated in Cells G27–G30.

SPARINGLY SOLUBLE SALTS OF WEAK ACIDS WITHOUT COMMON IONS

The program on Sheet 3 of the solubilities program does calculations for sparingly soluble salts of weak acids without common ions. We will talk later about a program to do calculations for salts of weak acids with common ions. Neither program accounts for ion pairing between different forms of phosphate and "inert" ions such as Na^+ and K^+.

SELECTED APPLICATIONS OF THE PROGRAM ON SHEET 3 OF THE SOLUBILITIES PROGRAM

The program on Sheet 3 of the solubilities program does calculations for unbuffered solutions with correction for ionic strength or buffered solutions without correction for ionic strength, both without common ions. We will talk first about some applications of the program and then discuss how the program does what it does.

To begin, open the solubility equilibria program and select Sheet 3. Then select File/Options/Formulas and confirm that Workbook Calculation is set to Automatic, there is a check mark in the *Enable iterative calculations* box, *Maximum Iterations* is set to 2,000 and that *Maximum Change* is set to 1E-35.

To do a set of calculations for zinc phosphate in otherwise pure water, copy the template in Cells A100–C114 on Sheet 3 and paste it into Cells A22–C36 using the Paste (P) option. Then set C22 to 1 and wait for H41 and H43 to change to "Solved." Press F9 one or more times if the iteration counter in H42 or H44 stops changing before H41 and H43 change to "Solved." Equilibrium hydrogen ion concentrations in F37 and G37 without and with correction for ionic strength are calculated in Cells F64 and F87. These hydrogen ion concentrations are used to calculate equilibrium concentrations of other reactants and products. Calculated quantities without and with correction for ionic strength are in Cells F26–F38.

Equilibrium Zn^{2+} and phosphate ion concentrations in Cells F27 and F32 are $[Zn^{2+}] = 9.87 \times 10^{-6}$ M and $[PO_4^{3-}] = 3.06 \times 10^{-9}$ M, respectively. These concentrations correspond to a reaction quotient of $Q_{sp} = [Zn^{2+}]^3[PO_4^{3-}]^2 = (9.87 \times 10^{-6})^3(3.06 \times 10^{-9}$ M$)^2 = 9.0 \times 10^{-33}$ which is the same as the solubility product constant. This supports the validity of the calculations.

To observe results calculated in a solution buffered at pH = 10, set C22 to 0, C35 to 10, and C34 to "Yes." Given that the buffer controls the hydrogen ion concentration at a fixed value and that the hydrogen ion concentration is used to calculate concentrations of other reactants, an iterative process isn't needed to find equilibrium concentrations in buffered solutions.

To repeat the processes for stannous phosphate, $Sn_3(PO_4)_2$, $K_{sp} = 1.4 \times 10^{-28}$, in an unbuffered solution, set C22 to 0, C25 to 1.4E-28, and C34 to No. Calculated results are in Cells F26–F38.

RATIONALE FOR THE PROGRAM FOR SALTS OF WEAK ACIDS

For a salt such as zinc(II) phosphate, concentrations of all reaction products depend directly or indirectly on the hydrogen ion concentration. For example, whereas the Zn^{2+} ion doesn't depend directly on the hydrogen ion concentration, it does depend on the phosphate ion concentration. Given that the phosphate ion concentration depends on the hydrogen ion concentration, the Zn^{2+} ion concentration also depends on the hydrogen ion concentration. For example, as shown by the reactions below, zinc phosphate will continue to dissolve until the Zn^{2+} concentration and the fully deprotonated form of phosphoric acid satisfy the solubility product constant.

$$Zn_3(PO_4)_2 \rightleftharpoons 3Zn^{2+} + 2PO_4^{3-} \qquad K_{sp} = \left[Zn^{2+}\right]^3 \left[PO_4^{3-}\right]^2$$

$$PO_4^{3-} \overset{H_2O}{\rightleftharpoons} HPO_4^{2-} + OH^- \overset{H_2O}{\rightleftharpoons} H_2PO_4^- + OH^- \overset{H_2O}{\rightleftharpoons} H_3PO_4 + OH^-$$

The program on Sheet 3 accounts for all these forms of phosphoric acid by solving a charge-balance equation to find the hydrogen ion concentration that corresponds to equilibrium concentrations of all reaction products, including Zn^{2+} and all the charged forms of phosphoric acid.

IONIC FORM OF A CHARGE-BALANCE EQUATION

I will use zinc phosphate, $Zn_3(PO_4)_2$, as an example. Having developed equations and a program to do calculations for zinc phosphate, I will show you how the equations and program can be adapted to any salt of a weak acid of the form, M_mB_b.

Let's begin by writing the reactions and identifying one each of the different ions as shown in boldface type below.

$$H_2O \rightleftharpoons \mathbf{H^+} + \mathbf{OH^-}$$

$$Zn_3(PO_4)_2 \rightleftharpoons 3\mathbf{Zn^{2+}} + 2\mathbf{PO_4^{3-}}$$

$$PO_4^{3-} + H_2O \rightleftharpoons \mathbf{HPO_4^{2-}} + OH^-$$

$$HPO_4^{2-} + H_2O \rightleftharpoons \mathbf{H_2PO_4^-} + OH^-$$

$$H_2PO_4^- + H_2O \rightleftharpoons H_3PO_4 + OH^-$$

Next, let's write the ionic form of a charge-balance equation as the algebraic sum of signed charges on the ions times equilibrium concentrations.

$$CBE = \left[H^+\right] - \left[OH^-\right] + 2\left[Zn^{2+}\right] - \left(\left[H_2PO_4^-\right] + 2\left[HPO_4^{2-}\right] + 3\left[PO_4^{3-}\right]\right) \qquad (7.9)$$

Multipliers for the ions are charges on the ions, e.g., $z_H = 1$, $z_{OH} = -1$, $z_{Zn} = 2$, $z_{H_2PO_4} = -1$, $z_{HPO_4} = -2$, and $z_{PO_4} = -3$.

Our task is to write a computational form of this equation in terms of one unknown, namely the hydrogen ion concentration.

EQUATIONS FOR DIFFERENT TERMS IN THE CHARGE-BALANCE EQUATION

Procedures and equations used for the different ions are discussed below.

The hydrogen ion concentration is calculated by using iterative procedures to solve the charge-balance equation. The hydroxide ion concentration is calculated as the autoprotolysis constant for water divided by the hydrogen ion concentration, i.e.,

$$\left[OH^-\right] = \frac{K_w}{\left[H^+\right]} \qquad (7.10)$$

As discussed in the acid-base chapters, concentrations of different forms of a polyprotic acid such as phosphoric acid can be calculated as products of concentration fractions times the total phosphate concentration.

As a reminder, the denominator term for fractions of a triprotic weak acid in different forms is

$$D_3 = \left[H^+\right]^3 + \kappa_1 \left[H^+\right]^2 + \kappa_2 \left[H^+\right] + \kappa_3$$

in which D_3 is the denominator term, κ_1, κ_2, and κ_3 are cumulative deprotonation constants and $[H^+]$ is the hydrogen ion concentration. As you may recall, cumulative deprotonation constants are products of successive deprotonation constants, i.e.,

$$\kappa_1 = K_{a1}; \qquad \kappa_2 = K_{a1}K_{a2}; \qquad \kappa_3 = K_{a1}K_{a2}K_{a3}$$

As you may also recall, successive fractions, α_0, α_1, α_2, and α_3 are successive denominator terms divided by the denominator, i.e.,

$$\alpha_0 = \frac{\kappa_0 \left[H^+\right]^3}{D_3}; \qquad \alpha_1 = \frac{\kappa_1 \left[H^+\right]^2}{D_3}; \qquad \alpha_2 = \frac{\kappa_2 \left[H^+\right]}{D_3}; \qquad \alpha_3 = \frac{\kappa_3}{D_3}$$

Equilibrium concentrations of forms of phosphate that have lost i = 1, 2, and 3 protons, respectively are as follows:

$$\left[H_2PO_4^-\right] = \alpha_1 C_{PO_4}; \qquad \left[HPO_4^{2-}\right] = \alpha_2 C_{PO_4}; \qquad \left[PO_4^{3-}\right] = \alpha_3 C_{PO_4}$$

in which α_1, α_2, and α_3 are fractions of phosphoric acid in forms that have lost 1, 2, and 3 protons, respectively, and C_{PO_4} is the total phosphate concentration.

The balancing coefficients in the solubility reaction and the equation for the solubility product constant are used to develop relationships between the Zn^{2+} concentration and total phosphate concentration, C_{PO_4}. The solubility reaction and relationship between the Zn^{2+} and total phosphate concentration are as follows:

$$Zn_3(PO_4)_2 \rightleftharpoons 3Zn^{2+} + 2PO_4^{3-} \qquad \left[Zn^{2+}\right] = \frac{3}{2}C_{PO_4} \qquad (7.11a)$$

in which 3 and 2 are the balancing coefficients, α_3 is the fraction of phosphoric acid in the form that has lost three protons, and C_{PO_4} is the total concentration of all forms of phosphate.

By using the coefficients in the balanced reaction and the equation for the solubility product constant, the Zn^{2+} concentration can be written in terms of the total phosphate concentration as follows:

$$K_{sp} = \left[Zn^{2+}\right]^3 \left[PO_4^{3-}\right]^2 = \left[Zn^{2+}\right]^3 \left(\alpha_3 C_{PO_4}\right)^2 \qquad \left[Zn^{2+}\right] = \left[\frac{K_{sp}}{\left(\alpha_3 C_{PO_4}\right)^2}\right]^{1/3} \qquad (7.11b)$$

After equating the right sides of Equations 7.11a and 7.11b and raising both sides of the resulting equation to the third power, the resulting equation can be rearranged into the following form:

$$C_{PO_4} = \left[\left(\frac{2}{3}\right)^3 \left(\frac{K_{sp}}{\alpha_3^2}\right)\right]^{1/(3+2)} \qquad (7.11c)$$

After substituting Equation 7.11c into Equation 7.11a, the Zn^{2+} concentration can be written as

$$\left[Zn^{2+} \right] = \frac{3}{2} C_{PO_4} = \frac{3}{2} \left[\left(\frac{2}{3} \right)^3 \left(\frac{K_{sp}}{\alpha_3^{\,2}} \right) \right]^{1/(3+2)} \tag{7.11d}$$

Concentrations of the three charged forms of phosphoric acid can be written as products of concentration fractions times the total phosphate concentration, i.e.,

$$\left(\alpha_1 C_{PO_4} + 2\alpha_2 C_{PO_4} + 3\alpha_3 C_{PO_4} \right) \overset{Eq.\ 7.11c}{=} \left(\alpha_1 + 2\alpha_2 + 3\alpha_3 \right) \left[\left(\frac{2}{3} \right)^3 \left(\frac{K_{sp}}{\alpha_3^{\,2}} \right) \right]^{1/(3+2)} \tag{7.11e}$$

Finally, the solubility is one half the total phosphate concentration, as expressed in Equation 7.11f:

$$S = \frac{1}{2} C_{PO_4} = \frac{1}{2} \left[\left(\frac{2}{3} \right)^3 \left(\frac{K_{sp}}{\alpha_3^{\,2}} \right) \right]^{1/(3+2)} \tag{7.11f}$$

We now have all the equations needed to write the charge-balance equation in terms of constants and quantities that depend on the hydrogen ion concentration.

COMPLETED CHARGE-BALANCE EQUATION

The charge-balance equation obtained by substituting Equations 7.10, 7.11d, and 7.11e into Equation 7.9 and factoring out the common term is

$$CBE = \left[H^+ \right] - \frac{K_w}{\left[H^+ \right]} + \left[2 \left(\frac{3}{2} \right) - \left(\alpha_1 + 2\alpha_2 + 3\alpha_3 \right) \right] \left[\left(\frac{2}{3} \right)^3 \left(\frac{K_{sp}}{\alpha_3^{\,2}} \right) \right]^{1/(3+2)} \tag{7.12a}$$

All terms in the charge-balance equation depend on constants and the hydrogen ion concentration.

For any acid of the form, H_nB, the program to be discussed is designed to set all fractions beyond α_n to zero. As examples, α_2 and α_3 will be set to zero for salts of monoprotic acids such as silver cyanide and α_3 will be set to zero for salts of diprotic acids such as calcium oxalate. As mentioned earlier, templates you can use to familiarize yourself with input information for different situations are included in Cells A100–O114 on Sheet 3 of the solubility program.

GENERALIZED CHARGE-BALANCE EQUATION

Although developed here using zinc phosphate as an example, Equation 7.12a can be used for any salt of the form, M_3B_2, in which the metal ion is a divalent ion, M^{2+}, and the base, B, is a trivalent ion, B^{3-}.

We can change Equation 7.12a to a generalized equation for any salt of the form, M_mB_b, of a metal ion with charge, z_m, with the fully deprotonated base of an uncharged acid with up to three acidic protons. As a starting point, let's write the solubility reaction in the following general form:

$$M_m B_b \rightleftharpoons mM^{z_m} + bB^{z_b}$$

Next, let's replace numbers in Equation 7.12a with the corresponding symbols, m, b, and z_m. The generalized equation is

$$CBE = \left[H^+\right] - \frac{K_w}{\left[H^+\right]} + \left[z_m\left(\frac{m}{b}\right) - \left(\alpha_1 + 2\alpha_2 + 3\alpha_3\right)\right]\left[\left(\frac{b}{m}\right)^m\left(\frac{K_{sp}}{\alpha_n^b}\right)\right]^{1/(m+b)} \quad (7.12b)$$

in which z_m is the charge on the metal ion, m and b are numbers of the metal ion and base per molecule of the salt, α_1–α_3 are fractions of forms of the parent acid, H_nB, that have lost 0–3 protons and α_n is the fraction of the form of the weak acid that has lost n protons.

Generalized equations for weak-acid and metal-ion concentrations and solubility are as follows:

$$C_B = \left[\left(\frac{b}{m}\right)^m\left(\frac{K_{sp}}{\alpha_n^2}\right)\right]^{1/(m+b)} \quad (7.13a)$$

$$\left[M^{z_m}\right] = \frac{m}{b}C_B \quad (7.13b)$$

$$S = \frac{1}{b}C_B \quad (7.13c)$$

in which C_B is the total concentration of all forms of the basic part of the solid, e.g., PO_4, b and m are coefficients associated with the base and metal ion in the solid, z_m is the charge on the metal ion, and S is the solubility. More specifically, for $Zn_3(PO_4)_2$, C_B is the concentration of all forms of phosphoric acid and $[M^{z_m}]$ is the Zn^{2+} concentration.

Concentrations of different forms of the basic form of the solid are calculated as products of concentration fractions times the total concentration of the basic form, i.e., $\alpha_i C_B$.

APPLICATIONS OF THE PROGRAM ON SHEET 3 TO OTHER SITUATIONS

Templates for salts with different numbers, m and b, of metal ions and weak bases are included in Cells E100–O114. You can use these templates to familiarize yourself with applications of salts with different stoichiometries, M_mB_b. To use each template, copy it and paste it into Cells A22–C36 and follow procedures described above for zinc and tin phosphates.

SPARINGLY SOLUBLE SALTS OF WEAK ACIDS WITH COMMON IONS

A program on Sheet 4 of the solubilities program does calculations for buffered solutions of salts of weak acids with common ions. At the time of this writing I haven't developed the mathematical treatment and programming for unbuffered solutions with common ions. In any event, the program described here uses a two-step process to calculate a solubility for each situation for which the ratio of the reaction quotient to the solubility product constant is equal to one, i.e., $Q_{sp}/K_{sp} = 1.0$.

Let's use zinc phosphate as an example to illustrate how the program works. The reaction, solubility product constant, and reaction quotient are as follows:

$$Zn_3(PO_4)_2 \rightleftharpoons 3Zn^{2+} + 2PO_4^{3-} \qquad K_{sp} = \left[Zn^{2+}\right]^3\left[PO_4^{3-}\right]^2 \qquad Q_{sp} = \langle Zn^{2+}\rangle^3\langle PO_4^{3-}\rangle^2$$

in which $[Zn^{2+}]$ and $[PO_4^{3-}]$ are equilibrium concentrations and $\langle Zn^{2+}\rangle$ and $\langle PO_4^{3-}\rangle$ are nonequilibrium concentrations.

Assuming that Zn^{2+} is the common ion, the ratio of the reaction quotient to the solubility product constant is as follows:

$$\frac{Q_{sp}}{K_{sp}} = \frac{\langle Zn^{2+}\rangle^3 \langle PO_4^{3-}\rangle^2}{K_{sp}} = \frac{(C_{Zn} + 3S)^3 \langle 2S\rangle^2}{K_{sp}}$$

in which C_{Zn} is the initial Zn^{2+} concentration, S is the zinc phosphate solubility, and K_{sp} is the solubility product constant. The program starts with a very small solubility, e. g., 10^{-30} M in F63 and increases it in 10-fold steps until the ratio of the reaction quotient to the solubility product constant in E63 is larger than one, i.e., $Q_{sp}/K_{sp} > 1$. The program then starts with a solubility in F64 equal to one tenth of that in F63 and increases it by small amounts until the ratio, Q_{sp}/K_{sp}, in E64 exceeds 1.0. The solubility for which the ratio, Q_{sp}/K_{sp}, first exceeds one is accepted as the equilibrium solubility.

Equilibrium Zn^{2+} and total phosphate concentrations are calculated as $[Zn^{2+}] = C_{Zn} + 3S$ and $C_{T,PO_4} = 2S$. The hydrogen ion concentration of the buffer is used to calculate fractions, $\alpha_0 - \alpha_3$, of phosphate in different forms in Cells F33–F36. Equilibrium concentrations of the different forms of phosphate in F29–F32 are calculated as products of concentration fractions times the total phosphate concentration.

Analogous considerations apply for phosphate as the common ion or for common ions of other sparingly soluble salts of weak acids. Templates for several situations are included in Cells A75–O94 on Sheet 4. Having pasted any of the templates into Cells A22–C41, you can use the template to do calculations with different combinations of the metal ion and/or the weak acid.

For the record, concentrations of different forms of weak acids are additive. In other words, if you add 0.010 M of one form of phosphate and 0.020 M of another form of phosphate, the program treats the common ions as 0.030 M phosphate. Also, if you include both the metal ion and one or more forms of an acid as common ions, the program will calculate and use the net concentration in Cell C53 or D53 for equilibrium calculations. For example, the net common ion concentration in a solution containing 0.030 M Zn^{2+} and 0.020 M total phosphate will be $C_{Net} = C_{Zn} - (3/2) C_{PO_4} = 0.030$ M $- (3/2)0.02$ M $= 0$. Therefore, the solution will be treated as a solution with no common ion.

SUMMARY

For computational purposes, sparingly soluble solids are divided into three groups consisting of salts of strong acids, hydroxides, and salts of weak acids. Procedures described for solving problems for each group involve solutions to exact equations using iterative spreadsheet programs.

The spreadsheet program for salts of strong acids finds solubilities that satisfy exact solubility product equations. For more soluble hydroxides, the hydroxides program finds the solubility for which the metal ion and hydroxide concentrations satisfy the solubility product constant. For less soluble hydroxides, the program finds the solubility for which the metal ion and hydroxide ion concentration satisfy both the solubility product constant and the autoprotolysis constant for water. One program for salts of weak acids can be used for both unbuffered and buffered solutions. Whereas the program accounts for effects of ionic strength in unbuffered solutions it doesn't account for effects of ionic strength in buffered solutions. A second program for salts of weak acids is designed to do calculations with or without common ions in buffered solutions without correction for ionic strength.

REFERENCES

1. P. Debye and E. Hückel, *Phys. Z.* **1923**, 24, 185–206.
2. H. E. Blayden and C. W. Davies, *J. Chem. Soc.*, **1930**, 949–956.
3. C. W. Davies, *J. Chem. Soc.*, **1938**, 2093–2098.
4. J. Kielland, *J. Am. Chem. Soc.*, **1937**, 59, 1675–1678.

8 Overview and Symbols for Complex Ion Reactions and Equilibria

This is the first of three chapters related to equilibrium calculations involving complex ions. This chapter describes a brief overview of types of complex-forming reactions to be discussed in more detail in subsequent chapters. It also includes a table of symbols to be used in subsequent chapters so that they will be available to you in one place. The symbols will be more meaningful when they are described in the context of specific applications in subsequent chapters.

A complex ion is formed when a metal ion shares one or more pairs of electrons with one or more donor ligands. In the absence of other complexing ligands, most metal ions in water exist as aquo complexes. For example, the usual symbol for silver ion in water is Ag^+. In reality, silver in the +1 oxidation state, Ag(I), exists as the free silver ion, Ag^+, and the aquo complexes $Ag(H_2O)^+$ and $Ag(H_2O)_2^+$. If we add sufficient ammonia to a solution containing silver in the +1 oxidation state, the ammonia will replace water in the aquo complexes to form mixed complexes and the silver/ammonia complex. The two complexes can be depicted as

$$H_2O: Ag: NH_3^+ \quad \text{and} \quad H_3N: Ag: NH_3^+$$

in which pairs of shared electrons are represented by of pairs of dots. Analogous reactions apply for aquo complexes of other metal ions and a variety of complexing ligands.

The complex of Ni^{2+} with ethylenediaminetetraacetic acid (EDTA) is an example of another group of complexes, commonly called chelates. Fully dissociated EDTA can be depicted as

$$\left(\ddot{O}OCCH_2 \right)_2 \ddot{N}CH_2CH_2\ddot{N} \left(CH_2CO\ddot{O} \right)_2^{4-}$$

in which pairs of dots represent sets of unpaired electrons.

When a fully dissociated EDTA ion forms a chelate with a metal ion such as Ni^{2+}, it does so by wrapping itself around the Ni^{2+} in a single step, thereby trapping the metal ion in what might be considered a "chemical cage" in which Ni^{2+} forms covalent bonds with six sets of the unpaired electrons in EDTA, i.e., one pair each in two nitrogen atoms and one pair each in four oxygen atoms in the acetate groups. As will be discussed in more detail later, ligands such as EDTA that wrap themselves around metal ions in this way are commonly called chelating ligands.

Complex ions and molecules are important in many aspects of our lives. For example, chloride ions in commercial cleaning products help to remove iron stains from bathroom fixtures by forming the dichloro complex of Fe^{3+}, $Fe(Cl_2)^+$, which dissolves in water. EDTA in cleaning products helps to remove calcium carbonate deposits from bathroom fixtures by forming the complex, $CaEDTA^{2-}$, which also dissolves in water.

Examples of naturally occurring complexes involved in life processes include hemoglobin and chlorophyll. Examples of synthetic complexes used in medicine include gramicidin, an antibiotic that facilitates the transport of ions across hydrophobic cell membranes, and *cis*-diamminedichloroPlatinum(II), one of a group of platinum compounds used in cancer therapy.

Whereas all these aspects of complexes are important, this and the other two chapters focus on equilibrium calculations and selected quantitative applications of complex ions. Complex ions are used in a variety of quantitative applications, including spectrophotometric determinations of metal ions based on absorbances of colored complexes, separations of metal ions based on extractions of the complexes of metal ions with a variety of ligands from water into organic solvents, and titrations of metal ions with ligands such as EDTA.

Some of these applications are discussed in more detail as we progress through this chapter. For now, let's talk about some of the reactants and reactions in which they are involved.

REACTANTS AND REACTIONS

This section describes some properties of metal ions, complexing ligands and complexes, also called coordination compounds.

METAL IONS

Let's talk first about two ways oxidation states of metal ions are represented herein using zinc in the +2 oxidation state as an example. Zinc in the +2 oxidation state forms ammonia complexes with up to four ammonias per zinc ion. The symbol, Zn(II), represents *all forms of zinc in the +2 oxidation state.* The symbol, Zn^{2+}, represents *the part of the zinc(II) that remains uncomplexed when the complexed-forming reactions reach equilibrium.* Symbols, $Zn(NH_3)^{2+}$, $Zn(NH_3)_2^{2+}$, $Zn(NH_3)_3^{2+}$, and $Zn(NH_3)_4^{2+}$ represent the complex ions. At equilibrium the total Zn(II) concentration, $C_{Zn(II)}$, will be distributed among the uncomplexed zinc ion, Zn^{2+}, and the four complexes. The total Zn(II) concentration will be equal to the sum of equilibrium concentrations of the uncomplexed Zn^{2+} ion and the four complex ions, i.e.,

$$C_{Zn(II)} = \left[Zn^{2+} \right] + \left[Zn(NH_3)^{2+} \right] + \left[Zn(NH_3)_2^{2+} \right] + \left[Zn(NH_3)_3^{2+} \right] + \left[Zn(NH_3)_4^{2+} \right]$$

Analogous symbols and relationships apply for other metal ions and complexes.

Whatever symbols are used to represent metal ions, complex ions involve the sharing of electron pairs between metal ions and ligands with one or more pairs of unshared electrons. The number of electron pairs a metal ion accepts in the formation of one or more complexes is called the *coordination number* and is represented herein by the symbol *CNum*. As examples, the usual coordination numbers for Ag(I), Zn(II), and Ni(II) are CNum = 2, 4, and 6, respectively, meaning that they can accept 2, 4, and 6 pairs of electrons, respectively.

LIGANDS

Compounds and ions that donate pairs of electrons to covalent bonds in complexes are called *ligands.* Nitrogen and oxygen are two of the more common ligands, i.e., electron donors.

One way to characterize a ligand is by the number of pairs of electrons it can share. For example, representing pairs of electrons by double dots in the formulae for ammonia, ethylenediamine, and EDTA below, it is observed that these three ligands have 1, 2, and 6 pairs of electrons, respectively, to share per molecule of the ligand.

Ammonia, L	Ethylenediamine, en	EDTA, Y^{4-}
$\ddot{N}H_3$	$H_2\ddot{N}CH_2CH_2\ddot{N}H_2$	$\left(\ddot{O}OCCH_2 \right)_2 \ddot{N}CH_2CH_2\ddot{N}\left(CH_2O\ddot{O} \right)_2^{4-}$

Based on numbers of electron pairs they can share, these three ligands are called *monodentate, bidentate,* and *hexadentate ligands,* respectively, meaning that they have one, two, and six pairs of unshared electrons each to share with a metal ion. The term, dentate, is derived from the Latin word, *dentatus*, for tooth. More generally, ligands with two or more pairs of electrons to share are called *polydentate ligands*.

As is common practice, ligands that form complexes with two or more ligands per metal ion such as ammonia and cyanide ion are represented herein by symbols L or L^- and fully dissociated EDTA is represented by the symbol, Y^{4-}. Using these symbols, the four complexes of Zn^{2+} with ammonia are represented as ZnL^{2+}, ZnL_2^{2+}, ZnL_3^{2+}, and ZnL_4^{2+}, and the Zn^{2+}/EDTA chelate is represented by ZnY^{2-}.

THE CHELATE EFFECT

The *chelate effect* is a term used to represent the increased stabilities associated with complexes that involve ring structures relative to complexes that don't involve ring structures.

When two donor atoms from a multidentate ligand bond with a metal ion, the resulting complex must necessarily be a ring structure involving the metal ion, the two donor atoms, and bridging groups between the donor atoms. For example, as shown earlier, each nitrogen atom in ethylenediamine has a pair of electrons it can share with a metal ion. Let's consider the situation when both nitrogen atoms in ethylenediamine share their two pairs of electrons with Ni^{2+}, to form the complex, $Ni(en)^{2+}$, in which en represents ethylenediamine. I suggest that you sketch the structure showing that it is a 5-membered ring consisting of Ni^{2+} bonded to two nitrogen atoms bridged by two CH_2 groups.

Given that complexes of Ni^{2+} with two ammonias and one ethylenediamine all involve bonding between Ni^{2+} and electron pairs in nitrogen, it will be informative to compare formation constants for the complexes. For the Ni^{2+}/ammonia complexes, the stepwise formation constants are $K_{f1} = 4.7 \times 10^2$ and $K_{f2} = 57$ corresponding to a cumulative or overall constant of $K_{f1} \times K_{f2} = 2.7 \times 10^4$. The formation constant for Ni^{2+} complexed with the two nitrogen atoms in ethylenediamine is $K_{f1} = 1.6 \times 10^7$. In other words, the formation constant for Ni^{2+} complexed with two nitrogen atoms in ethylenediamine is about 500 times larger than that for Ni^{2+} complexed with two nitrogen atoms in two ammonias. Although both complexes involve Ni(II) complexed with two nitrogen atoms, the ethylenediamine complex is much more stable than complexes with ammonia.

What happens if the number of ring structures is increased? Nickel(II) bound to two and three ethylenediamines, i.e., $Ni(en)_2^{2+}$ and $Ni(en)_3^{2+}$, contain two and three such ring structures each. Analogously, the EDTA chelate with Ni(II) is a three-dimensional structure involving six 5-membered rings. (See Figure 2 on the internet site, EDTA-Chemistry LibreTexts, for a depiction of the structure.)

Overall formation constants for Ni^{2+} bonded to (a) six nitrogen atoms in six ammonias, (b) six nitrogens in three ethylenediamines, and (c) two nitrogen atoms and four oxygen atoms in EDTA are compared below.

$Ni(NH_3)_6^{2+}$	$Ni(en)_3^{2+}$	$NiEDTA^{2-}$
$K_{f,Overall} = K_{f1} \times K_{f2} \ldots \times K_{f6}$	$K_{f,Overall} = K_{f1} \times K_{f2} \times K_{f3}$	K_f
1.0×10^8	2.1×10^{18}	4.3×10^{18}

Cumulative formation constants for the two chelates involving ring structures are about ten orders of magnitude larger than that for the ammonia complex that doesn't include a ring structure. This is called the chelate effect and the resulting complexes are called chelates.

In summary, chelates represent a special group of complexes with enhanced stabilities resulting from ring structures produced when two atoms from each of one or more ligands share electron pairs with one metal ion.

CHARGES ON REACTANTS AND PRODUCTS

Cadmium(II) forms complexes with ammonia and cyanide with one to four ligands per Cd(II). Accounting for the facts that ammonia is uncharged and cyanide ion is negatively charged, the ligands are represented by the symbols L and L^-, respectively. Accounting for these charges, Cd^{2+} and the four complexes with each metal ion are represented as

Ammonia complexes: Cd^{2+}, CdL^{2+}, CdL_2^{2+}, CdL_3^{2+}, and CdL_4^{2+}

Cyanide complexes: Cd^{2+}, CdL^+, CdL_2^0, CdL_3^{1-}, and CdL_4^{2-}

As you can see, charges on the complexes depend on charges on the metal ion and ligand as well as the number of ligands per metal ion.

Many treatments of complex ion equilibria ignore charges on reactants and products in order to simplify the treatments. However, given that iterative procedures to be described in subsequent chapters are based on charge-balance equations, we will include and account for charges on all reactants and products in this and subsequent chapters.

MULTISTEP AND ONE-STEP REACTIONS

Whereas some complexes and chelates form or dissociate in two or more steps, some chelates form or dissociate in a single step. Examples of multistep r*eactions* include complexes of Ag(I), Zn(II), and Ni(II) with ammonia. Using L for ammonia, the stepwise reactions are represented as follows:

$$Ag^+ + L \overset{K_{f1}}{\rightleftharpoons} AgL^+ + L \overset{K_{f2}}{\rightleftharpoons} AgL_2^+;$$

$$Zn^{2+} + L \overset{K_{f1}}{\rightleftharpoons} ZnL^{2+} + L \overset{K_{f2}}{\rightleftharpoons} ZnL_2^{2+} + L \overset{K_{f3}}{\rightleftharpoons} ZnL_3^{2+} + L \overset{K_{f4}}{\rightleftharpoons} ZnL_4^{2+}$$

$$Ni^{2+} + L \overset{K_{f1}}{\rightleftharpoons} NiL^{2+} + L \overset{K_{f2}}{\rightleftharpoons} NiL_2^{2+} + L \overset{K_{f3}}{\rightleftharpoons} NiL_3^{2+} + L \overset{K_{f4}}{\rightleftharpoons} NiL_4^{2+} + L \overset{K_{f5}}{\rightleftharpoons} NiL_5^{2+} + L \overset{K_{f6}}{\rightleftharpoons} NiL_6^{2+}$$

Ethylenediamine chelates of several metal ions form and dissociate in three steps.

The constants, K_{f1}–K_{f6}, in the Ni(II) complexes as well as analogous constants in other multistep reactions are called stepwise formation constants. We will have more to say about these reactions and constants later.

By contrast, EDTA chelates of Ag(I), Zn(II), and Ni(II) form or dissociate in one step each, i.e.,

$$Ag^+ + Y^{4-} \overset{K_{f,Y}}{\rightleftharpoons} AgY^{3-} \qquad Zn^{2+} + Y^{4-} \overset{K_{f,Y}}{\rightleftharpoons} ZnY^{2-} \qquad Ni^{2+} + Y^{4-} \overset{K_{f,Y}}{\rightleftharpoons} NiY^{2-}$$

in which Y^{4-} represents fully deprotonated EDTA.

For reasons discussed below, reactions such as those involving ammonia are called *multistep reactions* herein and reactions such as those involving EDTA are called *one-step reactions* herein.

Of all the differences among metal ions, ligands, and complexes, the one that has the greatest impact on the ways equilibrium calculations are done is that some complexes form or dissociate in two or more steps and others form or dissociate in one step. Therefore, equilibrium calculations in this text are divided into three groups, namely, multistep reactions, one-step reactions, and simultaneous multistep and one-step reactions. Discussions of one-step reactions herein focus exclusively on reactions of metal ions with EDTA to form metal-ion/EDTA chelates.

TABLE 8.1
Generic Symbols Used in This Discussion

Quantity	Ligand, e.g., NH$_4^+$, NH$_3$	EDTA	Metal ion	Buffer acid	Conj. base of buffer acid
Symbol	L	Y	M	HB	B

SOME GOALS AND RESULTING SYMBOLS

A goal of these chapters is to develop concepts, equations, procedures, and spreadsheet programs to solve problems involving a variety of metal ions, buffers, and ligands. To do this, I have used generic symbols in place of chemical symbols in both discussions and programs. Ignoring charges, it is common practice to represent ligands such as ammonia by the symbol, L, and to represent EDTA by Y. Still ignoring charges, metal ions and conjugate bases of buffer acids are represented by M and B, respectively.

These quantities and symbols used to represent them are summarized in Table 8.1.

Some reactants such as ammonia can be used either as part of a buffer, a ligand, or both. For example, when ammonium ion and ammonia are used as a buffer they are represented by HB$^+$ and B, respectively. When ammonia is used as a ligand, the two forms are represented as HL$^+$ and L, respectively. Ammonia is commonly used both as part of an ammonium-ion/ammonia buffer and as the ligand. In such cases, symbols HB$^+$ and B are used in discussions of ammonia in its role as a buffer and L is used for ammonia in its role as a ligand.

These symbols are used as subscripts to differentiate among quantities such as concentration fractions and diluted concentrations associated with buffers or ligands. For example, fractions of the ammonium ion/ammonia pair in the fully deprotonated forms when used as a buffer or a ligand are represented by $\alpha_{1,B}$ or $\alpha_{1,L}$, respectively, in which the subscripted 1 tells us this is the form of the acid that has lost one proton.

Extensions of these symbols to a variety of other quantities are summarized in Table 8.2. As noted earlier, these symbols are included here so that they will be available to you in one place. The symbols will be more meaningful when they are described in the context of specific applications in subsequent chapters.

I have attempted to identify all symbols as they are used in subsequent chapters. In case I fail to do so, I trust that you will refer to this table for help.

TABLE 8.2
Symbols Used in This and Subsequent Chapters

Symbol	Description
$K_{a,B}$, $K_{a,L}$	Deprotonation constants for acids associated with a buffer, B, and ligand, L.
Log $K_{f,j}$	Logarithm of stepwise formation constants for complexes with j ligands per metal ion.
C_0^0, C_1^0	Undiluted concentrations of forms of monoprotic acid-base pairs that have lost 0 and 1 proton, respectively.
C_0, C_1	Diluted concentrations of forms of monoprotic acid-base pairs that have lost 0 and 1 protons, respectively.
C_{HB}^0, C_B^0	Undiluted concentrations of the individual forms of an acid-base pair used as a buffer.
C_{HB}, C_B	Diluted concentrations of individual forms of an acid-base pair used as a buffer.
$C_{B,T}$	Total diluted concentration of the diluted concentrations of both forms of an acid-base pair used as a buffer.
[HB], [B]	Equilibrium concentrations of individual forms of an acid-base pair used as a buffer.
C_{HL}^0, C_L^0	Undiluted concentrations of individual forms of an acid-base pair used as a ligand.

(continued)

TABLE 8.2 (Continued)

Symbols Used in This and Subsequent Chapters

Symbol	Description
C_{HL}, C_L	Diluted concentrations of individual forms of an acid-base pair used as a ligand.
$C_{L,T}$, $C_{L,U}$	Diluted, $C_{L,T}$, and uncomplexed, $C_{L,U}$, concentrations of both forms of an acid-base pair used as a ligand.
[HL], [L]	Equilibrium concentrations of individual forms of an acid-base pair used as a ligand.
C_M^0, C_M, V_M	Undiluted concentration, C_M^0, diluted concentration, C_M, and volume, V_M, of a metal ion.
V_{HB}, V_B, V_L,	Volumes of protonated and unprotonated forms of the buffer, B, and the unprotonated form of ligand, L.
V_T	Total volume to which a solution is diluted.
n_{HB}, n_{HL}	Numbers of acidic protons on the protonated forms of the acid-base pairs used as the buffer and ligand, respectively.
z_{HB}, z_{HL}	Charges on the protonated forms of the acid-base pairs used as the buffer and ligand.
C_{Num}	Coordination number of the metal ion involved in a set of complex-formation reactions.
K_w	Autoprotolysis constant for water.
$D_{\alpha,B}$, $D_{\alpha,L}$	Denominator terms for fractions of the buffer and ligand in different forms.
$\alpha_{0,B}$, $\alpha_{1,B}$, $\alpha_{0,L}$, $\alpha_{1,L}$	Fractions of the buffer, B, and ligand, L, in the protonated and unprotonated forms, i.e., forms that have lost 0 and 1 protons, respectively.
κ_0–κ_6	Cumulative deprotonation constants for acids with 0 to 6 acidic protons.
CBE, C_H	Charge-balance equation, CBE; trial values, C_H, of the H^+ concentration in iterative calculations.
$C_{L,Start}$	Diluted ligand concentration at the start of the equilibration reactions.
β_0–β_6	Cumulative formation constants for complexes with 0–6 ligands per metal ion with $\beta_0 = 1$.
γ_0–γ_6	Fractions of the diluted metal ion concentration complexed with 0–6 ligands.
[M], [ML], … [ML$_6$]	Equilibrium concentrations of metal ions complexed with 0–6 ligands.
Sum j[ML$_j$]	Sum of products of numbers, j, of ligands per metal ion times equilibrium concentrations of complexes with j ligands per metal ion.
[H$_n$L]–[H$_{n-6}$L]	Equilibrium concentrations of forms of ligands that have lost 0–6 protons.
$K_{f,Y}$	Formation constants for metal-ion chelates with EDTA.
$K_{a1,Y}$–$K_{a6,Y}$	Stepwise deprotonation constants for EDTA.
$\kappa_{0,Y}$–$\kappa_{6,Y}$	Cumulative deprotonation constants for EDTA.
$C_{0,Y}^0$–$C_{6,Y}^0$	Initial concentrations of forms of protonated EDTA that have lost 0–6 protons.
$C_{0,Y}$–$C_{6,Y}$	Diluted concentrations of forms of protonated EDTA that have lost 0–6 protons.
[H$_6$Y^{2+}]–[Y^{4-}]	Equilibrium concentrations of forms of EDTA that have lost 0–6 protons.
$\alpha_{0,Y}$–$\alpha_{6,Y}$	Fractions of forms of EDTA that have lost 0–6 protons.

SUMMARY

We begin with a brief review of types of reactions to be discussed and continue with a more detailed discussion of reactants and reactions. We then describe the chelate effect as a term used to represent the increased stabilities associated with complexes that involve ring structures relative to complexes that don't involve ring structures. We then discuss charges on reactants and products that are used to write charge balance equations. We then differentiate between multistep and one-step reactions. Finally we discuss goals of and symbols used in subsequent chapters.

9 Equilibrium Calculations for Multistep Complex-Formation Reactions

Many complex-formation reactions approach equilibrium in two or more steps. As examples, reactions of Ag(I), Zn(II), and Ni(II) with ammonia approach equilibrium in two, four, and six steps, respectively. Reactions of ferrous iron, Fe(II), with 1,10-phenanthroline, also called o-phenanthroline or o-Phen for short, approach equilibrium in three steps. This chapter focuses on equilibrium calculations for reactions that approach equilibrium in two to six steps.

Using the Fe(II)/o-Phen reactions in acetate buffer as an example, we will talk briefly about approximate calculations and then extend the discussion to a process used to develop a charge-balance equation for the reactions. Having developed the charge-balance equation for this situation, we will talk about ways to extend the equation to other situations, including reactions of uncharged and charged monoprotic ligands to complexes with up to six ligands per metal ion. We will then talk about an iterative program to solve the charge-balance equation for equilibrium concentrations of metal ions and their complexes as well as other reactants and products with acid-base properties. Finally, we will talk about some practical applications of the concepts and program.

We will assume that anions such as acetate ion are added as sodium salts and cations except those that form insoluble chlorides are added to solutions as chloride salts. We will assume that ions such as Ag(I), Hg(I), and Pb(II) that form sparingly soluble chloride salts would be added as nitrate salts.

Spreadsheet programs discussed in this chapter are available on the website http://crcpress.com/9781138367227.

REACTIONS INVOLVING FERROUS IRON AND o-PHENANTHROLINE

o-phenanthroline, or o-Phen for short, is an uncharged conjugate base of a singly charged monoprotic weak acid with a deprotonation constant, $K_{a,L} = 1.38 \times 10^{-5}$. Using the symbol, L, for the ligand, o-Phen, as is common practice, the acid-base reaction is

$$HL^+ \overset{K_{a,L}}{\rightleftharpoons} H^+ + L \qquad K_{a,L} = 1.38 \times 10^{-5}$$

Ferrous iron, Fe^{2+}, reacts in three steps with o-Phen to form three complexes with one, two, and three o-Phens per Fe(II) atom. The reactions are as follows:

$$Fe^{2+} + L \overset{K_{f1}}{\rightleftharpoons} FeL^2 + L \overset{K_{f2}}{\rightleftharpoons} FeL_2^{2+} + L \overset{K_{f3}}{\rightleftharpoons} \underset{\lambda_{Max} \cong 506 \text{ nm}}{\overset{\text{Red}}{FeL_3^{2+}}}$$

in which K_{f1}, K_{f2}, and K_{f3} are stepwise formation constants.

As shown in the reactions, the complex with three o-Phens per Fe(II), FeL_3^{2+}, is red with an absorption maximum near 506 nm and a molar absorptivity of $a = 1.1 \times 10^4$ L/mol-cm.[1] The absorbance of FeL_3^{2+} at a wavelength near 506 nm is often used to quantify Fe(II) concentrations in a variety of sample types such as iron tablets and blood.

Reactants Used to Adapt the Treatment to Metal Ions
and Ligands Other than Fe(II) and o-Phen

Detailed procedures for determinations of Fe(II) concentrations in a variety of samples are available on the internet. Calibration solutions containing known Fe(II) concentrations are usually prepared in sulfuric acid solutions using ferrous ammonium sulfate, $Fe(NH_4)_2(SO_4)_2$. When mixed with a solution containing sodium acetate, sulfuric acid and acetate react to form an acetic acid/acetate buffer.

I began by writing an iterative program based on these reactants. Whereas the program worked well, it wasn't easy to adapt it to other combinations of metal ions and ligands. I then developed a program based on the use of ferrous chloride in place of ferrous ammonium sulfate and a buffer prepared using acetic acid and sodium acetate in place of sulfuric acid and sodium acetate.

The two programs gave virtually identical equilibrium concentrations of reactants and products in common between the two sets of reactions for each of several sets of starting conditions. Therefore, given that concepts, equations, and a program based on ferrous chloride as the source of Fe(II) and a buffer based on acetic acid and sodium acetate are easily extended to other metal-ion/ligand reactions, I have chosen to use this latter set of reactions in subsequent discussions.

For the record, hydroxylamine hydrochloride is usually included with the other reactants to reduce any Fe(III) that might be formed to Fe(II). The reaction is:

$$2Fe^{3+} + 2NH_2OH \cdot HCL + 2OH^- \rightarrow 2Fe^{2+} + N_2 + 4H_2O + H^+ + Cl^-$$

I have assumed in subsequent discussions that all the iron is in the +2 oxidation state and have not included the foregoing reaction in the charge-balance equation.

An Example

Example 9.1 is used as the basis for our initial discussions of the Fe(II)/o-Phen reactions.

> **Example 9.1 Calculate equilibrium FeL_3^{2+} concentrations in solutions prepared by adding 10.0 mL each, of solutions containing (A) 1.0×10^{-4} M, (B) 2.5×10^{-4}, and (C) 5×10^{-4} M ferrous chloride to each of three 100-mL volumetric flasks followed by 25.0 mL of 0.014 M o-Phen and 10.0 mL of a solution containing 1.0 M each of acetic acid and sodium acetate and diluting each solution to 100 mL with water.**
>
> **For acetic acid: $K_{a,B} = 1.8 \times 10^{-5}$; for protonated o-Phen HL^+: $K_{a,L} = 1.38 \times 10^{-5}$; for FeL_j^{2+}: $\log K_{f1} = 5.86$, $\log K_{f2} = 5.25$, and $\log K_{f3} = 10.03$.**

We will talk about both approximate and iterative procedures to solve this example after we have discussed some background information relevant to both options.

BACKGROUND INFORMATION

Let's begin with descriptions and symbols for different measures of concentration.

Concentration Symbols

Calculations for Example 9.1 and other analogous problems involve three "types" of concentrations. The three "types" of concentrations are (a) *undiluted concentrations* before solutions are mixed, (b) *diluted concentrations* after solutions are mixed but before any reactions occur, and (c) *equilibrium concentrations* after all reactions are complete. For any reactant, X, undiluted concentrations are represented by C_X^0; diluted concentrations are represented by C_X and equilibrium concentrations are represented by [X]. Symbols for selected reactants are summarized in Table 9.1.

TABLE 9.1

Summary of Symbols Used for Undiluted, Diluted, and Equilibrium Concentrations of Acid-Base Pairs Used as Buffers and Complexing Ligands in This Section

Undiluted Concentrations				Diluted Concentrations			
Buffer Pair		Ligand Pair[a]		Buffer		Ligand Pair[a]	
HB	B^-	HL^+	L	HB	B	HL^+	L
C_{HB}^0	C_B^0	C_{HL}^0	C_L^0	C_{HB}	C_B	C_{HL}	C_L
Equilibrium Concentrations							
Acid-Base Pairs Used as Buffers				Acid-Base Pairs Used as Ligands			
Buffer acid		Buffer base		Ligand acid		Ligand	
[HB]		$[B^-]$		$[HL^+]$		[L]	

[a] It is assumed throughout this chapter that none of the protonated forms of the ligand is added initially, i.e., $C_{HL}^0 = C_{HL} = 0$.

For Example 9.1, acetic acid and acetate ion are the *buffer acid* and *buffer base*, respectively, and protonated o-Phen and unprotonated o-Phen are the *ligand acid* and *ligand base*, respectively. As noted in the table, it is assumed throughout this chapter that none of the ligand acid is added to solutions initially. However, some of the ligand will be converted to the ligand acid at equilibrium.

Analogous symbols are used for other reactants and products.

DILUTED REACTANT CONCENTRATIONS

Diluted concentrations of reactants are used in equilibrium calculations. For any volume, V_X, of a reactant, X, the diluted concentration is the volume of the reactant times the undiluted concentration, C_X^0, divided by the total volume, V_T, to which the solution is diluted, as expressed in Equation 9.1:

$$C_X = \frac{V_X C_X^0}{V_T} \qquad (9.1)$$

For example, the diluted ligand concentration in a solution prepared by diluting 25.0 mL of 0.014 M o-Phen to 100 mL is

$$C_L = \frac{V_L C_L^0}{V_T} = \frac{(25.0\,\text{mL})(0.0140\text{M})}{100\,\text{mL}} = 3.5 \times 10^{-3}\,\text{M}$$

Analogous calculations are used for diluted concentrations of other reactants, including buffer components.

APPROXIMATE HYDROGEN ION CONCENTRATION

The dissociation constant for acetic acid can be used in the rearranged form below to approximate the hydrogen ion concentration, as expressed in Equation 9.2:

$$K_{a,HB} = \frac{\left[H^+\right]\left[B^-\right]}{[HB]} \quad \Rightarrow \quad \left[H^+\right] = K_{a,HB}\frac{[HB]}{\left[B^-\right]} \cong K_{a,HB}\frac{C_{HB}}{C_B} \qquad (9.2)$$

Substituting $K_{a,HB} = 1.8 \times 10^{-5}$ and $C_{HB} = C_B = 0.10$ M gives the same approximate value of the hydrogen ion concentration $[H^+] \cong 1.8 \times 10^{-5}$ M, as the concentrations, 1.0 M each, in the example. The approximate option is based on an assumption that the H^+ concentration is controlled exclusively by the buffer. This simplification doesn't account for effects of o-Phen, a weak base, on the hydrogen ion concentration.

FRACTIONS OF MONOPROTIC WEAK ACIDS IN DIFFERENT FORMS

For any monoprotic weak base used as a ligand, fractions in forms that have lost $i = 0$ and 1 proton are expressed in Equations 9.3a and 9.3b:

$$\alpha_{0,L} = \frac{\left[H^+\right]}{\left[H^+\right] + K_{a,L}} \qquad (9.3a)$$

$$\alpha_{1,L} = \frac{K_{a,L}}{\left[H^+\right] + K_{a,L}} \qquad (9.3b)$$

I will trust you to write the analogous equations for acid-base pairs used as buffers by replacing subscripted L's with B's.

RELATIONSHIPS AMONG STEPWISE FORMATION CONSTANTS AND EQUILIBRIUM CONCENTRATIONS

Stepwise reactions for the formation of the Fe(II)/o-Phen complexes are repeated here for your convenience.

$$Fe^{2+} + L \overset{K_{f1}}{\rightleftharpoons} FeL^{2+} + L \overset{K_{f2}}{\rightleftharpoons} FeL_2^{2+} + L \overset{K_{f3}}{\rightleftharpoons} FeL_3^{2+}$$

Relationships among stepwise formation constants and equilibrium concentrations are written in the usual way, i.e., as equilibrium concentrations of the products divided by the product of reactants, as expressed in Equations 9.4a–9.4c:

$$K_{f1} = \frac{\left[FeL^{2+}\right]}{\left[Fe^{2+}\right][L]} \quad (9.4a) \qquad K_{f2} = \frac{\left[FeL_2^{2+}\right]}{\left[FeL^{2+}\right][L]} \quad (9.4b) \qquad K_{f3} = \frac{\left[FeL_3^{2+}\right]}{\left[FeL_2^{2+}\right][L]} \quad (9.4c)$$

in which [L] is the equilibrium ligand concentration, i.e., the o-Phen concentration.

CUMULATIVE FORMATION CONSTANTS

Cumulative formation constants for multistep complexes are products of successive constants, $K_{f,j}$. Cumulative formation constants for steps corresponding to *j ligands per metal ion* are commonly represented by symbols, β_j. It is common practice to tabulate logarithms of stepwise constants, i.e., log K_{fj}, rather than the constants themselves.

Calculating Cumulative Formation Constants

One approach to calculating cumulative constants is to calculate the stepwise constants as $10^{\log K_{fj}}$ and then to calculate cumulative constants as products of stepwise constants. However, a more efficient way is to calculate cumulative constants as 10 raised to the power of the sum of logarithms of stepwise constants, as expressed in Equation 9.5:

$$\beta_j = K_{f1}K_{f2}\ldots K_{fj} = 10^{\left(\log K_{f1} + \log K_{f2} + \cdots + \log K_{fj}\right)} \qquad (9.5)$$

in which β_j is the cumulative constant corresponding to formation of the complex, FeL_j^{2+} and K_{fj} is the stepwise constant for the complex, FeL_j^{2+}. For example, given log $K_{f1} = 5.86$, log $K_{f2} = 5.25$, and log $K_{f3} = 10.03$ for the Fe(II)/o-Phen complexes, the cumulative constants for first and third complexes, FeL^{2+} and FeL_3^{2+} are

$$\beta_1 = 10^{(5.86)} = 7.24 \times 10^5 \quad \text{and} \quad \beta_3 = 10^{(5.86+5.25+10.03)} = 1.38 \times 10^{21}$$

I will trust you to show that $\beta_2 = 1.29 \times 10^{11}$.

Given that β_3 is much larger than β_1 and β_2 and the o-Phen concentration in the example is significantly larger than the Fe(II) concentration, it is reasonable to expect that most of the Fe(II) will be converted to the FeL_3^{2+} complex. We will test this assumption when we begin to do calculations later.

Relationships among Cumulative Constants and Equilibrium Concentrations

Equations for the cumulative formation constants in terms of concentrations are obtained as products of equations for the stepwise constants. For example, the equation for β_3 for the Fe(II)/o-Phen complexes in terms of equilibrium concentrations is obtained as the product of equations for the three stepwise constants, i.e.,

$$\beta_3 = K_{f1}K_{f2}K_{f3} = \frac{\left[FeL^{2+}\right]}{\left[Fe^{2+}\right][L]} \frac{\left[FeL_2^{2+}\right]}{\left[FeL^{2+}\right][L]} \frac{\left[FeL_3^{2+}\right]}{\left[FeL_2^{2+}\right][L]} = \frac{\left[FeL_3^{2+}\right]}{\left[Fe^{2+}\right][L]^3}$$

I will trust you to use similar procedures to confirm Equations 9.6a–9.6c, summarized below, for cumulative constants, β_1 and β_2, along with that for β_3.

$$\beta_1 = \frac{\left[FeL^{2+}\right]}{\left[Fe^{2+}\right][L]} \quad (9.6a) \qquad \beta_2 = \frac{\left[FeL_2^{2+}\right]}{\left[Fe^{2+}\right][L]^2} \quad (9.6b) \qquad \beta_3 = \frac{\left[FeL_3^{2+}\right]}{\left[Fe^{2+}\right][L]^3} \quad (9.6c)$$

These observations can be generalized. Ignoring charges for metal ions and ligands, the relationship between the j'th cumulative constant and equilibrium concentration can be written as follows in Equation 9.6d:

$$\beta_j = \frac{[ML_j]}{[M][L]^j} \tag{9.6d}$$

in which β_j is the cumulative constant, $[ML_j]$ is the equilibrium concentration of the metal ion complexed with j ligands and $[M]$ and $[L]$ are equilibrium metal-ion and ligand concentrations.

As was done for weak acids in earlier chapters, cumulative formation constants are used to develop equations for fractions of the Fe(II) in different forms, i.e., Fe^{2+}, FeL^{2+}, FeL_2^{2+}, and FeL_3^{2+}.

FRACTIONS OF Fe(II) IN DIFFERENT FORMS

Fractions of metal ions complexed with different numbers, j, of ligands are represented herein by symbols, γ_j, to differentiate them from fractions, α_i, of weak acids in different forms. As examples, fractions of Fe(II) complexed with 0–3 o-Phens per Fe(II) ion are represented herein by the symbols, γ_0, γ_1, γ_2, and γ_3.

Basis for Concentration Fractions

Example 9.2 illustrates a procedure used to develop an equation for the fraction, γ_0, for the form of Fe(II) complexed with zero ammonias, i.e., Fe^{2+}. Having worked through the example, I will show you how the resulting equation can be used to write fractions for the other Fe(II)/oPhen complexes as well as complexes with different numbers of ligands per metal ion.

Example 9.2 Given a solution containing Fe(II) and 1,10-phenanthroline, develop an equation for the fraction, γ_0, of Fe(II) present as Fe^{2+}.

Step 1: Write the reactions and cumulative formation constants, $\beta_1-\beta_3$.

$$Fe^{2+} + L \overset{\beta_1}{\rightleftharpoons} FeL^{2+} + L \overset{\beta_2}{\rightleftharpoons} FeL_2^{2+} + L \overset{\beta_3}{\rightleftharpoons} FeL_3^{2+} + L$$

Step 2: Write the equation for the fraction of Fe(II) present as Fe^{2+} in terms of equilibrium concentrations of all forms of Fe(II) and the total Fe(II) concentration, $C_{T,Fe}$.

$$\gamma_0 = \frac{\left[Fe^{2+}\right]}{C_{T,Fe}} = \frac{\left[Fe^{2+}\right]}{\left[Fe^{2+}\right] + \left[FeL^{2+}\right] + \left[FeL_2^{2+}\right] + \left[FeL_3^{2+}\right]}$$

Step 3: Write equations for the cumulative formation constants.

$$\beta_1 = \frac{\left[FeL^{2+}\right]}{\left[Fe^{2+}\right][L]}, \ \beta_2 = \frac{\left[FeL_2^{2+}\right]}{\left[Fe^{2+}\right][L]^2}, \ \beta_3 = \frac{\left[FeL_3^{2+}\right]}{\left[Fe^{2+}\right][L]^3}$$

Step 4: Rearrange equations for cumulative constants to express each complex ion concentration as the product of a function, $\beta_i[L]^j$, times the equilibrium Fe^{2+} concentration.

$$\left[FeL^{2+}\right] = \beta_1[L]\left[Fe^{2+}\right], \left[FeL_2^{2+}\right] = \beta_2[L]^2\left[Fe^{2+}\right], \left[FeL_3^{2+}\right] = \beta_3[L]^3\left[Fe^{2+}\right]$$

Step 5: Substitute equations from Step 4 into the equation in Step 2 and cancel $[Fe^{2+}]$ terms in the numerator and denominator (see Equation 9.7a).

$$\gamma_0 = \frac{\left[Fe^{2+}\right]}{\left(1+\beta_1[L]+\beta_2[L]^2+\beta_3[L]^3\right)\left[Fe^{2+}\right]} = \frac{1}{1+\beta_1[L]+\beta_2[L]^2+\beta_3[L]^3} = \frac{1}{D_{Fe}} \equiv \frac{1}{D_3} \quad (9.7a)$$

Step 6: Write the denominator term for any complex with 3 ligands per metal ion (see Equation 9.7b).

$$D_{3,L} = 1+\beta_1[L]+\beta_2[L]^2+\beta_3[L]^3 = \sum_{j=0}^{3} \beta_j[L]^j \quad (9.7b)$$

Briefly, the procedure for any fraction is to (a) write the equation for the fraction in terms of equilibrium concentrations of the metal complexes, (b) to rearrange the cumulative constants in order to set equilibrium concentrations of other reactants proportional to the concentration of the form of the metal ion of interest as in Step 4, and (c) to substitute these equations into the equation for the fraction, and (d) to simplify the resulting equation.

You can test your ability to apply the process by doing Exercise 9.1.

EXERCISE 9.1

Using procedures similar to those in Example 9.2, show that the equation for the fraction, γ_1, of Fe(II) in the form, FeL^{2+}, can be written in terms of cumulative constants and the ligand concentration as follows:

$$\gamma_1 = \frac{\beta_1[L]}{1+\beta_1[L]+\beta_2[L]^2+\beta_3[L]^3} = \frac{\beta_1[L]}{D_3}$$

Hints: Write the fractional equation as the concentration, $[FeL^{2+}]$, divided by the sum of the equilibrium concentrations of all forms of Fe(II), rearrange equations for cumulative constants to express all denominator terms as multipliers times $[FeL^{2+}]$, substitute these equations into the fractional equation, cancel $[FeL^{2+}]$ terms in the numerator and denominator and multiply top and bottom by $\beta_1[L]$.

As an example,

$$\left[Fe^{2+}\right] = \frac{1}{\beta_1[L]}\left[FeL^{2+}\right]$$

Whether or not you do this exercise, notice that (a) the denominator term is the same for both fractions and (b) the numerator terms for the first and second fractions, γ_0 and γ_1, are the same as the first and second denominator terms. These observations can be generalized. Denominator terms are the same for all fractions of complexes with the same numbers of ligands per metal ion and numerator terms for successive fractions are successive denominator terms.

Although developed here for Fe(II)/o-Phen complexes, the foregoing equations and observations apply for all complexes with 0–3 ligands per metal ion.

Complexes with Fewer and More Ligands per Metal Ion

We could use procedures similar to that illustrated above to develop equations for complexes with fewer and more ligands per metal ion. However, it is easier to use patterns in equations for the Fe(II)/o-Phen complexes to write denominator terms for complexes with fewer and more ligands per metal ion. Using patterns in Equation 9.7b, denominator terms for complexes with fewer and more ligands per metal ion can be written by dropping and adding terms of the form, $\beta_j[L]^j$.

As examples, denominator terms, D_2, D_4, and D_6, for complexes with up to 2, 4, and 6 ligands per metal ion written by dropping and adding $\beta_j[L]^j$ terms in Equation 9.7b are expressed in Equations 9.8a–9.8c:

$$D_2 = 1 + \beta_1[L] + \beta_2[L]^2 \tag{9.8a}$$

$$D_4 = 1 + \beta_1[L] + \beta_2[L]^2 + \beta_3[L]^3 + \beta_4[L]^4 \tag{9.8b}$$

$$D_6 = 1 + \beta_1[L] + \beta_2[L]^2 + \beta_3[L]^3 + \beta_4[L]^4 + \beta_5[L]^5 + \beta_6[L]^6 \tag{9.8c}$$

I will trust you to identify terms dropped from and added to that for D_3 to obtain these equations.

As noted earlier, successive fractions are successive denominator terms divided by the denominator.

DISTRIBUTION PLOTS

By analogy with the way concentration fractions for weak acids were used to prepare distribution plots for weak acids, concentration fractions for complex ions can be used to prepare distribution plots for complex ions in different forms. Distribution plots for complex ions are plotted herein as concentration fractions vs. the ligand concentration, $[L]$, on a logarithmic scale.

A program on Sheet 6 of the complex ions program is designed to plot high resolution distribution plots for the Fe(II)/o-Phen reaction. A program on Sheet 7 of the same program is designed to plot distribution plots for a variety of reactions, including low resolution plots for the Fe(II)/o-Phen reaction. The program on Sheet 7 is more versatile than that on Sheet 6 but gives discontinuities in plots for the Fe(II)/o-Phen reaction at points at which rates of change in the o-Phen concentration in Column J increase abruptly.

To access the program, open the complex ions program from the programs website http://crcpress. com/9781138367227 and select Sheet 6 for high resolution plots for the Fe(II)/o-Phen reaction or

Sheet 7 for other reactions. Instructions for using the programs are in Rows 1–11 in Columns A–I on each sheet. Plots prepared using the programs can be used to help us visualize how different variables influence fractions of metal ions in different forms.

A template for the Fe(II)/o-Phen reaction is in Cells A51–G53 on Sheet 6. Templates for three sets of reactions, including the Fe(II)/o-Phen reaction are in Cells A51–G67 on Sheet 7 of the of the complex ions program.

Complexes of Fe(II) with o-Phen

As a reminder, Fe(II) reacts with o-phenanthroline in three steps to produce the colored form, FeL_3^{2+}, the absorbance of which is measured and used to quantify ferrous iron concentrations, i.e.,

$$Fe^{2+} + L \overset{K_{f1}}{\rightleftharpoons} FeL^{2+} + L \overset{K_{f2}}{\rightleftharpoons} FeL_2^{2+} + L \overset{K_{f3}}{\rightleftharpoons} \underbrace{\overset{Red}{\overbrace{FeL_3^{2+}}}}_{\lambda_{Max} \cong 506 \text{ nm}}$$

For successful determinations of Fe(II) concentrations using this reaction it is important that most of the Fe(II) is converted the FeL_3^{2+} complex.

Figure 9.1 is a plot of the fractions of ferrous iron in the three forms with j = 0, 1, 2, and 3 o-Phens per Fe(II) in a solution with a hydrogen ion concentration of 1.8×10^{-5} M and a diluted o-Phen concentration of 3.5×10^{-3} M.

As reminders, the stepwise formation constants for the three steps in the reactions are $K_{f1} = 7.2 \times 10^5$, $K_{f2} = 1.8 \times 10^5$, and $K_{f3} = 1.1 \times 10^{10}$. These constants tell us that driving forces for the first two steps are quite large but the driving force for the third step is about five orders of magnitude larger than for either of the first two steps.

The solid line in the figure shows us that most of the Fe(II) is in the Fe^{2+} form at the lowest o-Phen concentrations and decreases toward zero as the o-Phen concentration increases. The dotted line tells us that small amounts of Fe(II) are converted to the FeL^{2+} form at intermediate

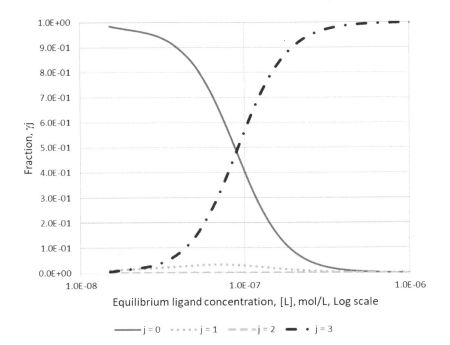

FIGURE 9.1 Distribution plots for Fe(II) reacting with o-phenanthroline in an acetate buffer with a diluted o-Phen concentration of $C_L = 3.5 \times 10^{-3}$ M a hydrogen ion concentration of $[H^+] = 1.8 \times 10^{-5}$ M.

o-Phen concentrations. The fraction of the Fe(II) in the FeL_2^{2+} remains very small for all o-Phen concentrations. The reason is that the formation constant for FeL_3^{2+} is so much larger than that for FeL_2^{2+} that virtually all of the Fe(II) is converted to FeL_3^{2+} for o-Phen concentrations at which FeL_2^{2+} would form.

The dash-dot line tells us that virtually all of the Fe(II) is converted to the colored FeL_3^{2+} form for o-Phen concentrations larger than about 1.0×10^{-6} M. Complete conversion of the Fe(II) to the colored form favors successful applications of the reaction for colorimetric determination of Fe(II) concentrations. Therefore, this plot helps to identify a set of conditions for which the Fe(II)/o-Phen reaction can be used for the photometric determination of Fe(II).

Distribution plots can also be used to help us visualize how different variables influence behaviors of complex formation reactions. For example, the figure in Cells A11–H31 on Sheet 6 of the complex ions program is designed to help us visualize how different variables influence distributions of Fe(II) among the different forms. For example, what do you think would happen to the distribution plots if we were to set logarithms of all three constants to 5.0? Try it.

Given that all constants have the same value, you should see is a set of distribution plots consisting of one descending plot corresponding to the Fe^{2+}, two intermediate peaks corresponding to the FeL^{2+} and FeL_2^{2+} complexes and an ascending sigmoid corresponding to the FeL_3^{2+} complex. If you do this, and I hope you will, restore the input section to its original condition by copying the template in A51–G53 and pasting it into A35–G37 using the Paste (P) option.

What do you think would happen if we were to set logarithms of K_{f1} and K_{f2} in Cells B35 and C35 successively to 1.0 each? Try it and see. What you should see is the crossing points for the pair of sigmoid plots moving to successively larger values of the o-Phen concentration. Can you explain this behavior? I'll try. As we decrease values of constants, K_{f1} and K_{f2}, we also decrease the cumulative constant, $\kappa_3 = K_{f1}K_{f2}K_{f3}$, for the formation of FeL_3^{2+}. As the cumulative formation constant is decreased, increasingly larger o-Phen concentrations are needed to force the complex formation reaction to completion.

There is much more we could do with these plots. However, the foregoing changes should be sufficient to illustrate ways that distribution plots can be used to visualize effects of experimental variables on complex formation reactions.

If you make the foregoing changes, don't forget to restore the input section of the program to the original version. Logarithms of the formation constants for the Fe(II)/o-Phen complexes are in Cells B33–D33 for your convenience.

Unless stated otherwise, the remainder of our discussion of multiligand complexes involves the program on Sheet 7 of the complex ion equilibria program.

Complexes of Zn(II) with Ammonia

Zinc(II) reacts with ammonia in four steps to produce complexes with one to four ligands per zinc ion, i.e.,

$$Zn^{2+} + L \underset{\rightleftharpoons}{\overset{K_{f1}}{}} ZnL^{2+} + L \underset{\rightleftharpoons}{\overset{K_{f2}}{}} ZnL_2^{2+} + L \underset{\rightleftharpoons}{\overset{K_{f3}}{}} ZnL_3^{2+} + L \underset{\rightleftharpoons}{\overset{K_{f4}}{}} ZnL_4^{2+}$$

Stepwise formation constants for the four steps in the reactions are included below.

$$K_{f1} = 1.6 \times 10^2 \qquad K_{f2} = 1.9 \times 10^2 \qquad K_{f3} = 2.3 \times 10^2 \qquad K_{f4} = 1.1 \times 10^2$$

These constants tell us that driving forces for the four steps are about the same. Based on what we observed for the Fe(II)/o-Phen complexes with approximately equal constants, we would expect a plot with one descending sigmoid corresponding to Zn^{2+}, three peaks corresponding to ZnL^{2+}, ZnL_2^{2+}, and ZnL_3^{2+}, respectively, and one ascending sigmoid corresponding to ZnL_4^{2+}.

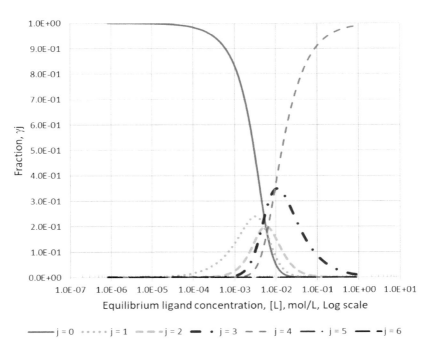

FIGURE 9.2 Distribution plots for Zn(II) reacting with 0.10 M ammonia in a solution with a hydrogen ion concentration of 1.0×10^{-10} M.

To observe the plots, select Sheet 7 of the complex ions program; copy the template in Cells A58–G60 and paste it into Cells A35–I37 using the Paste (P) option. Figure 9.2 is the distribution plot you should see.

The plots confirm the shapes suggested above, namely, one descending sigmoid corresponding to Zn^{2+}, three peaks corresponding to ZnL^{2+}, ZnL_2^{2+}, and ZnL_3^{2+}, respectively, and one ascending sigmoid corresponding to ZnL_4^{2+}.

What do you think would happen to the plots if we were to set log K_{f2} to some larger value, e.g., log $K_{f2} = 7$? I would expect the peak for ZnL_2^{2+} to be more dominant in the figure.

To test this expectation, copy the template for the Zn(II)/ammonia complexes in A58–G60, paste it into Cells A35–G37 and observe the interactive plot in Cells K7–R30. Then set C35 to log $K_{f2} = 7$, corresponding $K_{f2} = 1.0E7$, and reexamine the interactive plots. The plot for j = 2 representing ZnL_2^{2+} will have become more dominant in the figure and the plot j = 1representing ZnL^{2+} will have disappeared under the plot for the ZnL_2^{2+} complex.

Complexes of Ni(II) with Ammonia

Nickel(II) reacts with ammonia in six steps to produce complexes with 1–6 ligands per nickel ion, i.e.,

$$NI^{2+} + L \underset{}{\overset{K_{f1}}{\rightleftharpoons}} NiL^{2+} + L \underset{}{\overset{K_{f2}}{\rightleftharpoons}} NiL_2^{2+} + L \underset{}{\overset{K_{f3}}{\rightleftharpoons}} NiL_3^{2+} + L \underset{}{\overset{K_{f4}}{\rightleftharpoons}} NiL_4^{2+} + L \underset{}{\overset{K_{f5}}{\rightleftharpoons}} NiL_5^{2+} + L \underset{}{\overset{K_{f6}}{\rightleftharpoons}} NiL_6^{2+}$$

A template used to plot distribution plots for Ni(II) complexes with ammonia at pH = 10.0 is in Cells A65–G67 on Sheet 7 of the complex ions program. To use the template, copy it and paste it into Cells A35–G37 and observe the interactive plot in Cells K7–R30. A more permanent version of the plot is in Cells AL7–AS7 on Sheet 7.

The plot consists of one descending sigmoid corresponding to the uncomplexed Ni(II), 5 peaks corresponding to Ni(II) with j = 1–5 ammonias and one partial ascending sigmoid corresponding to Ni(II) complexed with j = six ammonias. As you can see, because of the small value of K_{f6} an unusually large ammonia concentration would be necessary to convert all the Ni(II) to the complex with six ammonias.

You can observe effects of variables such as the hydrogen ion concentration or formation constants on the plot by changing the quantities in Cells A35–G37. For example, you can observe effects of hydrogen in concentrations by changing the hydrogen ion concentration in A37.

Now, let's return to our discussion of the complexes of Fe(II) with o-phenanthroline.

UNCOMPLEXED o-Phen CONCENTRATION

As shown in earlier equations, the equilibrium o-Phen concentration, [L], is used to calculate fractions of Fe(II) in different forms. As also shown in earlier reactions, some of the o-Phen concentration is used to form the Fe(II)/o-Phen complexes. The part of the o-Phen concentration not complexed with a metal ion is called the uncomplexed ligand concentration and represented by the symbol, $C_{L,U}$, herein.

On the one hand, it is assumed in conventional approximations that the uncomplexed o-Phen concentration is the same as the diluted concentration, i.e., $C_{L,U} \cong C_L$. On the other hand, the iterative program to be discussed later calculates the uncomplexed ligand concentration and uses it to calculate the equilibrium o-Phen concentration.

Accounting for the fact that the three Fe(II)/o-Phen complexes contain 1, 2, and 3 o-Phens, respectively, the uncomplexed ligand concentration is expressed in Equation 9.9a:

$$C_{L,U} = C_L - \left(\left[FeL^{2+} \right] + 2 \left[FeL_2^{2+} \right] + 3 \left[FeL_3^{2+} \right] \right) \tag{9.9a}$$

in which $C_{L,U}$ and C_L are the uncomplexed and diluted ligand concentrations and multipliers, 1–3, are the numbers of ligands per metal ion. Analogous equations can be written for metal-ion/ligand complexes with fewer and more ligands per metal ion. For example, Equation 9.9b shows complexes with six ligands per metal ion:

$$C_{L,U} = C_L - \left([ML] + 2[ML_2] + 3[ML_3] + 4[ML_4] + 5[ML_5] + 6[ML_6] \right) \tag{9.9b}$$

Analogous terms for complexes with fewer than six ligands per metal ion can be written by dropping terms from right to left. The program to be described later uses the coordination number in E42 to set all terms beyond those corresponding to the coordination number to zero (Exercise 9.2).

EXERCISE 9.2

Using concentrations below calculated using an iterative program to solve Part C of Example 9.1A, show that the uncomplexed o-Phen concentration at equilibrium is $C_{L,U} = 3.2 \times 10^{-3}$ M.

$$C_L = 3.5 \times 10^{-3} M, \left[Fe^{2+} \right] = 1.6 \times 10^{-17} M, \left[FeL^2 \right]^+ = 2.0 \times 10^{-14} M, \left[FeL_2^{2+} \right]$$
$$= 5.7 \times 10^{-12} M, \text{ and } \left[FeL_3^{2+} \right] = 1.0 \times 10^{-4} M$$

The uncomplexed o-Phen concentration, $C_{L,U} = 3.2 \times 10^{-3}$ M, differs by about 9% from the diluted concentration, $C_L = 3.5 \times 10^{-3}$ M.

Equilibrium o-Phen Concentration

For approximate calculations, the equilibrium ligand concentration is approximated as the fraction, $\alpha_{1,L}$, of o-Phen in the fully dissociated form times the diluted o-Phen concentration, as expressed in Equation 9.10a:

$$[L] \cong \alpha_{1,L} C_L \cong \frac{K_{a,L}}{[H^+] + K_{a,L}} C_L \qquad \text{Conventional approximations} \qquad (9.10a)$$

in which $\alpha_{1,L}$ is the fraction of o-Phen in the fully dissociated form, C_L is the diluted o-Phen concentration, $K_{a,L}$ is the deprotonation constant for protonated o-Phen and $[H^+]$ is the equilibrium hydrogen ion concentration.

For iterative calculations, the equilibrium ligand concentration is calculated as the fraction, $\alpha_{1,L}$, of o-Phen in the fully dissociated form times the uncomplexed o-Phen concentration, as expressed in Equation 9.10b:

$$[L] = \alpha_{1,L} C_{L,U} = \frac{K_{a,L}}{[H^+] + K_{a,L}} C_{L,U} \qquad \text{Iterative option} \qquad (9.10b)$$

in which α_{1L} is the fraction of o-Phen in the fully deprotonated form, $C_{L,U}$ is the uncomplexed o-Phen concentration and other symbols are as described above.

EXERCISE 9.3

Given $K_{a,HL} = 1.38 \times 10^{-5}$, $[H^+] \cong 1.8 \times 10^{-5}$ M and $C_{L,U} \cong 3.2 \times 10^{-3}$ M for Part C of Example 9.1A, show that the equilibrium o-Phen concentration is approximated as $[L] \cong 1.4 \times 10^{-3}$ M, i.e., about 43% of the diluted o-Phen concentration.

Fe(II) Concentration from Ferrous Chloride

Given that ferrous chloride, $FeCl_2$, is a salt, it is reasonable to assume that it will dissolve and dissociate completely in water. Assuming this to be the case, the reaction is:

$$FeCl_2 \rightarrow Fe^{2+} + 2Cl^-$$

It follows that each molecule of the salt will produce one ferrous ion and two chloride ions.

Volumes, undiluted concentrations and diluted concentrations of ferrous chloride and Fe(II) are represented herein by the symbols, V_{Fe}, C_{Fe}^0, and C_{Fe}, respectively. Diluted ferrous chloride and Fe(II) concentrations are calculated as the product of the volume, V_{Fe}, times the undiluted concentration, C_{Fe}^0, divide by the total volume, as expressed in Equation 9.11a:

$$C_{Fe} = \frac{V_{Fe} C_{Fe}^0}{V_T} \qquad (9.11a)$$

in which C_{Fe} is the diluted Fe(II) concentration, V_{Fe} is the volume of the ferrous chloride solution, C_{Fe}^0 is the undiluted ferrous chloride concentration, and V_T is the total volume.

Based on the coefficient of chloride in ferrous chloride, the diluted chloride ion concentration is two times the Fe(II) concentration, as expressed in Equation 9.11b:

$$C_{Cl} = 2C_{Fe} = 2\frac{V_{Fe} C_{Fe}^0}{V_T} \qquad (9.11b)$$

CONCENTRATIONS OF DIFFERENT FORMS OF Fe(II)

Assuming that the only reactions involving Fe(II) are the complex-forming reactions, the equilibrium concentration of each form of Fe(II) is the fraction in that form times the diluted Fe(II) concentration.

As examples, concentrations of the Fe(II)/o-Phen complexes with 0–3 o-Phens per Fe(II) ion can be written as expressed in Equations 9.12a–9.12d:

$$\left[Fe^{2+}\right] = \gamma_0 C_{Fe} \qquad (9.12a)$$

$$\left[FeL^{2+}\right] = \gamma_1 C_{Fe} \qquad (9.12b)$$

$$\left[FeL_2^{2+}\right] = \gamma_2 C_{Fe} \qquad (9.12c)$$

$$\left[FeL_3^{2+}\right] = \gamma_3 C_{Fe} \qquad (9.12d)$$

in which $\gamma_0 - \gamma_3$ are fractions of the diluted Fe(II) concentrations complexed with 0–3 o-Phens and C_{Fe} is the diluted Fe(II) concentration.

Patterns in these equations can be used to write the equation for the concentration of any complex of any metal ion with any number of ligands per metal ion, as expressed in Equation 9.13:

$$\left[ML_j\right] = \gamma_j C_M \quad \text{for } 0 \le j \le C_{Num} \qquad (9.13)$$

in which ML_j is the concentration of a metal ion complexed with j ligands, γ_j, is the fraction of the metal ion concentration complexed with j ligands, C_M is the diluted concentration of the metal ion, and C_{Num} is the coordination number of the metal ion.

CALCULATION OPTIONS

My primary focus in this chapter is on an iterative procedure to solve problems such as Example 9.1A. However, for sake of completeness, I have also included an approximate procedure for the example. Given that the iterative procedure eliminates the need for simplifying assumptions on which approximate calculations are based, it gives smaller errors over wider ranges of conditions than the approximate option for many situations. Therefore, your time may be better spent by skipping the discussion of approximate calculations and going directly to the discussion of the iterative procedure.

APPROXIMATE CALCULATIONS

Approximate calculations involving multistep complex-formation reactions are based on assumptions that (a) concentrations of all reactants except the metal ion are approximately the same at the beginning and end of the equilibration process and (b) that the hydrogen ion concentration is controlled exclusively by the buffer. These assumptions can be used for an approximate solution to Example 9.1A.

PART A OF EXAMPLE 9.1

I will guide you through a stepwise solution to Part A of the example and then describe a program that you can use to simplify some of the calculations.

Example 9.1, Part A Calculate the equilibrium FeL_3^{2+} concentration in a solution prepared by adding 10.0 mL of a solution containing 1.0×10^{-4} M ferrous chloride to a 100-mL volumetric flask followed by 25.0 mL of 0.014 M o-Phen and 10.0 mL of a solution containing 1.0 M each of acetic acid and sodium

acetate and diluting the solution to 100 mL with water. For acetic acid:
$K_{a,B} = 1.8 \times 10^{-5}$; for protonated o-Phen, HL^+: $K_{a,L} = 1.38 \times 10^{-5}$;
for FeL_j^{2+}: log K_{f1} = 5.86, log K_{f2} = 5.25, and log K_{f3} = 10.03.

Step 1: Calculate diluted concentrations of acetic acid, acetate ion, o-Phen, and Fe(II).

$$C_{HB} = C_B = \frac{10(1.0)}{100} = 0.10\,M$$

$$C_L = \frac{25(0.014)}{100} = 3.5 \times 10^{-3}\,M$$

$$C_{Fe} = \frac{10(1.0 \times 10^{-4})}{100} = 1.0 \times 10^{-5}\,M$$

Step 2: Approximate the hydrogen ion concentration using the deprotonation constant, $K_{a,B}$, for acetic acid and the diluted concentrations, C_{HB} and C_B, of acetic acid and acetate ion.

$$\left[H^+\right] \cong K_{a,B}\frac{C_{HB}}{C_B} \cong 1.8 \times 10^{-5}\frac{0.10}{0.10} \cong 1.8 \times 10^{-5}\,M; \qquad pH = -\log\left[H^+\right] \cong 4.74$$

Step 3: Calculate the fraction of diluted o-Phen in the fully dissociated form using the dissociation constant, $K_{a,L}$, of protonated o-Phen and the approximate H^+ concentration.

$$\alpha_{1L} = \frac{K_{a,L}}{([H^+] + K_{a;L})} \cong \frac{1.38 \times 10^{-5}}{1.8 \times 10^{-5} + 1.38 \times 10^{-5}} \cong 0.434$$

Step 4: Calculate the equilibrium o-Phen concentration using the fraction, $\alpha_{1,L}$, of o-Phen in the deprotonated form and diluted concentration, C_L, of o-Phen.

$$[L] = \alpha_{1L}C_L \cong (0.434)(3.5 \times 10^{-3}\,M) \cong 1.52 \times 10^{-3}\,M$$

Step 5: Calculate values of cumulative formation constants, β_1–β_3, for the FeL_j^{2+} complexes using sums of logarithms of stepwise formation constants, K_{f1}–K_{f3}.

$$\beta_1 = 10^{5.86} = 7.24 \times 10^5 \qquad \beta_2 = 10^{(5.86+5.25)} = 1.29 \times 10^{11} \qquad \beta_3 = 10^{(5.86+5.25+10.03)} = 1.38 \times 10^{21}$$

Step 6: Calculate the denominator term for fractions of iron in different forms using the ligand concentration, [L], from Step 4 and cumulative formation constants, β_j, from Step 5.

$$D_\gamma = 1 + \beta_1[L] + \beta_2[L]^2 + \beta_3[L]^3 \cong 4.84 \times 10^{12}$$

Step 7: Calculate the fraction of the diluted Fe(II) concentration complexed with j = 3 o-Phens using quantities identified and calculated above.

$$\gamma_3 = \frac{\beta_3 L^3}{D_{Fe}} = \frac{(1.38 \times 10^{21})(1.52 \times 10^{-3})^3}{4.84 \times 10^{12}} \cong 1.00$$

Step 8: Calculate the equilibrium FeL_3^{2+} concentration as the product of the fraction, γ_3, in that form times the diluted Fe(II) concentration, C_{Fe}.

$$\left[FeL_3^{2+}\right] = \gamma_3 C_{Fe} \cong (1.00)(1.0 \times 10^{-5}\,M) \cong 1.0 \times 10^{-5}\,M$$

As a reminder, Parts B and C of Example 9.1 involved ferrous iron concentrations of $C_{Fe} = 2.5 \times 10^{-4}$ M and 5.0×10^{-4} M, respectively. I will trust you to show that the FeL_3^{2+} concentrations for Parts B and C of the example are 2.5×10^{-5} M and 5.0×10^{-5} M, respectively.

A requirement for a successful determination of Fe(II) concentrations using the Fe(II)/o-Phen reaction is that most of the Fe(II) concentration be converted to the colored complex, FeL_3^{2+}. As shown in Step 7, the fraction of Fe(II) complexed with three o-Phens is $\gamma_3 \cong 1.00$ corresponding to virtually 100% of Fe(II) being converted to the colored form. There are two primary reasons for this. First, the diluted o-Phen concentration, $C_L = 3.5 \times 10^{-3}$ M, is much larger than the diluted Fe(II) concentration, $C_{Fe} = 1.0 \times 10^{-5}$ M. Second, the overall formation constant for the colored form, $\beta_3 = 1.4 \times 10^{21}$, is ten orders of magnitude larger than the next largest constant, $\beta_2 = 1.3 \times 10^{11}$.

These factors combine to ensure that virtually all of the Fe(II) is converted to the colored form, FeL_3^{2+}. We will talk later about a program on Sheet 3 of the complex ions program used to compare concentrations calculated using approximate and iterative procedures.

A TIME-SAVING PROGRAM FOR APPROXIMATE CALCULATIONS

A program on Sheet 2 of the complex ions program can be used to calculate a variety of quantities, including fractions and concentrations of weak acids and complexes in different forms. A template for Part A of Example 9.1 is included in Cells B75–E89 on Sheet 2 of the program. Instructions for using the template are included with it on Sheet 2.

I suggest that you use the template and program to do Exercise 9.4.

EXERCISE 9.4

Use the program on Sheet 2 of the complex ions program to confirm fractions below rounded to two significant figures. Also use the fractions with diluted concentrations included in the table to calculate approximate equilibrium concentrations for both forms of acetic acid and o-phenanthroline and the four forms of Fe(II) for conditions in Part A of Example 9.1:

$$pH = -\log[H^+] \cong -\log(1.8 \times 10^{-5}) \cong 4.74$$

	Buffer		Ligand		Fe(II)			
Concentration fractions	$\alpha_{0,B}$, I30	$\alpha_{1,B}$, I31	$\alpha_{0,L}$, G20	$\alpha_{1,L}$, G21	γ_0, I21	γ_1, I22	γ_2, I23	γ_3, I24
	0.50	0.50	0.57	0.43	2.1E-13	2.3E-10	6.2E-8	1.00
Diluted C's	$C_{T,B} = 0.020$ M		$C_{T,L} = 3.50E\text{-}3$ M		CT,Fe = 1.0E-5 M			
Equilibrium concentrations	[HB]	[B$^-$]	[HL$^+$]	[L]	[Fe^{2+}]	[FeL^{2+}]	[FeL$_2^{2+}$]	[FeL$_3^{2+}$]
	0.010	0.010	2.0E-5	1.5E-5	2.1E-18	2.3E-15	6.2E-13	1.00E-5

The equilibrium concentration of each reactant and product is calculated as the product of the corresponding fraction times the total concentration of all forms of the reactant. For example, the Fe^{2+} concentration is calculated as $[Fe^{2+}] = \gamma_0 C_{T,Fe} \cong (2.1 \times 10^{-13})(1.0 \times 10^{-5}$ M$) \cong 2.1 \times 10^{-18}$ M. I will trust you to confirm the other concentrations using similar procedures.

On the one hand, concentrations of Fe^{2+}, FeL^{2+}, and FeL_2^{2+} are very small. On the other hand, procedures used to calculate the concentrations apply to other situations for which concentrations of two or more complex ions represent significant fractions of the total metal ion concentration.

Sheet 3 of the complex ions program includes a program to compare concentrations calculated using conventional approximation and iterative procedures. We will talk in more detail about that program later. For now, let's talk about potential sources of error in the approximate concentrations.

SOURCES OF ERROR IN THE APPROXIMATION PROCEDURE

As a reminder, the approximate calculations are based on assumptions that (a) the o-Phen concentration is about the same at the beginning and end of the equilibration reactions and (b) that the hydrogen ion concentration is controlled exclusively by the acetate buffer. On the one hand, given that the diluted o-Phen concentration, $C_L = 3.5 \times 10^{-3}$ M, is significantly larger than the largest Fe(II) concentrations, $C_{Fe} = 5.0 \times 10^{-5}$ M, the assumption that the uncomplexed o-Phen concentration is approximately the same at the start and end of the equilibration process is valid. On the other hand, given that the approximation procedure does not account for the fact that o-Phen is a weak base, the assumption that the hydrogen ion concentration is controlled exclusively by the acetate buffer may not be valid.

For example, o-Phen, being a weak base, will react with water to produce hydroxide ion, i.e.,

$$L + H_2O \rightleftharpoons HL^+ + OH^-$$

This could make the solution more basic. For example, reduction of the starting acetic acid and acetate concentrations from 1.0 M to 0.10 M would result in concentration errors much larger than the usual target of 5% for approximate calculations.

This is the reason for the relatively large starting acetic acid and acetate ion concentrations. The larger acetic acid and acetate ion concentrations are used to ensure that the hydrogen ion concentration and pH are controlled by the buffer and are not influenced to a significant extent by the basic properties of o-phenanthroline.

BASIS FOR AN ITERATIVE PROGRAM

Except for situations involving ions such as Ag(I), Hg(I), and Pb(II) that form chloride precipitates, we will assume that reactants are added as sodium and chloride salts. Nitrate salts can be used for ions that form chloride precipitates.

The program to be described here is on Sheet 3 of the complex ions program. Given that all cell numbers in the remainder of this chapter refer to Sheet 3 of the program this would be a good time to open the program and have it available as we discuss the basis for it.

The iterative program is based on the following facts:

a. Equilibrium concentrations of all reactants and products except the sodium ion and chloride ion depend on the hydrogen ion concentration.
b. A charge-balance equation expressed in terms of diluted reactant concentrations, equilibrium constants, and the hydrogen ion concentration can be solved for equilibrium concentrations of all reactants and products except Na^+ and Cl^-.
c. Sodium ion and chloride ion concentrations are proportional to diluted concentrations of salts from which they are derived.

Let's talk next about the charge-balance equation for Part A of Example 9.1 and then talk about ways to extend that equation to other situations.

Example 9.1 is repeated here for your convenience.

Example 9.1 (repeated) Calculate equilibrium FeL_3^{2+} concentrations in solutions prepared by adding 10.0 mL each, of solutions containing (A) 1.0×10^{-4} M, (B) 2.5×10^{-4}, and (C) 5.0×10^{-4} M ferrous chloride to each of three 100 mL volumetric flasks followed by 25.0 mL of 0.014 M o-Phen and 10.0 mL each of a solution containing 1.0 M each of acetic acid and sodium acetate and diluting each solution to 100 mL with water. For acetic acid: $K_{a,B} = 1.8 \times 10^{-5}$; for protonated o-Phen HL^+: $K_{a,L} = 1.38 \times 10^{-5}$; for FeL_j^{2+}: log $K_{f1} = 5.86$, log $K_{f2} = 5.25$, and log $K_{f3} = 10.03$.

TABLE 9.2

Reactions Used to Obtain the Charge-Balance Equation for Example 9.1 with One Each of the Different Ions Is in Bold Type. (Conjugate Bases of Acids Corresponding the Buffer and Ligand Are Represented by Symbols, B^- and L, Respectively.)

Reactants	Reactions
Water	$H_2O \overset{K_w}{\rightleftharpoons} \mathbf{H^+} + \mathbf{OH^-}$
Sodium salt, NaB, of the acetate ion, B^-	$NaB \rightarrow \mathbf{Na^+} + \mathbf{B^-}$
Equilibration reaction of acetic acid, HB, and acetate, B^-	$HB \overset{K_{a,B}}{\rightleftharpoons} H^+ + B^-$
Ferrous chloride	$FeCl_2 \rightarrow \mathbf{Fe^{2+}} + 2\mathbf{Cl^-}$
Equilibration reaction of protonated o-Phen and o-Phen	$\mathbf{HL^+} \overset{K_{a,HL}}{\rightleftharpoons} H^+ + L$
Stepwise Fe(II)/o-Phen reactions	$\mathbf{Fe^{2+}} + L \overset{K_{f1}}{\rightleftharpoons} \mathbf{FeL^{2+}} + L \overset{K_{f2}}{\rightleftharpoons} \mathbf{FeL_2^{2+}} + L \overset{K_{f3}}{\rightleftharpoons} \mathbf{FeL_3^{2+}}$

The first step in writing the charge-balance equation for any situation is to identify all ions associated with the situation. The best way to do this is to write the reactions associated with the problem and to identify *one each* of the different ions.

REACTIONS ASSOCIATED WITH EXAMPLE 9.1

Reactions and symbols associated with reactants and products for Example 9.1 are summarized in Table 9.2 with *one each of the different ions in bold type*.

Ions in bold type are the ions used to write the ionic form of the charge-balance equation.

IONIC FORM OF THE CHARGE-BALANCE EQUATION

As reminders, (a) the contribution of each ion to the charge-balance equation is the signed charge on the ion times the equilibrium concentration of the ion and (b) the charge-balance equation is obtained herein by setting the algebraic sum of contributions of all ions to zero.

Equation 9.14a is the *ionic form* of the charge-balance equation for reactions in Table 9.2 written as the algebraic sum of products of signed charges on the ions times equilibrium concentrations. The equation is:

$$CBE = \left[H^+\right] - \left[OH^-\right] + \left[Na^+\right] - \left[B^-\right] + \left[HL^+\right] - \left[Cl^-\right]$$
$$+ z_{Fe}\left(\left[Fe^{2+}\right] + \left[FeL^{2+}\right] + \left[FeL_2^{2+}\right] + \left[FeL_3^{2+}\right]\right) \tag{9.14a}$$

in which z_{Fe} is the charge on Fe^{2+} and the complexes, HL^+ is the protonated form of o-Phen and other symbols have their usual meanings. All complexes have charges of 2+ because ferrous iron has a charge of 2+, i.e., $z_{Fe} = 2$, and the ligand, o-Phen, is uncharged. I will talk later about situations involving singly charged ligands such as cyanide, CN^-. For now I will trust you to confirm that terms for all charged reactants and products are included and that each term is the charge on the ion times its equilibrium concentration.

TABLE 9.3

Summary of Equations and Constants Used to Calculate Ionic Concentrations of Reactants and Products in the Charge-Balance Equation

Equation	Explanation
$\left[H^+\right]$	H^+ concentration that causes the CBE to change from $-$ to $+$.
$\left[OH^-\right] = \dfrac{K_w}{\left[H^+\right]}$	K_w is the autoprotolysis constant for water.
$\left[Na^+\right] = C_{NaB}$	C_{NaB}: Diluted sodium acetate concentration.
$\left[B^-\right] = \alpha_{1,B} C_{T,B} = \dfrac{K_{a,HB}}{\left[H^+\right] + K_{a,HB}} C_{T,B}$	$\alpha_{1,B}$: Fraction of the buffer in the fully dissociated form; $K_{a,HB}$: Deprotonation constant for the buffer acid; $C_{T,B}$: Total buffer concentration, i.e., $C_{T,B} = C_{HB} + C_B$.
$\left[HL^+\right] = \alpha_{0,L} C_{L,U} = \dfrac{\left[H^+\right]}{\left[H^+\right] + K_{a,HL}} C_{L,U}$	$\alpha_{0,L}$: Fraction of the ligand acid in the fully protonated form; $K_{a,HL}$: Deprotonation constant for the ligand acid, e.g., protonated o-Phen, $C_{L,U}$: Uncomplexed ligand concentration.
$\left[Cl^-\right] = z_{Fe} C_{Fe} = 2 C_{Fe}{}^a$	z_{Fe}: Charge on Fe(II)
$\left[Fe^{2+}\right] = \gamma_0 C_{Fe}$	γ_0: Fraction of Fe(II) in the form, Fe^{2+}
$\left[FeL^{2+}\right] = \gamma_1 C_{Fe}$	γ_1: Fraction of Fe(II) in the form, FeL^{2+}
$\left[FeL_2^{2+}\right] = \gamma_2 C_{Fe}$	γ_2: Fraction of Fe(II) in the form, $FeL_2{}^{2+}$
$\left[FeL_3^{2+}\right] = \gamma_3 C_{Fe}$	γ_3: Fraction of Fe(II) in the form, $FeL_3{}^{2+}$.

^{a} C_{Fe} is the diluted $FeCl_2$ concentration in this and the following four equations.

EXPRESSING IONIC CONCENTRATIONS IN TERMS OF EXPERIMENTAL QUANTITIES

Equations used to express the ionic concentrations in terms of experimental quantities are summarized in Table 9.3.

Most of the terms are discussed above and are not discussed further here. However, equations for the Na^+ and Cl^- concentrations merit some discussion.

Assuming that the Na^+ and Cl^- salts dissociate completely and that Na^+ and Cl^- ions aren't involved in subsequent reactions, it follows that their equilibrium concentrations are proportional to the diluted concentrations of compounds from which they are derived.

CHARGE-BALANCE EQUATION IN TERMS OF EXPERIMENTAL QUANTITIES AND CONSTANTS

Relationships in Table 9.3 are used to write Equation 9.14b in terms of the hydrogen ion concentration, concentration fractions, diluted concentrations of reactants, and the uncomplexed o-Phen concentration, $C_{L,U}$.

$$CBE = \left[H^+\right] - \overbrace{\frac{K_w}{\left[H^+\right]}}^{\left[OH^-\right]} + \overbrace{C_{NaB}}^{\left[Na^+\right]} - \overbrace{\alpha_{1,B} C_{T,B}}^{\overbrace{\left[B^-\right]}^{Buffer\ term}} + \overbrace{\alpha_{0,HL} C_{L,U}}^{\overbrace{\left[HL^+\right]}^{Ligand\ term}} - \overbrace{z_{Fe} C_{Fe}}^{\overbrace{}^{\left[Cl^-\right]from\ FeCl_2}}$$

$$+ z_{Fe} (\overbrace{\gamma_0 C_{Fe}}^{\left[Fe^{2+}\right]} + \overbrace{\gamma_1 C_{Fe}}^{\left[FeL^{2+}\right]} + \overbrace{\gamma_2 C_{Fe}}^{\left[FeL_2^{2+}\right]} + \overbrace{\gamma_3 C_{Fe}}^{\left[FeL_3^{2+}\right]})$$

$$\text{Complex ion concentrations}$$

(9.14b)

The equation is annotated to identify ions represented by the different terms included in Table 9.3.

If we were to limit our attention to this relatively simple situation, the next step would be to express quantities in Equation 9.14b in terms of cells in a spreadsheet program. However, it will save time and effort to talk first about ways we can extend Equation 9.14b to other types of situations and then associate terms with cells in a spreadsheet program.

EXTENSIONS OF THE CHARGE-BALANCE EQUATION FOR THE Fe(II)/o-Phen REACTIONS TO OTHER SITUATIONS

The charge-balance equation described above, like any other equation developed for a specific situation, is limited to a narrow range of conditions. Let's use the equation as a starting point to identify changes needed to extend the equation to a wider range of situations. More specifically, let's talk about ways to extend the equation to situations involving (a) buffers prepared using uncharged or singly charged monoprotic weak acids, (b) ligands consisting of conjugate bases of uncharged or singly charged monoprotic acids, (c) complexes with charges resulting from uncharged **or** singly charged ligands, and (d) complexes with two to six ligands per metal ion.

Let's begin with an equation used to adapt the charge-balance equation to buffers prepared using uncharged or charged monoprotic acids with charges, $z_{HB} = 0$ or $z_{HB} = 1$, in Cell D42.

BUFFER TERM

The buffer term in Equation 9.14b applies only to the acid-base pair of an uncharged buffer acid, e.g., acetic acid. Dissociation reactions for sodium acetate, NaAc, and ammonium chloride, NH_4Cl, are used below to illustrate the basis for an equation to differentiate between contributions of buffers prepared using monoprotic uncharged and charged acids. Differences among the ions are illustrated in bold type.

Uncharged acid, HAc, D42 = 0 Charged acid, NH_4^+, D42 = 1

$NaAc \rightarrow \mathbf{Na^+ + Ac^-}$ $NH_4Cl \rightarrow \mathbf{Cl^- + NH_4^+}$

We will assume that (a) the salts dissociate completely, (b) Na^+ and Cl^- do not react further, and (c) acetate ion and ammonium ion concentrations can be calculated as fractions in the deprotonated form for acetate and in the protonated form for ammonium ion times total concentrations of both forms of each acid. Accounting for charges on the ions, an ionic form of the buffer term is expressed in Equation 9.15a:

$$\text{Ionic form of buffer term} = \text{IF}\left(D42 = 0, \left[Na^+\right] - \left[Ac^-\right]\right), \text{IF}\left(D42 = 1, -\left[Cl^-\right] + \left[NH_4^+\right]\right) \quad (9.15a)$$

Equating sodium-ion and chloride-ion concentrations from buffer components to concentrations of the salts, C_{NaAc} and C_{NH_4Cl}, and equating acetate-ion and ammonium-ion concentrations to concentration fractions times the total concentration of both forms of each conjugate acid/base pair, an equation representing contributions to the charge-balance equation of ions from buffers prepared using a monoprotic uncharged or charged acid is

$$\text{Buffer term} = \text{IF}\left(D42 = 0, C_{NaAc} - \alpha_{1,Ac}C_{HAc,T}\right), \text{IF}\left(D42 = 1, -C_{NH_4Cl} + \alpha_{0,NH_4}C_{NH_4,T}\right) \quad (9.15b)$$

in which D42 is the charge on the acid, C_{NaAc} and C_{NH_4Cl} are diluted concentrations of the salts, $\alpha_{1,HAc}$ and α_{0,NH_4} are fractions of acetic acid and ammonia in the deprotonated and protonated forms, respectively, and $C_{HAc,T}$ and $C_{NH_4,T}$ are total concentrations of both forms of each acid.

Equation 9.15b can be generalized to any monoprotic acid-base pair, HB/B^- or HB^+/B, used as the buffer by replacing Ac with B and NH_4^+ with HB^+.

LIGAND TERM

Whereas an analogous procedure could be used to develop an equation for a ligand term, it is easier to write the equation by analogy with Equation 9.15b by substituting cyanide ion for the acetate ion. The only conceptual difference is that the uncomplexed ligand concentration, $C_{L,U}$, is used in place of the total concentrations of the two forms of each acid. The equation is

$$\text{Ligand term} = IF\left(D43 = 0, C_{NaCN} - \alpha_{1,HCN}C_{CN,U}, IF(D43 = 1, -C_{NH4Cl} + \alpha_{0,NH4}C_{NH4,U}\right) \quad (9.15c)$$

in which D43 is the charge on the ligand acid, $C_{CN,U}$ and $C_{NH4,U}$ are uncomplexed concentrations of both forms of each acid and other terms are as described above.

Equation 9.15c can be generalized to any monoprotic acid-base pair, HL/L^- or HL^+/L, used as the ligand by replacing CN^- with L^- and NH_4^+ with HL^+.

Buffer terms and ligand terms are in Columns BK and BL of the program on Sheet 3 of the complex ions program.

CHARGES ON COMPLEXES

As described above, charges on Fe(II) and the Fe(II)/o-Phen complexes were all +2 because charges on Fe(II) and o-Phen are +2 and 0, respectively. Charges on complexes involving negatively charged ligands such as cyanide will depend on the charge on the metal ion and numbers of ligands per metal ion.

Given a complex of the form, ML_j^z, with j ligands per metal ion, the charge on the complex can be calculated as

$$z_j = z_M + jz_L \quad (9.15d)$$

in which z_j is the charge on the complex, z_M is the positive charge on the metal ion, j is the number of ligands per metal ion and z_L is the signed charge on the ligand, usually 0 or -1 for monoprotic ligands. For example, the charge on the complex, CdL_4, of Cd(II) with four cyanide ions is $z_j = z_4 = z_M + j(z_L) = 2 + 4(-1) = -2$. It follows that the complex can be written as CdL_4^{2-}.

EXERCISE 9.5

Show that charges on forms of Cd(II) complexed with $j = 0$–4 cyanide ions are Cd^{2+}, CdL^{1+}, CdL_2^{0}, CdL_3^{1-}, and CdL_4^{2-}.

Whereas it is easy to calculate the charges on the ions by hand it is tedious to do so. Equation 9.15d is used in Cells C55–I55 on Sheet 3 of the complex ions program to calculate charges on metal ions and complexes for uncharged ligands such as ammonia or negatively charged ligands such as the cyanide ion.

COMPLEX ION TERMS

Contributions of complex ions to the charge-balance equation are sums of products of charges on the ions times equilibrium concentrations of the complexes. For complexes with up to six ligands per metal ion, the complex ion term is

$$\overset{C55}{z_0}[M] + \overset{D55}{z_1}[ML] + \overset{E55}{z_2}[ML_2] + \overset{F55}{z_3}[ML_3] + \overset{F55}{z_4}[ML_4] + \overset{H55}{z_5}[ML_5] + \overset{I55}{z_6}[ML_6] \quad (9.15e)$$

in which z_0–z_6 are charges, z_j, on ions with j = 0–6 ligands per metal ion and C55–I55 are cells in which charges are calculated.

Equation 9.15e can be written in the following compact forms:

$$Sum\left[(C\$55:I\$55)*([M]:[ML_6])\right]=Sum\left[(C\$55:I\$55)*(\gamma_0:\gamma_6)\,C_M\right] \qquad (9.15f)$$

Symbols are as described earlier. Calculated fractions, γ_0–γ_6, are in columns AT–AZ, C_M is in Column BC and concentrations, [M]–[ML_6], are in Columns BD–BJ of the program on Sheet 3 of the complex ions program.

CHLORIDE ION FROM THE METAL CHLORIDE

The chloride ion concentration from the metal ion will be the charge on the metal ion times the metal ion concentration, as expressed in Equation 9.15g:

$$\left[Cl^-\right]_{MCl_z}=z_M C_M \qquad (9.15g)$$

in which z_M is the charge on the metal ion and C_M is the diluted concentration of the metal ion.

COMPLEXES WITH FEWER THAN SIX LIGANDS PER METAL ION

The program adapts Equation 9.15f to situations involving fewer than six ligands per metal ion by using the coordination number, C_{Num}, in If/Then statements to set to zero all terms beyond that for the coordination number entered into Cell E42 by the user. For example, for complexes such as the Zn(II) complexes of ammonia with up to four ammonias per Zn(II), multipliers H55 and I55 corresponding to charges, $z_j = 5$ and $z_j = 6$, would be set to zero by the program, thereby limiting calculations to those for ions with 0–4 ligands per metal ion.

EXPANDED CHARGE-BALANCE EQUATION

The charge-balance equation obtained by combining all the terms above is expressed in Equation 9.16a:

$$CBE=\left[H^+\right]-\overbrace{\frac{K_w}{\left[H^+\right]}}^{\left[OH^-\right]}+Eq.\ 9.15b+Eq.\ 9.15c-\overbrace{z_M*C_M}^{\left[Cl^-\right]}+Sum\overbrace{\left[z_j*(\gamma_0:\gamma_6)C_M\right]}^{z_j*[ML_j]} \qquad (9.16a)$$

in which z_M and C_M are the charge and concentration of the metal ion, z_j is the charge on the j'th form of the complex and γ_0–γ_6 are fractions of the metal ion complexed with 0–6 ligands.

The complete charge-balance equation is written below in terms of cells and columns on Sheet 3 in which different quantities are entered or calculated.

$$CBE=AG30-\overbrace{\frac{H42}{AG30}}^{[OH^-]}+\overbrace{BK}^{Eq.\ 9.15b}+\overbrace{BL}^{Eq.\ 9.15c}-\overbrace{F\$42}^{z_M}*\overbrace{BC}^{C_M}+Sum\ [(C\$55:I\$55)*\overbrace{(AT:AZ)BC]}^{(\gamma_0:\gamma_6)C_M} \qquad (9.16b)$$

As examples, whereas F\$42 and C\$55:I\$55 refer to specific cells, BK, BL, BC, and AT–AZ refer to columns in which various quantities in the annotated equation are calculated.

A SPREADSHEET PROGRAM

Let's begin with the rationale for an iterative program to solve the charge-balance equation for equilibrium concentrations of reactants and products.

Rationale for an Iterative Program to Solve the Charge-Balance Equation

As discussed earlier, sodium ion and chloride ion concentrations are calculated using diluted concentrations of the salts that produce them. Equilibrium concentrations of all other reactants and products can be expressed in terms of the hydrogen ion concentration. Therefore, the charge-balance equation can be solved by finding the hydrogen ion concentration that satisfies the equation and using it to calculate equilibrium concentrations of other reactants and products.

An Unexpected Problem and a Way to Solve It

My first attempt to implement a spreadsheet program to solve Equation 9.16b was to use the built-in iterative feature of Excel in the same way that it was used to solve equations for acid-base equilibria. Whereas the resulting program worked well for small ratios of metal-ion and ligand concentrations it failed for larger ratios of these concentrations. The problem was traced to the way the program calculated the uncomplexed ligand concentration.

To solve that problem, I then developed a multistep program that combines the iterative feature of Excel with a successive approximations approach. The successive approximations program not only gives the same results as the simpler program for small ratios of metal ion to ligand concentrations but also extends the useful range to ratios of these concentrations at least 10-fold larger than the largest ratios for which the simpler program gave results. I will focus here on the successive approximations program.

Basis for the Successive Approximations Program

Iterative calculations for the successive approximations program are in Cells AE23–BN47 on Sheet 3 of the complex ions program. My purpose here is to help you understand how the successive approximations program solves charge-balance equations for complex ion equilibria. Believing that the discussion of the basis for the successive approximations program will be more meaningful if you have observed the iterative part of the program in operation, I will guide you through a process used to observe a key part of the iterative calculations as the program solves Example 9.3.

> **Example 9.3 Given a solution containing 0.020 M Ni(II) and 0.025 M ammonia buffered near $[H^+] \cong 5 \times 10^{-8}$ M with 0.075 M protonated TRIS and 0.0125 M unprotonated TRIS, calculate equilibrium concentrations of Ni^{2+}, the complex ions, NiL^{2+}–NiL_6^{2+}, the hydrogen ion, the hydroxide ion, and uncomplexed ammonia. For protonated TRIS: $K_{a,B} = 8.4 \times 10^{-9}$; for the ammonium ion: $K_{a,L} = 5.7 \times 10^{-10}$; for the Ni(II)/ammonia complexes: log $K_{f1} = 2.72$, log $K_{f2} = 2.17$, log $K_{f3} = 1.66$, log $K_{f4} = 1.12$, log $K_{f5} = 0.67$, log $K_{f6} = -0.03$.**

Solving the Example

If you haven't already done so, open the complex ions program, select Sheet 3, copy the template in A94–I109 and paste it into A29–I44 using the Paste (P) option. Then, with C29 set to 0, set the Ni(II) concentration in B39 to 0.020, set C29 to 1, scroll down until Cells A46–I46 are on the screen, left click anywhere on the screen and wait until E46 changes from Solving to Solved and the iteration counter in H46 stops changing. If the iteration counter in H46 should stop changing before E46 changes to Solved for this or any other problem, press F9 one or more times until E46 changes to solved and the iteration counter stops changing at the same time.

To observe part of the program in operation, set C29 to 0 and left click anywhere on the sheet. Then set C29 to 1 and, before left clicking anywhere on the screen, scroll to the right and up or down until Cells AF25–AH47 are on the screen. Then left click anywhere on the screen and watch changes in Columns AF–AH as the program progresses through the several steps. Cells in Column AG should converge to a hydrogen ion concentration of $[H^+] = 1.56 \times 10^{-8}$ M in steps 10–20. By scrolling to the right to Columns AT–AZ and BD–BJ, you can see that the program will have converged on fixed values of the fractions, γ_0–γ_6, and concentrations of Ni^{2+}–NiL_6^{2+}. If you would like to observe any part of the process in more detail, you can rerun the program one or more times by setting C29 to 0 and 1 and scrolling to the area you want to observe before left clicking on the screen.

Iterative Calculations

As stored on the programs website http://crcpress.com/9781138367227, the program does the iterative calculations in 20 steps in Cells AE23–BN47 on Sheet 3 of the complex ions program. Whereas the program converges on equilibrium conditions in ten of fewer steps for most situations, I chose to err on the side of reliability over shorter convergence times by using a larger number of steps.

Each step in the program uses an iterative process similar to that used in the acid-base programs to resolve the hydrogen ion concentration to the number of significant figures in I29, usually three. For example, with F29 set to 0.5 and I29 set to 3, the algorithm in AG27 starts with a hydrogen ion concentration equal to 50% of the H^+ concentration of the buffer and increases it by amounts equal to $10^{-I29} = 10^{-3}$ times the same H^+ concentration until the charge-balance equation in AH27 changes from – to +.

The hydrogen ion concentration that causes the – to + change in AH27 is accepted as a first estimate of the equilibrium concentration for conditions in that row and is used with equations discussed earlier to calculate equilibrium concentrations of all other reactants and products except the Na^+ and Cl^- concentrations which are calculated as diluted concentrations of sodium and chloride salts of negatively charged and positively charged ions used to prepare solutions.

Stepwise Approach to the Successive Approximations Program

Calculations in each step of the program are done with a fixed value of the uncomplexed ligand concentration, $C_{L,U}$, in Column AQ of each row. Calculations in Row 27 are done with the uncomplexed ligand concentration in AQ27 set equal to the diluted ligand concentration, i.e., $C_{L,U} = C_L$. When C29 is set to 1, the iterative program calculates concentrations of other reactants and products corresponding to $C_{L,U} = C_L$. Complex ion concentrations calculated in BE–BJ for this set of conditions are used with an Equation in BM27 to calculate an updated estimate of the uncomplexed ligand concentration.

Calculations in Row 28 are done with the uncomplexed ligand concentration in AQ28 set to the concentration calculated in BM27 in the first step. Concentrations of complex ions calculated for this new set of conditions are used to calculate another estimate of the uncomplexed ligand concentration in BM28. Estimates of the uncomplexed ligand concentration in BM27 and BM28 are then used to calculate an average value of $C_{L,U}$ in BN28.

Calculations in Row 29 and all remaining rows are as described for Rows 27 and 28 except that calculations in all rows beyond Row 28 are done using average values of the uncomplexed ligand concentration calculated in the two preceding steps. The use of average values of uncomplexed ligand concentrations speeds the convergence process by decreasing differences between the uncomplexed concentrations in consecutive steps.

Whereas other details of the program are interesting to some, myself included, a more detailed discussion of the program is beyond the scope of this chapter.

CALCULATED QUANTITIES

Concentration fractions and concentrations for Example 9.3 calculated using the iterative program and the approximation procedure discussed earlier are summarized in Cells J39–Q53 of the program on Sheet 3 of the complex ions program. As you can see, there are relatively large errors associated with the approximate results for this example. We will talk later about reasons for the errors. For now, let's talk about some applications of the iterative program.

SELECTED APPLICATIONS OF THE ITERATIVE PROGRAM

We will talk here about two applications of the successive approximations program. One application is to determine how effective the Fe(II)/o-Phen reaction is at converting Fe(II) to a colored form that can be used for a colorimetric determination of Fe(II) concentrations. The other is to predict whether or not a metal hydroxide will precipitate from a solution containing ammonia at relatively high pH.

COLORIMETRIC DETERMINATION OF FERROUS IRON

As discussed earlier, the reaction of Fe(II) with o-phenanthroline to form the red colored complex, FeL_3^{2+}, is often used for the colorimetric determination of Fe(II). As a reminder, reactions of Fe(II) and o-Phen are as follows:

$$Fe^{2+} + L \underset{}{\overset{K_{f1}}{\rightleftharpoons}} FeL^{2+} + L \underset{}{\overset{K_{f2}}{\rightleftharpoons}} FeL_2^{2+} + L \underset{}{\overset{K_{f3}}{\rightleftharpoons}} \overbrace{FeL_3^{2+}}^{Red}$$
$$\lambda_{Max} \cong 506 \text{ nm}$$

in which K_{f1}, K_{f2}, and K_{f3} are stepwise formation constants.

As shown in the reactions, the complex with three o-Phens per Fe(II), FeL_3^{2+}, is red with an absorption maximum near 506 nm and a molar absorptivity of a $\cong 1.1 \times 10^4$ L/mol-cm.[1] The absorbance of FeL_3^{2+} at a wavelength near 506 nm is used to quantify Fe(II) concentrations in a variety of sample types, including iron tablets and iron in blood. For the record, these reactions and calculations have more than academic interest to me. Because of a severe loss of blood following a medical procedure during the preparation of this text, it was necessary for me to have iron infusions to restore my hemoglobin.

In any event, for a successful determination the Fe(II) concentration, it is important that conditions be such that most of the Fe(II) is converted to the colored complex. Stated differently, it is important that the fraction, γ_3, be very close to 1.0 corresponding to virtually 100% conversion of Fe(II) to the colored form and that γ_0–γ_2 be very small.

Let's talk about how you can use the program to solve Example 9.1 which is repeated here for your convenience.

> **Example 9.1 Calculate equilibrium FeL_3^{2+} concentrations in solutions prepared by adding 10.0 mL each, of solutions containing (A) 1.0×10^{-4} M, (B) 2.5×10^{-4}, and (C) 5×10^{-4} M ferrous chloride to each of three 100-mL volumetric flasks followed by 25.0 mL of 0.014 M o-Phen and 10.0 mL of a solution containing 1.0 M each of acetic acid and sodium acetate and diluting each solution to 100 mL with water.**
>
> **For acetic acid: $K_{a,B} = 1.8 \times 10^{-5}$; for protonated o-Phen HL^+: $K_{a,L} = 1.38 \times 10^{-5}$; for FeL_j^{2+}: log $K_{f1} = 5.86$, log $K_{f2} = 5.25$, and log $K_{f3} = 10.03$.**

To solve Part A of the example, open the complex ions program, select Sheet 3, copy the template in Cells A73–I88 and paste it into Cells A29–I44 using the Paste (P) option. Then set C29 to 1 and wait until E46 changes from "Solving" to "Solved." If the iteration counter in H46 stops changing before E46 changes to "Solved," press F9 and wait until E46 changes to solved. Repeat this process if the iteration counter stops changing before E46 changes to "Solved."

Now scroll to N51 and N52 and note that the fractions, γ_3, calculated using the approximate and iterative options are both 1.00 corresponding to virtually 100% conversion of Fe(II) to the colored complex, FeL_3^{2+}. In other words, both the approximate and iterative options predict that virtually all of the Fe(II) is converted to the colored form as is necessary for a successful determination.

To solve Parts B and C of the example, set C29 to zero, set B39 to the desired Fe(II) concentration, 2.5E-4 or 5.0E-4, reset C29 to 1 and follow the procedure described for Part A.

As noted earlier, conditions in this example were selected such that the approximation procedure would give valid results. The iterative procedure gives valid results for conditions for which the approximation procedure fails.

WILL A METAL HYDROXIDE PRECIPITATE FROM SOLUTION?

For reasons to be discussed in the next chapter, it is advantageous to do EDTA titrations of some metal ions at pH's near 10. A problem is that several metal ions tend to precipitate from solution as metal hydroxides at high pH. It is common practice to use ammonia to prevent precipitation of metal hydroxide by forming metal-ion/ammonia complexes. Example 9.4 expresses the problem in the form of a question that we can answer with the aid of the complex ions program.

Example 9.4 Will Ni(OH)₂ precipitate from a solution containing 0.0010 M Ni(II) and 0.025 M ammonia buffered with 0.075 M protonated TRIS, Tris(hydroxymethyl) amino-methane, and 0.0125 M unprotonated TRIS?

For protonated TRIS: $K_{a,B} = 8.4 \times 10^{-9}$; for ammonium ion: $K_{a,L} = 5.7 \times 10^{-10}$; for Ni(OH)₂, $K_{sp} = 2.0 \times 10^{-15}$; for the Ni(II)/ammonia complexes: log $K_{f1} = 2.72$, log $K_{f2} = 2.17$, log $K_{f3} = 1.66$, log $K_{f4} = 1.12$, log $K_{f5} = 0.67$, log $K_{f6} = -0.03$.

Rationale for Using Ammonia to Prevent Precipitation of Metal Hydroxides

The solubility reaction for Ni(OH)₂ and the solubility-product constant are as follows:

$$Ni(OH)_2 \rightleftharpoons Ni^{2+} + 2OH^- \quad K_{sp} = \left[Ni^{2+} \right]\left[OH^- \right]^2 = 2.0 \times 10^{-15}$$

Nickel hydroxide will precipitate if the product, $[Ni^{2+}][OH^-]^2$, exceeds the solubility-product constant, i.e., if $[Ni^{2+}][OH^-]^2 > K_{sp}$.

Representing ammonia by the symbol, L, the complex formation reactions for Ni(II) and ammonia are as follows:

$$Ni^{2+} + L \overset{K_{f1}}{\rightleftharpoons} NiL^{2+} + L \overset{K_{f2}}{\rightleftharpoons} NiL_2^{2+} + L \overset{K_{f3}}{\rightleftharpoons} NiL_3^{2+} + L \overset{K_{f4}}{\rightleftharpoons} NiL_4^{2+} + L \overset{K_{f5}}{\rightleftharpoons} NiL_5^{2+} + L \overset{K_{f6}}{\rightleftharpoons} NiL_6^{2+}$$

Given that some of the Ni(II) is converted to the ammonia complexes, the Ni^{2+} concentration will be less than it would be in the absence of ammonia. The question is, will the Ni^{2+} concentration be small enough to prevent precipitation. We can use the complex ions program to help us answer this question by calculating equilibrium Ni^{2+} and OH^- concentrations.

Adapting the Successive Approximations Program to Example 9.4

To use the successive approximations program to help solve Example 9.4, open it into an unused Excel file and select Sheet 3. Then copy the template in Cells A94–I109 and paste it into Cells A39–I44 using the Paste (P) option. Then set the Ni(II) concentration in B39 from 0.020 M to 1.0E-3 and set C29 to 1, left click somewhere on the screen and wait for Cell E46 to change from "Solving" to "Solved." Press F9 one or more times if the iteration counter in H46 stops changing before E46 changes to "Solved." After E46 changes to "Solved," scroll to Cells K40 and R46 and read the Ni^{2+}

concentration from K40, $[Ni^{2+}] = 5.97E\text{-}4\,M$, and the OH^- concentration from R46, $[OH^-] = 8.23E\text{-}7\,M$. Then calculate the ion product, $[Ni^{2+}][OH^-]^2$, and compare it to the solubility product constant for nickel hydroxide as shown below.

$$\left[Ni^{2+}\right]\left[OH^-\right]^2 = \left(5.97 \times 10^{-4}\,M\right)\left(8.23 \times 10^{-7}\,M\right)^2 = 4.0 \times 10^{-16} < K_{sp} = 2.0 \times 10^{-15}$$

Given that the concentration product is less than the solubility product we predict that nickel hydroxide will not precipitate from solution.

COMPARISONS OF CONCENTRATIONS BASED ON APPROXIMATE AND ITERATIVE OPTIONS

As we've worked through this chapter, we've noted significant differences between results calculated using the approximate and iterative options for some situations. Let's use results for Example 9.4 to try to understand reasons for the differences.

The first seven columns of Table 9.4 include results for the seven forms of Ni(II) calculated using the approximate and iterative options for Example 9.4.

Notice first that percentage differences between concentrations are the same as differences between concentration fractions.

There are two primary sources of the differences between the two options. On the one hand, the approximate option is based on assumptions that (a) the ligand concentrations at the start and end of the equilibration process are about the same and (b) the hydrogen ion concentration is controlled exclusively by the buffer. On the other hand, the iterative procedure accounts for changes in these quantities that occur during the equilibration process.

Regarding the assumption that the ligand concentrations at the beginning and end of the equilibration process are the same, notice that the uncomplexed ligand concentrations calculated using the approximate and iterative options in N45 and N46 differ by less than 2%, i.e., 0.0250 M vs. 0.0245 M. Given that the uncomplexed ligand concentration for the approximate option is assumed to be the same as the diluted concentration, i.e., $C_{L,U} \cong C_L$, it follows that the first assumption is satisfied. This conclusion is supported by the fact that the diluted ligand concentration, $C_L = 0.0250$ M, is about 25 times larger than the diluted Ni(II) concentration, $C_{Ni} = 1.0 \times 10^{-3}$ M.

Regarding the assumption that the hydrogen ion concentration is controlled exclusively by the buffer, results show that the hydrogen ion concentration calculated using the iterative option, $[H^+]_{Iter} = 1.2 \times 10^{-8}\,M$, is about 24% of that of the buffer, i.e., $[H^+]_{Iter}/[H^+]_{Buf} = 1.2 \times 10^{-8}$ M/5.04×10^{-8} M = 0.24.

TABLE 9.4
Concentrations Calculated Using Approximation and Iterative Options for Example 9.4

	Selected Concentrations and Percentage Differences									
	$[Ni^{2+}]$	$[NiL^{2+}]$	$[NiL_2^{2+}]$	$[NiL_3^{2+}]$	$[NiL_4^{2+}]$	$[NiL_5^{2+}]$	$[NiL_6^{2+}]$	$[H^+]$	$C_{L,U}$	$[L]$
Approx.	8.7E-4	1.3E-4	5.3E-6	6.7E-8	2.5E-10	3.3E-13	8.5E-17	5.04E-8	0.0250	2.8E-4
Iterative	6.0E-4	3.4E-4	5.6E-5	2.8E-6	4.1E-8	2.1E-10	2.2E-13	1.21E-8	0.0245	1.1E-3
Diff., %	45	−63	−91	−98	−99	−100	−100	310	1.9	−75
	Selected Fractions and Percentage Differences									
Fractions	γ_0	γ_1	γ_2	γ_3	γ_4	γ_5	γ_6			
Approx.	8.7E-01	1.3E-01	5.3E-03	6.7E-05	2.5E-07	3.2E-10	8.5E-14			
Iterative	6.0E-01	3.4E-01	5.6E-02	2.8E-03	4.1E-05	2.1E-07	2.2E-10			
Diff., %	45	−63	−91	−98	−99	−100	−100			

Is there a logical reason why the hydrogen ion concentration calculated using the iterative option is different than that of the buffer? Yes there is.

Remember, the ligand in this example is ammonia. As you know, ammonia is a weak base. For example, as shown below, ammonia will react with hydrogen ions to produce the ammonium ion, i.e.,

$$NH_3 + H^+ \rightleftharpoons NH_4^+$$

It is assumed in the approximate procedure that the ammonia concentration remains constant during the reaction. Therefore, changes in the ammonia concentration will result in errors in approximate results. Whereas the iterative program accounts for such changes, the approximate option doesn't.

The acid-base program discussed in an earlier chapter was used as an independent test of the H^+ concentration calculated using the iterative option. Using values, $K_{a,B} = 8.4E-9$, $K_{a,L} = 5.7E-10$, $C_{HB} = 0.075$, $C_B = 1.25E-2$, and $C_L = 2.45E-2$ in the acid-base equilibria program gave $[H^+] = 1.21E-8$ which is the same as that calculated using the complex ions program. This result doesn't prove the accuracy of the complex ions program but it does support it.

So, assuming that the H^+ concentration calculated using the iterative program is correct, how does this effect calculated concentrations of different forms of Ni(II). It effects the ammonia concentration which in turn effects calculated fractions of Ni(II) in different forms which effect concentrations of Ni(II) in different forms. As shown in the last column of Table 9.4, the ammonia concentration calculated using the iterative option, $[L] = 1.1 \times 10^{-3}$ M, is about four times that assumed for the approximate calculations. This results from the fact that the hydrogen ion concentration calculated using the iterative option is significantly less than that calculated using the approximate option. As discussed earlier, fractions of Ni(II) in different forms complexed with j ammonias are proportional to ammonia concentration to the j'th power, i.e., $\gamma_j \propto [L]^j$.

The iterative option accounts for changes that are assumed to remain constant in the approximate option.

SUMMARY

This chapter describes concepts, equations, and procedures associated with the use of an iterative program based on a charge-balance equation to solve problems involving complex formation reactions that approach equilibrium in two or more steps. The three-step reaction of ferrous iron with o-phenanthroline, o-Phen, to form complexes with one to three o-Phens per Fe(II) is used as a starting point to show how selected equations and reactions can be used to develop a charge-balance equation for situations involving two-step and three-step reactions. Equations include relationships between diluted ligand concentrations and uncomplexed ligand concentrations, relationships between stepwise formation constants and cumulative formation constants, relationships between cumulative constants and equilibrium concentrations of reactants and products, and equations for fractions of metal ions complexed with two to six ligands per metal ions.

Having developed the background equations, it is shown how reactions associated with the Fe(II)/o-Phen complexes can be used to develop a charge-balance equation in terms of equilibrium concentrations of reactants and products. That equation is then used to develop a charge-balance equation in terms of the hydrogen ion concentration, concentration fractions, and diluted concentrations of reactants and products. Then it is shown how the charge-balance equation for this simpler situation can be extended to more complicated situations involving buffers prepared using uncharged or charged monoprotic weak acids, ligands consisting of conjugate bases of uncharged or charged monoprotic weak acids, and complexes with up to six ligands per metal ion.

It is then shown how a multistep, iteratively assisted, successive approximations program can be used to solve the charge-balance equation for larger ratios of metal-ion to ligand concentrations than is possible with a simpler program based exclusively on the iterative feature of Excel. It is then shown how the successive approximations program can be adapted for two practical problems. One problem involves calculations of concentrations of the colored complex formed between Fe(II) and three o-Phens for different starting conditions. The other problem involves calculations needed to show that complexes of Ni(II) with ammonia can be used to prevent precipitation of nickel hydroxide at a relatively high pH.

Finally, reasons for errors associated with approximation procedures are discussed.

REFERENCE

1. L. G. Hargis, *Analytical Chemistry: Principles and Techniques*, Prentice Hall: Englewood Cliffs, NJ, 1988, pp. 416–417.

10 Equilibrium Calculations for Metal-Ion/EDTA Reactions

This chapter is concerned primarily with equilibrium calculations for chelate-forming reactions with emphasis on metal-ion/EDTA reactions. More specifically, it focuses on an iterative approach to calculations that eliminates the need for simplifying assumptions commonly used for conventional approximations. The iterative procedure is based on the fact that concentrations of most reactants and products except alkali metal ions and halogen or nitrate ions added as salts of weak acids and weak bases depend on the hydrogen ion concentration. This makes it possible to express equilibrium concentrations of reactive components in terms of a charge-balance equation in terms of the hydrogen ion concentration.

Calculations for each set of reactants are done by using an iterative procedure to solve the charge-balance equation for the hydrogen ion concentration that satisfies other equilibria in each solution. The hydrogen ion concentration calculated in this way is used to calculate fractions of different reactants in different forms that are then used by the program to calculate equilibrium concentrations of reactants and products. For reasons to be discussed later, concentrations of metal-ion/EDTA complexes are not included in the charge-balance equation for results described herein.

The iterative procedure is based on the fact that the charge-balance equation is less than zero for hydrogen ion concentrations less than the equilibrium concentration and larger than zero for hydrogen ion concentrations larger than the equilibrium concentration. As with analogous programs described in acid-base chapters, the iterative procedure is implemented by starting with a hydrogen ion concentration less than the equilibrium concentration and increasing it by small amounts until the sign of the charge-balance equation changes from − to +. The hydrogen ion concentration that causes the − to + change for each situation is accepted as the equilibrium concentration and is used to calculate concentrations of other reactants and products.

The program is applied to calculations for metal ions such as Zn(II) and Ni(II) that form ammonia complexes as well as ions such as Ca(II) and Mg(II) that don't form ammonia complexes. To save time and space, we will treat calculations for the latter group of ions as special cases of calculations for ions that form ammonia complexes. We will compare results calculated using both the iterative and conventional options to illustrate types of errors introduced by the simplifying assumptions associated with approximations procedures.

On the one hand, approximation errors for pM values, i.e, the negative logarithm of the metal ion concentration, are usually less than about 10% meaning that there are relatively small differences between titration curves prepared using iterative and approximate options. On the other hand, concentration errors are quite large for many situations. Finally, whereas iterative calculations can be adapted to other applications of chelate-forming reactions, we will use calculations associated with titrations of metal ions with EDTA as illustrative examples.

Spreadsheet programs discussed in this chapter are available on the programs website http://crcpress.com/9781138367227. Programs to be discussed are on Sheets 4, 5, and 9 of the complex ions program. We will talk first about some applications and talk later about the basis for the programs.

OVERVIEW OF EDTA CHELATES

Ethylenediaminetetraacetic acid, EDTA, is usually added as the disodium salt, $Na_2H_2Y{\cdot}2H_2O$, of the form of EDTA that has lost four protons. Representing EDTA by Y as is common practice, the dissociation reaction for the disodium salt is as follows:

$$Na_2H_2Y \cdot 2H_2O \rightarrow 2Na^{2+} + H_2Y^{2-} + 2H_2O$$

The single left-to-right arrow signifies that the salt dissociates completely. The ionic form of EDTA produced in this reaction, H_2Y^{2-}, participates in the stepwise equilibration reactions shown below.

$$H_6Y^{2+} \underset{\pm H^+}{\overset{K_{a1}}{\rightleftharpoons}} H_5Y^+ \underset{\pm H^+}{\overset{K_{a2}}{\rightleftharpoons}} H_4Y \underset{\pm H^+}{\overset{K_{a3}}{\rightleftharpoons}} H_3Y^- \underset{\pm H^+}{\overset{K_{a4}}{\rightleftharpoons}} H_2Y^{2-} \underset{-H^+}{\overset{K_{a5}}{\rightleftharpoons}} HY^{3-} \underset{-H^+}{\overset{K_{a6}}{\rightleftharpoons}} Y^{4-}$$

It is the fully deprotonated form of EDTA, Y^{4-}, that reacts with metal ions. For example, the reaction of Zn(II) and EDTA is:

$$Zn^{2+} + Y^{4-} \rightleftharpoons ZnY^{2-}$$

The formation of the desired form of EDTA, Y^{4-}, is favored by small hydrogen ion concentrations or high pH.

SOME PRACTICAL PROBLEMS INVOLVING COMPETING REACTIONS

Based on the foregoing discussion, it is desirable to implement each reaction at a pH high enough to ensure that a significant fraction of EDTA is in the fully deprotonated form. However, the use of EDTA at high pH is complicated by the fact that most metal ions of interest form hydroxide precipitates at high pH, e.g.,

$$Zn^{2+} + 2OH^- \rightleftharpoons Zn(OH)_2$$

If permitted to form, hydroxide precipitates of metal ions would interfere with the metal-ion/EDTA reactions by removing the metal ions from solution. Therefore, for each metal ion of interest, it is necessary to select a hydrogen ion concentration that is low enough that a significant fraction of the EDTA is in the fully deprotonated form but high enough that a hydroxide precipitate doesn't form.

As discussed in the preceding chapter, ammonia forms complexes with several metal ions in which we will be interested. For example, Zn(II) reacts with ammonia in four steps to form complexes with one to four ammonias per Zn(II). Using the symbol, L, for ammonia as is common practice, reactions of Zn(II) with ammonia can be written as follows:

$$Zn^{2+} + L \overset{K_{f1}}{\rightleftharpoons} ZnL^{2+} + L \overset{K_{f2}}{\rightleftharpoons} ZnL_2^{2+} + L \overset{K_{f3}}{\rightleftharpoons} ZnL_3^{2+} + L \overset{K_{f4}}{\rightleftharpoons} ZnL_4^{2+}$$

in which L is ammonia and K_{f1}–K_{f4} are stepwise formation constants.

As also discussed in the preceding chapter, chemists have learned that ammonia can be used as an *auxiliary ligand* to help prevent precipitation of metal hydroxides by forming complexes that reduce concentrations of the free ions such Zn^{2+} below values that exceed solubility-product constants. My experience with the programs to be discussed is that they will function satisfactorily if the ratio of the ammonia and metal-ion concentrations in C38 and B39 in the programs on Sheets 4 and 5 is at least 4:1; i.e., $C_L^0/C_M^0 \geq 4$, but may fail if the ratio is significantly smaller than this.

Obviously, the inclusion of ammonia doesn't help for metal ions that don't form ammonia complexes. The only option for ions such as Ca(II) and Mg(II) that don't form complexes with ammonia is to limit metal-ion and hydroxide concentrations to values that don't exceed solubility-product constants. This is made easier by the fact that hydroxides of these latter ions tend to be more soluble than hydroxides of metal ions that do form ammonia complexes.

A consequence of these facts for equilibrium calculations is that *metal ions are commonly divided into groups that do and don't form ammonia complexes.* We could treat equilibria for these two groups of metal ions independently as is usually done. However, there is a simpler option. For reasons described below, we can save time and effort by focusing first on ions such as Zn(II) and Ni(II) that form ammonia complexes and then adapting what we have learned for these ions to calculations for ions that don't form ammonia complexes.

As a reminder from the preceding chapter, fractions of metal ions in forms with up to six ligands per metal ion are represented herein by symbols γ_0–γ_6. For ions such as Zn(II) and Ni(II) that form ammonia complexes, fractions, γ_0–γ_6, of the ions in different forms are all less than or equal to zero, i.e., γ_1–$\gamma_6 \leq 0$. For ions such as Ca(II) and Mg(II) that don't form ammonia complexes, the fraction, γ_0, of the metal ion in the uncomplexed form is equal to 1 and all other fractions are equal to zero, i.e., $\gamma_0 = 1$ and γ_1–$\gamma_6 = 0$. As we will see later, these differences make it very easy to adapt concepts, equations, and procedures developed for ions that form ammonia complexes to ions that don't form ammonia complexes.

AN AID TO PLOTTING TITRATION CURVES

Titration curves for metal-ion/EDTA titrations are plots of negative logarithms of metal ion concentrations, $-\log[M^{z+}]$, vs. the EDTA volume, V_Y. By analogy with the way pH is defined, negative logarithms of a metal ion concentration are represented by the symbol pM. As examples, $pCd = -\log[Cd^{2+}]$ and $pCa = -\log[Ca^{2+}]$ in which $[Cd^{2+}]$ and $[Ca^{2+}]$ are equilibrium concentrations of the metal ions.

We will talk later about effects of variables such as metal-ion concentrations, EDTA concentrations, ammonia concentrations, etc. on shapes of titration curves. To be able to visualize effects of the different values of each variable on shapes of titration curves it will be helpful to be able to plot titration curves for different values of each variable on one figure.

Plotting up to Six Titration Curves on One Figure

One part of the iterative program on Sheet 5 of the complex ions program is designed to plot up to six titration curves on each figure. I will talk about the overall program later. For now, I will guide you through a process used to plot titration curves for six solutions containing different Cd(II) concentrations titrated with a fixed EDTA concentration on one figure. Having used Example 10.1 to show you how to plot six titration curves on one figure, I will then describe a minor change to the process for figures with fewer than six plots.

Example 10.1 Given solutions buffered at pH ≅ 10 using 0.036 M ammonium chloride and 0.20 M ammonia, plot titration curves for titrations of 25.0-mL volumes of solutions containing (A) 3.0×10^{-3}, (B) 4.0×10^{-3}, (C) 5.0×10^{-3}, (D) 6.0×10^{-3}, (E) 7.0×10^{-3}, and (F) 8.0×10^{-3} M Cd(II) with 5.0×10^{-3} M EDTA.

A template for this example is in Cells A80–I98 on Sheet 5 of the complex ions program. To use the template, begin by opening the complex ions program and selecting Sheet 5. Then copy the template in Cells A80–I98 on Sheet 5 and paste it into Cells A26–I44 on that sheet using the Paste (P) option. Then set F37 to 5.0E-3 for the EDTA concentration and I26 to 1 to tell the program where to store the EDTA volume and pCd values for Plot 1. Then follow the sequence in Table 10.1 starting with Plot 1 to plot titration curves for the six Cd(II) concentrations.

TABLE 10.1

Sequence of Cell Settings to Plot Titration Curves for Six Cd(II) Concentrations Titrated with 5.0×10^{-3} M EDTA on One Plot

	Plot 1	Plot 2	Plot 3	Plot 4	Plot 5	Plot 6
C29	0	0	0	0	0	0
B39	3.0E-3	4.0E-3	5.0E-3	6.0E-3	7.0E-3	8.0E-3
C29	1	1	1	1	1	1
Wait for D46 to change from "Solving" to "Solved" and then change I26 to values below[a]						
I26	2	3	4	5	6	6.5
Number Sequence for Four Plots on One Figure						
I26	2	3	4	4.5		

[a] Press F9 one or more times if the iteration counter in G46 stops changing before D46 changes to "Solved."

After completing the sequence in each column, (a) scroll to the figure in Cells U4–AA25 to see the plot for the last concentration for which calculations are complete, (b) scroll to Cells AC2–AN25 to view volume and calculated pCd values for which sequences in Table 10.1 are complete, and (c) scroll to Cells AP5–AV25 to view plots for all concentrations for which sequences in Table 10.1 are complete.

Sequences for Fewer than Six Plots on One Figure

To plot one to five titration curves on a single figure, the number in I26 after pM vs. volume results for the last group have been plotted should be 1.5, 2.5, … 5.5. This prevents the program from plotting duplicates of results for the last plot on the figure. As an example, the number sequence in I16 for four plots on one figure is included in the last row of Table 10.1. Notice that the last number is 4.5 rather than 5.

Saving Numerical Results and Figures for Later Use

Results in the storage cells in AC2–AN25 and the figure in Cells AP5–AV25 are temporary in the sense that they change any time the sequence in I26 is changed. To save any set of numerical results more permanently, copy the results in AC2–AN25 and paste them into a convenient location using the "123" paste option. It will usually be necessary to reformat the data set. To save a figure more permanently after it is formatted to your satisfaction, copy it and paste into a convenient location using the Picture (U) option. Some features of figures saved using the Picture (U) option can be changed by right-clicking on the figure and following the menu options. I have saved several sets of numerical results and corresponding figures on Sheet 8 of the complex ions program.

Return to Results for Example 10.1

In any event, the titration curves for Example 10.1 are included in Figure 10.1A. All plots exhibit the expected sigmoid shapes being relatively flat before and after equivalence points with rapid increases in pM values vs. EDTA volumes near equivalence points. Rapid increases in pCd values just before to just after the equivalence point result from the fact that virtually all of the Cd(II) is being converted to the CdY^{2-} chelate and Cd^{2+} concentrations are being reduced to very small values.

Equivalence-point volumes are used in the usual way to calculate Cd(II) concentrations. For example, using the equivalence-point volume of 20.0 mL for Plot 2, we can calculate

$$C_{Cd}{}^0 = \frac{V_{Y,Eq}}{V_{Cd}} C_Y{}^0 = \frac{20.0 \text{ mL}}{25.0 \text{ mL}} 5.0 \times 10^{-3} = 4.0 \times 10^{-3} \text{ M}$$

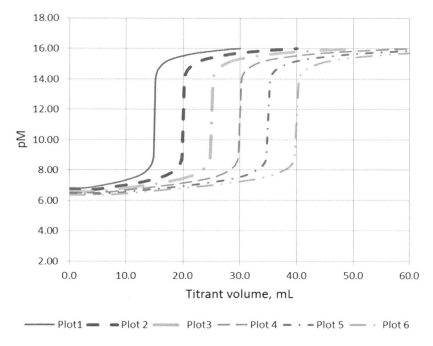

FIGURE 10.1A Titration curves for cadmium(II) titrated with EDTA in solutions buffered near pH = 10 with 0.036 M ammonium chloride and 0.20 M ammonia.

Plots 1–6: 25.0 mL of 3.0×10^{-3}, 4.0×10^{-3}, 5.0×10^{-3}, 6.0×10^{-3}, 7.0×10^{-3}, and 8.0×10^{-3} M Cd(II) titrated with 5.0×10^{-3} M EDTA.

Symbols C_{Cd}^0 and C_Y^0 are initial concentrations of Cd(II) and EDTA, and $V_{Y,Eq}$ and V_{Cd} are the equivalence-point volume of EDTA and the volume of Cd(II).

An Exercise

Figure 10.1A illustrates effects on titration curves of different Cd(II) concentrations titrated the same EDTA concentration. We can get additional information on effects of Cd(II) concentrations on shapes of titration curves by plotting titration curves for a wider range of Cd(II) concentrations with EDTA concentrations equal to each Cd(II) concentration. Exercise 10.1 is designed to do this.

EXERCISE 10.1

Given solutions buffered at pH ≅ 10 using 0.036 M ammonium chloride and 0.20 M ammonia, plot titration curves for titrations of 25.0-mL volumes of solutions containing (A) 1.0×10^{-4}, (B) 1.0×10^{-3}, (C) 0.010, (D) 0.020, (E) 0.030, and (F) 0.050 M Cd(II) with solutions containing EDTA concentrations equal to each Cd(II) concentration.

The template used for Example 10.1 on Sheet 5 and the sequence in Table 10.1 can be used for this exercise. The only changes are to set Cells F37 and B39 to the EDTA and Cd(II) concentrations for each part of the exercise.

Titration curves you should obtain are plotted in Figure 10.1B.

Plots 1–6 represent increasing EDTA and Cd(II) concentrations. Equivalence-point volumes are the same for all plots because each Cd(II) concentration is titrated with an equal EDTA concentration. These plots illustrate more clearly than Figure 10.1A how pCd values increase with increasing Cd(II) concentrations. The larger the change in pM values, the more easily end-point volumes can be detected reliably.

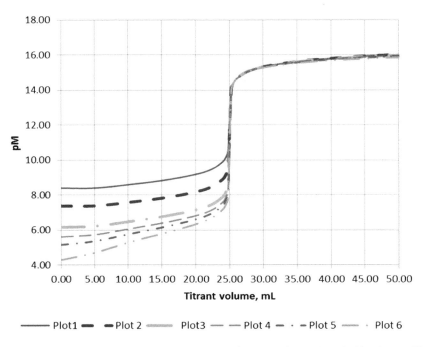

FIGURE 10.1B   Titration curves for cadmium(II) titrated with EDTA in solutions buffered near pH = 10 with 0.036 M ammonium chloride and 0.20 M ammonia.

BACKGROUND INFORMATION

The purpose of this part is to describe some background information associated with calculations involving metal-ion/EDTA equilibria. Example 10.2 is used as the basis for our discussion. We will solve the example later. What we learn from this example can be extended to reactions of other metal ions with EDTA.

> **Example 10.2 Given solutions buffered near pH = 10.0 with 0.036 M ammonium chloride and 0.20 M ammonia, describe background information needed to calculate equilibrium ZnY^{2-}, Zn^{2+}, and Y^{4-} concentrations and pZn after adding (A) 15.0 mL, (B) 25.0 mL, and (C) 35.0 mL of 1.00×10^{-3} M EDTA to 25.0 mL of a solution containing 1.00×10^{-3} M Zn(II). Constants needed to solve this example are provided as needed in subsequent discussions.**

REACTIONS AND FORMATION CONSTANTS

The reaction of primary interest is the reaction of Zn^{2+} with the fully deprotonated form of EDTA to form the chelate. The reaction and formation constant are as shown in Equation 10.1:

$$Zn^{2+} + Y^{4-} \rightleftharpoons ZnY^{2-} \qquad K_{f,Y} = \frac{\left[ZnY^{2-}\right]}{\left[Zn^{2+}\right]\left[Y^{4-}\right]} \tag{10.1}$$

in which $K_{f,Y}$ is the formation constant for the formation of the Zn(II)/EDTA chelate, ZnY^{2-}, and $[Zn^{2+}]$ and $[Y^{4-}]$ are equilibrium concentrations of Zn^{2+} and fully deprotonated EDTA, respectively.

There are differences between ways equilibrium constants for acid-base reactions and chelate-formation reactions are tabulated. Whereas deprotonation constants are tabulated for acid-base reactions, logarithms of formation constants are tabulated for chelate-formation reactions. For example, the logarithm of the formation constant, $K_{f,Y}$, for the Zn(II)/EDTA reaction is log $K_{f,Y} = 16.5$. This corresponds to a formation constant, $K_{f,Y} = 10^{16.5} = 3.2 \times 10^{16}$.

UNCHELATED Zn(II) AND EDTA CONCENTRATIONS

Given that the parts of metal-ion and EDTA concentrations that remain after the chelate-formation reaction has reached equilibrium are used in a variety of calculations, it is desirable to have terms and symbols for these quantities. The parts of the metal-ion and EDTA concentrations that are not chelated with each other are called unchelated metal-ion and EDTA concentrations herein and are represented herein by symbols, $C_{M,U}$ and $C_{Y,U}$, respectively, the subscripted U signifying that these are unchelated concentrations.

We will talk later about more detailed computational forms of equations for unchelated concentrations. For now it will suffice to describe the unchelated concentrations as differences between the diluted concentrations and the chelate concentration. As examples, unchelated Zn(II) and EDTA concentrations are represented in Equations 10.2a and 10.2b:

$$C_{Zn,U} = C_{Zn} - \left[ZnY^{2-} \right] \qquad (10.2a) \qquad\qquad C_{Y,U} = C_Y - \left[ZnY^{2-} \right] \qquad (10.2b)$$

in which $C_{Zn,U}$ and C_{Zn} are the unchelated and diluted Zn(II) concentrations, respectively, $C_{Y,U}$ and C_Y are the unchelated and diluted EDTA concentrations, respectively, and $[ZnY^{2-}]$ is the equilibrium concentration of the Zn(II)/EDTA chelate.

EQUILIBRIUM Zn^{2+} AND Y^{4-} CONCENTRATIONS

However unchelated Zn(II) and EDTA concentrations are calculated, equilibrium Zn^{2+} and Y^{4-} concentrations are products of the fractions in each form times the unchelated concentrations, as expressed in Equations 10.3a and 10.3b:

$$\left[Zn^{2+} \right] = \gamma_0 C_{Zn,U} \qquad (10.3a) \qquad\qquad \left[Y^{4-} \right] = \alpha_{6,Y} C_{Y,U} \qquad (10.3b)$$

in which $C_{Zn,U}$ and $C_{Y,U}$ are unchelated Zn(II) and EDTA concentrations and γ_0 and $\alpha_{6,Y}$ are fractions of the unchelated concentrations present as Zn^{2+} and Y^{4-}, respectively (see Exercise 10.2).

EXERCISE 10.2
Given a solution containing 1.00×10^{-3} M unchelated Zn(II) for conditions such that $\gamma_0 = 7.7 \times 10^{-7}$, show that the equilibrium Zn^{2+} concentration is $[Zn^{2+}] = 7.7 \times 10^{-10}$ M.

Returning to Example 10.2, we will frame the discussion in the context of a titration of Zn(II) with EDTA. However, concepts, equations, and procedures discussed in connection with this example can be adapted to other applications of such reactions.

EQUIVALENCE-POINT VOLUME

The volume of EDTA at the equivalence point is the Zn(II) volume times the initial Zn(II) concentration, $C_{Zn}{}^0 = 1.0 \times 10^{-3}$ M, divided by initial EDTA concentration, $C_Y{}^0 = 1.0 \times 10^{-3}$ M, i.e.,

$$V_{Eq} = \frac{V_{Zn} C_{Zn}^0}{C_Y^0} = \frac{(25.0 \text{ mL})(1.0 \times 10^{-3} \text{ M})}{(1.0 \times 10^{-3} \text{ M})} = 25.0 \text{ mL}$$

in which V_{Eq} is the equivalence-point volume of EDTA, V_{Zn} is the volume of Zn(II), and $C_{Zn}{}^0$ and $C_Y{}^0$ are initial Zn(II) and EDTA concentrations. It follows that Parts A–C of Example 10.2 involving EDTA volumes of 15.0 mL, 25.0 mL, and 35.0 mL represent situations before, at, and after the equivalence point.

Excess and Limiting Reactants

Before the equivalence point, the diluted Zn(II) concentration will be larger than the EDTA concentration, i.e., $C_{Zn} > C_Y$. As such, the Zn(II) is called the excess reactant and the EDTA is called the limiting reactant because the EDTA concentration will limit the chelate concentration that can be produced. Given the 1 to 1 stoichiometry of the Zn(II)/EDTA reaction, diluted Zn(II) and EDTA concentrations will be the same at the equivalence point and therefore both are limiting reactants. It follows that EDTA and Zn(II) are the excess and limiting reactants, respectively, after the equivalence point.

Salts of EDTA and Zn(II)

Let's assume that Zn(II) is added as the chloride salt, $ZnCl_2$, and, as is common practice, that EDTA is added as the disodium salt, $Na_2H_2Y \cdot 2H_2O$. Let's also assume that both salts dissociate completely, i.e.,

$$ZnCl_2 \rightarrow Zn^{2+} + 2Cl^- \qquad Na_2H_2Y \cdot 2H_2O \rightarrow 2Na^{2+} + H_2Y^{2-} + 2H_2O$$

The single left-to-right arrows signify that the salts dissociate completely.

Diluted Zn(II) and EDTA Concentrations

Diluted concentrations of reactants are needed for most calculations. Diluted concentrations are products of reactant volumes times starting concentrations divided by sums of volumes of Zn(II) and EDTA, as expressed in Equations 10.4a and 10.4b:

$$C_{Zn} = \frac{V_{Zn}C_{Zn}^0}{V_{Zn} + V_Y} \qquad (10.4a) \qquad\qquad C_Y = \frac{V_Y C_Y^0}{V_{Zn} + V_Y} \qquad (10.4b)$$

in which C_{Zn} and C_Y are diluted Zn(II) and EDTA concentrations, $C_{Zn}{}^0$ and $C_Y{}^0$ are initial Zn(II) and EDTA concentrations, and V_{Zn} and V_Y are volumes of Zn(II) and EDTA (see Exercise 10.3).

EXERCISE 10.3

Show that diluted Zn(II) and EDTA concentrations for Part A of Example 10.2 are $C_{Zn} = 6.25 \times 10^{-4}$ M and $C_Y = 3.75 \times 10^{-4}$ M. (Part A involves addition of 15.0 mL of 1.0×10^{-3} M EDTA to 25.0 mL of a solution containing 1.0×10^{-3} M Zn(II).)

Acid-Base Reactions of EDTA

Fully protonated EDTA is a doubly charged hexaprotic acid, H_6Y^{2+}. It follows that the H_2Y^{2-} produced by dissociation of the sodium salt, $Na_2H_2Y \cdot H_2O$, will be involved in equilibration reactions involving H^+ and the other six forms of EDTA, as expressed in Equation 10.5:

$$H_6Y^{2+} \underset{\pm H^+}{\overset{K_{a1}}{\rightleftharpoons}} H_5Y^+ \underset{\pm H^+}{\overset{K_{a2}}{\rightleftharpoons}} H_4Y \underset{\pm H^+}{\overset{K_{a3}}{\rightleftharpoons}} H_3Y^- \underset{\pm H^+}{\overset{K_{a4}}{\rightleftharpoons}} \mathbf{H_2Y^{2-}} \underset{-H^+}{\overset{K_{a5}}{\rightleftharpoons}} HY^{3-} \underset{-H^+}{\overset{K_{a6}}{\rightleftharpoons}} Y^{4-} \qquad (10.5)$$

H_2Y^{2-} is in bold type to emphasize the fact that it is the form added to the solution.

TABLE 10.2

Deprotonation Constants for Fully Protonated EDTA

K_{a1}	K_{a2}	K_{a3}	K_{a4}	K_{a5}	K_{a6}
1.0	3.2×10^{-2}	1.0×10^{-2}	2.2×10^{-3}	6.9×10^{-7}	5.8×10^{-11}

Stepwise deprotonation constants for the different forms of EDTA are summarized in Table 10.2. Fractions of EDTA in forms that have lost $i = 0-6$ protons are represented herein by the symbols, $\alpha_{0Y}-\alpha_{6Y}$, in which the subscripted numbers 0–6 represent numbers of protons lost and subscripted Y's are used to differentiate the EDTA fractions from fractions of other acids.

The denominator term for fractions of EDTA in different forms is as shown in Equation 10.6a:

$$D_{6Y} = \left[H^+\right]^6 + \kappa_1\left[H^+\right]^5 + \kappa_2\left[H^+\right]^4 + \kappa_3\left[H^+\right]^3 + \kappa_4\left[H^+\right]^2 + \kappa_5\left[H^+\right] + \kappa_6 \qquad (10.6a)$$

in which $[H^+]$ is the hydrogen ion concentration and $\kappa_1-\kappa_6$ are cumulative deprotonation constants for EDTA. Cumulative deprotonation constants are products of successive stepwise constants. Successive fractions, $\alpha_{0Y}-\alpha_{6Y}$, are successive denominator terms divided by the denominator. For example, the fraction, α_{6Y}, of EDTA in the fully deprotonated form is expressed in Equation 10.6b:

$$\alpha_{6,Y} = \frac{\kappa_6}{D_Y} \qquad (10.6b)$$

Example 10.3 illustrates a calculation of the fraction, $\alpha_{6,Y}$.

Example 10.3 Calculate the fraction, $\alpha_{6,Y}$, of EDTA in the fully deprotonated form at pH = 10.0. For EDTA: $K_{a1} = 1.0$, $K_{a2} = 3.2 \times 10^{-2}$, $K_{a3} = 1.0 \times 10^{-2}$, $K_{a4} = 2.2 \times 10^{-3}$, $K_{a5} = 6.9 \times 10^{-7}$, $K_{a6} = 5.8 \times 10^{-11}$.

Step 1: Calculate the hydrogen ion concentration.

$$\left[H^+\right] = 10^{-pH} = 1.0 \times 10^{-10}$$

Step 2: Calculate successive cumulative formation constants for EDTA as products of successive constants. (I'll trust you to confirm the numbers below.)

$$\kappa_1 = K_{a1} = 1.0, \kappa_2 = 3.2 \times 10^{-2}, \kappa_3 = 3.2 \times 10^{-4}, \kappa_4 = 7.0 \times 10^{-7}, \kappa_5 = 4.9 \times 10^{-13}, \kappa_6 = 2.82 \times 10^{-23}$$

Step 3: Calculate the denominator term for all fractions.

$$D_{6,Y} = \left[H^+\right]^6 + \kappa_1\left[H^+\right]^5 + \kappa_2\left[H^+\right]^4 + \kappa_3\left[H^+\right]^3 + \kappa_4\left[H^+\right]^2 + \kappa_5\left[H^+\right] + \kappa_6 = 7.68 \times 10^{-23}$$

Step 4: Calculate the fraction in the fully dissociated form.

$$\alpha_{6,Y} = \frac{\kappa_6}{D_{6,Y}} = \frac{2.82 \times 10^{-23}}{7.68 \times 10^{-23}} = 0.367$$

In other words, 36.7% of the EDTA is in the fully deprotonated form, Y^{4-}, at pH = 10.0.

The fraction, $\alpha_{6,Y} = 0.367$, is used later to develop a relationship for quantities called conditional constants that account for effects of hydrogen ion concentration on metal-ion/EDTA reactions. For now, let's talk about one of several templates that can save you time and effort for such calculations.

Time-Saving Templates

Whereas the calculations in Example 10.3 and other analogous examples are straightforward, they are tedious and time consuming. To help you devote less time punching numbers into a calculator and devote more time and effort to more general concepts, I have included a program for these and other analogous calculations on Sheet 2 of the complex ions program. I have also included templates for selected problems to help you familiarize yourself with the program.

A template for Example 10.3 is included in Cells B56–E67 of the program on Sheet 2. To use the template, copy it and paste it into Cells B19–E30 using the Paste (P) option. The fraction, $\alpha_{6Y} = 0.367$, is calculated in Cell G30.

On the one hand, this and other templates don't free you of the responsibility to understand the concepts involved. On the other hand, if you use them wisely they can give you more time to devote to understanding the concepts associated with the calculations.

Effects of Hydrogen Ion Concentrations on the Fraction, α_{6Y}, of EDTA in the Fully Deprotonated Form

For most applications of metal-ion/EDTA reactions, it is desirable that most of the limiting reactant(s) be converted to the chelate. A modified and rearranged form of the formation constant equation for the reaction can give some insight into the relationship between the fraction, α_{6Y}, of EDTA in the fully deprotonated form and the chelate concentration. Using equations described above, the formation constant for the Zn(II)/EDTA chelate can be written and rearranged into the forms shown in Equation 10.7:

$$K_{f,Y} = \frac{\left[ZnY^{2-}\right]}{\left[Zn^{2+}\right]\left[Y^{4-}\right]} \Rightarrow \left[ZnY^{2-}\right] = K_{f,Y}\left[Zn^{2+}\right]\left[Y^{4-}\right] \overset{\text{Eq. 10.3b}}{=} \alpha_{6,Y}K_{f,Y}\left[Zn^{2+}\right]C_{Y,U} \qquad (10.7)$$

in which $\alpha_{6,Y}$ is the fraction of EDTA in the fully deprotonated form, $C_{Y,U}$ is the unchelated EDTA concentration, and other symbols have their usual meanings.

The rearranged equation tells us that the chelate concentration, $[ZnY^{2-}]$, is proportional to the fraction, $\alpha_{6,Y}$, of EDTA in the fully deprotonated form. Given that the fraction of EDTA in the fully deprotonated form depends on the hydrogen ion concentration, degrees of completion of reactions of EDTA with metal ions also depend on the hydrogen ion concentration.

The program on Sheet 2 of the complex ions program was used to calculate percentages, $100\alpha_{Y,6}$, of EDTA in the fully deprotonated form by using the template in Cells B56–E67 on Sheet 2, and varying the pH in Cell C20. Results are plotted in Figure 10.2.

As you can see, virtually all of the EDTA is in the fully deprotonated form for very small hydrogen ion concentrations, high pH's, and decreases to very small percentages for larger hydrogen ion concentrations, low pH's. If you look closely at the point for $[H^+] = 1.0 \times 10^{-10}$ M, you will see that the percentage is about 37% as calculated in Example 10.3.

More generally, given that it is the fully deprotonated form of EDTA that reacts with metal ions, low hydrogen ion concentrations or high pH's favor formation of metal ion/EDTA chelates.

Effects of Hydrogen Ion Concentration on Titration Curves for Ca(II)

Given that fully deprotonated EDTA, Y^{4-}, is the form of EDTA that reacts with metal ions and that the Y^{4-} concentration is influenced by the hydrogen ion concentration, it is reasonable to assume that hydrogen ion concentration will influence shapes of titration curves for metal ions. Figure 10.3 illustrates effects of hydrogen ion concentration on titrations of 50 mL of 0.050 M Ca(II) with 0.050 M EDTA.

As you can see, changes in pCa at the equivalence point increase with decreasing hydrogen ion concentrations or increasing pH. The primary reason for increases in pCa with decreasing hydrogen

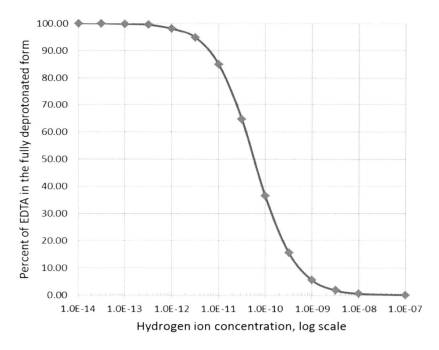

FIGURE 10.2 Effects of hydrogen ion concentrations on the percent of EDTA in the fully deprotonated form, Y^{4-}.

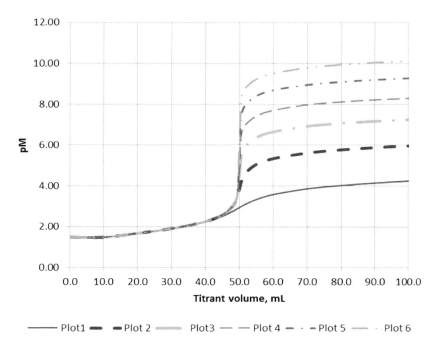

FIGURE 10.3 Effects of hydrogen ion concentrations on titration curves for 50 mL of 0.050 M Ca(II) titrated with 0.050 M EDTA.

Plots 1–6. pH = 5, 6, 7, 8, 9, and 10.

ion concentrations is that concentrations of the fully deprotonated form of EDTA increase with decreasing hydrogen ion concentrations.

A template for these calculations is in Cells A106–I124 on Sheet 5 of the complex ions program. Hydrogen ion concentrations are varied from 1×10^{-5} to 1×10^{-10} by setting $K_{a0,L}$ in B33 from 1E-5 to 1E-10 in 10-fold steps. Instructions for using the template are as discussed earlier. For the record, concentrations of other forms of the simulated acids with constants, $K_{a0,L}$, are such that hydrogen ion concentrations are equal to values of the deprotonation constant, i.e., $[H^+] = K_{a0,L}$ in B33.

Given the foregoing discussion, you may wonder why we wouldn't always use a hydrogen ion concentration for which virtually all of the EDTA is in the fully deprotonated form. Continue reading.

COMPETING REACTIONS, FORMATION OF METAL HYDROXIDES

The reason we can't use metal-ion/EDTA reactions at sufficiently low hydrogen ion concentrations, high pH's, that all of the EDTA is in the Y^{4-} form is that metal ions of interest form hydroxide precipitates at very low hydrogen ion concentrations (or high pH's). As shown below, fully deprotonated EDTA competes with hydroxide ion for Zn(II).

$$Zn^{2+} + Y^{4-} \rightleftharpoons ZnY^{2-} \quad Zn^{2+} + 2OH^- \rightleftharpoons Zn(OH)_2 \quad K_{sp} = \left[Zn^{2+} \right]\left[OH^- \right]^2$$

Given that the hydroxide precipitate would interfere with the EDTA reaction by removing Zn(II) from solution, we want to help EDTA win the competition for Zn(II). Given that concentrations of both deprotonated EDTA and hydroxide ion depend on the hydrogen ion concentration, we may be able to use the hydrogen ion concentration to help EDTA win the competition by preventing precipitation of zinc hydroxide.

To understand how to help EDTA win the competition for Zn(II), it would be helpful to have a way to calculate Zn(II) concentrations above which a precipitate will form.

The part of the Zn(II) concentration that will be available to react with hydroxide ion is the part not chelated with EDTA, i.e., the unchelated Zn(II) concentration, $C_{Zn,U}$. In solutions without a complexing ligand, all of the unchelated Zn(II) concentration is in the Zn^{2+} form, i.e., $C_{Zn,U} = [Zn^{2+}]$.

To prevent precipitation of $Zn(OH)_2$ it is necessary that the Zn^{2+} concentration satisfy the inequality, $[Zn^{2+}] \leq K_{sp}/[OH^-]^2$. After substituting $[OH^-] = K_w/[H^+]$ into this inequality, it can be written as shown in Equation 10.8:

$$\left[Zn^{2+} \right] = C_{Zn,U} \leq \frac{K_{sp}}{\left[OH^- \right]^2} \leq K_{sp}\left(\frac{\left[H^+ \right]}{K_w} \right)^2 \tag{10.8}$$

As an example, given a solubility-product constant, $K_{sp} = 1.2 \times 10^{-17}$, for zinc hydroxide and $K_w = 1.0 \times 10^{-14}$, it follows that the maximum unchelated Zn(II) concentration in solution with $[H^+] = 1.0 \times 10^{-10}$ without forming zinc hydroxide is

$$\left[Zn^{2+} \right] \leq K_{sp}\left(\frac{\left[H^+ \right]}{K_w} \right)^2 \leq 1.2 \times 10^{-17}\left(\frac{1.0 \times 10^{-10}}{1.0 \times 10^{-14}} \right)^2 \leq 1.2 \times 10^{-9}\ M$$

This Zn^{2+} concentration is much too small to be titrated with EDTA (see Exercise 10.4).

EXERCISE 10.4

A more commonly tabulated value of the solubility-product constant for zinc hydroxide is $K_{sp} = 3.0 \times 10^{-16}$. Show that the upper limit of the solubility of unchelated Zn(II) at pH = 10 is $[Zn^{2+}] \leq 3.0 \times 10^{-8}\ M$.

This latter concentration, 3.0×10^{-8}, although significantly larger than that calculated using the smaller K_{sp}, is still much too small to be titrated with EDTA. So, what do we do?

Fortunately, chemists have devised a way to use ammonia as an *auxiliary ligand* to increase *unchelated* concentrations of Zn(II) and other metal ions in solutions with low hydrogen ion concentrations, high pH's.

Use of Ammonia as an Auxiliary Ligand to Increase Solubilities of Metal Hydroxides

Ammonia or any other ligand used to increase the solubility of metal hydroxides is commonly called an *auxiliary ligand*. Let's talk about the basis for the use of ammonia to increase the solubility of zinc hydroxide.

How Ammonia Helps to Increase Zn(II) Concentrations for which Zinc Hydroxide Won't Precipitate from Solution

Zinc(II) reacts with ammonia in four steps to form complexes with up to four ammonias per zinc ion. Representing ammonia by L as is common practice, stepwise reactions for the formation of the Zn(II)/ammonia complexes are as follows:

$$Zn^{2+} + L \overset{K_{f1}}{\rightleftharpoons} ZnL^{2+} + L \overset{K_{f2}}{\rightleftharpoons} ZnL_2^{2+} + L \overset{K_{f3}}{\rightleftharpoons} ZnL_3^{2+} + L \overset{K_{f4}}{\rightleftharpoons} ZnL_4^{2+}$$

Given a fixed Zn(II) concentration, the formation of the Zn(II)/ammonia complexes will reduce the Zn^{2+} concentration. If the formation of the complexes reduces the Zn^{2+} concentration to or below that for which the inequality, $[Zn^{2+}] < K_{sp}/[OH^-]^2$, is satisfied, then zinc hydroxide will not precipitate from solution.

As shown earlier, the Zn^{2+} concentration is the fraction, γ_0, of unchelated Zn(II) concentration in the Zn^{2+} form times the unchelated Zn(II) concentration, i.e.,

$$\left[Zn^{2+} \right] = \gamma_0 C_{Zn,U}$$

As discussed in the preceding chapter, the fraction, γ_0, is as shown in Equation 10.9:

$$\gamma_0 = \frac{1}{1 + \beta_1 [L] + \beta_2 [L]^2 + \beta_3 [L]^3 + \beta_4 [L]^4} = \frac{1}{D_{\gamma 4}} \tag{10.9}$$

in which [L] is the equilibrium ammonia concentration and β_1–β_4 are cumulative formation constants for the formation of the Zn(II)/ammonia complexes and $D\gamma_4$ is a short-hand representation of the denominator term for complexes of a metal ion with a coordination number of four.

Example 10.4A illustrates an application of this equation.

Example 10.4A Calculate the fraction, γ_0, of Zn(II) in the free form, Zn^{2+}, in a solution buffered at pH = 10.0 with 0.036 M ammonium ion and 0.20 M ammonia. For ammonium ion: $K_{a,L} = 5.7 \times 10^{-10}$; for the Zn(II)/ammonia complexes: log $K_{f1} = 2.21$, log $K_{f2} = 2.29$, log $K_{f3} = 2.36$, log $K_{f4} = 2.03$; $[H^+] = 10^{-pH} = 1.0 \times 10^{-10}$ M

Step 1: Calculate the equilibrium ammonia concentration.

$$[L] = \alpha_{1L} C_{L,T} = \frac{K_a}{\left[H^+ \right] + K_a} \left(C_{0L} + C_{1L} \right) = 0.200$$

Step 2: Calculate the cumulative formation constants for the Zn(II)/ammonia complexes using the relationship, $10^{Sum(K_{f,j})}$.

$$\beta_1 = 1.62 \times 10^2, \ \beta_2 = 3.16 \times 10^4, \ \beta_3 = 7.24 \times 10^6, \ \beta_4 = 7.76 \times 10^8$$

Step 3: Calculate the denominator term for the fractions of Zn(II) complexed with 0–4 ammonias.

$$D_{\gamma,4} = 1 + \beta_1[L] + \beta_2[L]^2 + \beta_3[L]^3 + \beta_4[L]^4 = 1.32 \times 10^6$$

Step 4: Calculate the fraction, γ_0.

$$\gamma_0 = \frac{1}{1 + \beta_1[L] + \beta_2[L]^2 + \beta_3[L]^3 + \beta_4[L]^4} = 7.6 \times 10^{-7}$$

The calculations are straightforward but tedious. The template in Cells B56–E67 on Sheet 2 of the complex ions program can be used to assist with the calculations.

Maximum Unchelated Zn(II) Concentration without Precipitation of Zinc Hydroxide

If there is no EDTA in a solution, then none of the Zn(II) will be chelated with EDTA and the Zn^{2+} concentration will be the same as the unchelated concentration, i.e., $[Zn^{2+}] = C_{Zn,U}$. Example 10.4B illustrates a process used to predict whether a metal hydroxide will or won't precipitate from a solution containing ammonia.

> **Example 10.4B Will zinc hydroxide precipitate from a solution containing $C_{Zn,U} = 1.0 \times 10^{-3}$ unchelated Zn(II) and [L] = 0.20 M ammonia buffered at pH = 10.0. For zinc hydroxide: $K_{sp} = 1.2 \times 10^{-17}$; from Example 10.4A: $\gamma_0 = 7.6 \times 10^{-7}$.**
>
> *Step 1:* Calculate the hydrogen ion and hydroxide ion concentrations in a solution at pH = 10.0.
>
> $$[H^+] = 10^{-pH} = 10^{-10} M, \ [OH^-] = K_w / [H^+] = 10^{-4} M$$
>
> *Step 2:* Calculate the Zn^{2+} concentration.
>
> $$[Zn^{2+}] \cong \gamma_0 C_{Zn} = (7.6 \times 10^{-7})(1.0 \times 10^{-3}) = 7.6 \times 10^{-10} M$$
>
> *Step 3:* Compare the concentration product to the solubility-product constant.
>
> $$[Zn^{2+}][OH^-]^2 = (7.6 \times 10^{-10} M)(1.0 \times 10^{-4} M)^2 = (7.6 \times 10^{-18}) < K_{sp} = 1.2 \times 10^{-17}$$
>
> Conclusion: Zinc hydroxide won't precipitate from solution.

On the one hand, this more conventional approach to such calculations is valid. On the other hand, it is quite easy to develop a way to combine all these calculations into one equation that is convenient for use with calculators and computers.

After substituting $[OH^-] = K_w/[H^+]$ and $[Zn^{2+}] = \gamma_0 C_{Zn,U}$ into the inequality, $[Zn^{2+}][OH^-]^2 < K_{sp}$, the result can be rearranged into the form

$$C_{Zn,U} \leq \frac{K_{sp}}{\gamma_0} \left(\frac{[H^+]}{K_w} \right)^2 \tag{10.10a}$$

in which $C_{Zn,U}$ is the unchelated Zn(II) concentration, K_{sp} is the solubility-product constant, γ_0 is the fraction of Zn(II) present as Zn^{2+} and $[H^+]$ and K_w have their usual significance.

Example 10.4C illustrates an application of Equation 10.10a.

> **Example 10.4C For conditions in Example 10.4A, what is the maximum unchelated Zn(II) concentration for which zinc hydroxide won't form? Conditions in Example 10.4A: $K_{sp} = 1.2 \times 10^{-17}$, $[H^+] = 1.0 \times 10^{-10}$ M, $\gamma_0 = 7.6 \times 10^{-7}$.**
>
> $$C_{Zn,U} \leq \frac{K_{sp}}{\gamma_0}\left(\frac{[H^+]}{K_w}\right)^2 \leq \frac{1.2 \times 10^{-17}}{7.6 \times 10^{-7}}\left(\frac{1.0 \times 10^{-10}}{1 \times 10^{-14}}\right)^2 \leq 1.6 \times 10^{-3}\,M$$

I will trust you to rearrange Equation 10.10a into a form that can be used to calculate the minimum hydrogen ion concentration needed to prevent precipitation of a metal hydroxide.

The more commonly quoted solubility-product constant for zinc hydroxide is 3.0×10^{-16} (see Exercise 10.5).

EXERCISE 10.5

Assuming conditions in Example 10.4A and solubility-product constant for zinc hydroxide of $K_{sp} = 3.0 \times 10^{-16}$, show that the upper limit of the unchelated Zn(II) concentration for which zinc hydroxide won't precipitate from solution is $C_{Zn,U} \leq 0.038$ M.

Given uncertainties associated with this and other constants, your primary take-home messages from this and other such calculations are the concepts and procedures rather than the numerical results. It would be necessary to do experiments to resolve the differences. To quote a famous analytical chemist, I. M. Kolthoff, "Theory guides, experiment decides."

In any event, whereas Equation 10.10a applies equally well for other doubly charged metal ions such as Cu^{2+} and Ni^{2+}, some modifications are needed for metal ions with different charges. For any metal ion, M^z, with a charge z, Equation 10.10a can be written as Equation 10.10b:

$$C_M \leq \frac{K_{sp}}{\gamma_0}\left(\frac{[H^+]}{K_w}\right)^z \tag{10.10b}$$

As examples, $z = 1$ for Ag(I) and $z = 3$ for Al(III).

For any metal ion that reacts with ammonia, the inclusion of ammonia will increase the solubility of the metal hydroxide. As often happens, any improvement in one area results in a compromise in another. The use of ammonia to prevent precipitation of metal hydroxides is no exception.

Effects of Ammonia Concentrations on Titration Curves for Cd(II)

Figure 10.4 illustrates effects of ammonia concentrations, 0.020 M to 0.22 M, on titration curves for 25.0 mL of 1×10^{-3} M Cd(II) titrated with 1×10^{-3} M EDTA at pH = 10.

As shown by Plots 1–6, increases in ammonia concentration for fixed Cd(II) and EDTA concentrations decrease the range of pM changes at the equivalence point. Such changes could decrease the reliability with which the end-point volume and metal ion concentration can be resolved.

For any who may want to plot the curves themselves, a template for the calculations is in A132–I150 on Sheet 5 of the complex ions program. After the template is pasted into Cells A27–I44, ammonia concentrations are varied in Cell C38 without changing contents of other cells. A special feature of the template is a program in Cell B38 that calculates ammonium ion concentrations needed to maintain $[H^+] \cong 1.0 \times 10^{-10}$ M, pH $\cong 10.0$, for any ammonia concentration entered into C38.

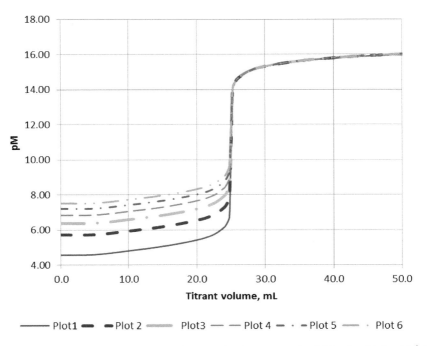

FIGURE 10.4 Effects of ammonia concentrations on titration curves for 25.0 mL of 1.0×10^{-3} M Cd(II) titrated with 1.0×10^{-3} M EDTA buffered near pH = 10 with ammonium chloride and ammonia.

Plots 1–6. = 0.02, 0.06, 0.10, 0.14, and 0.22 M ammonia with ammonium chloride concentrations adjusted to give pH ≅ 10.

CONDITIONAL CONSTANTS

Conditional constants are constants that account for effects of solution conditions on behaviors of reactions of interest. Let's begin with effects of ammonia concentrations and hydrogen ion concentrations on the behavior of the Zn(II)/EDTA reaction.

As a starting point, let's write the reaction and equation for the formation constant for the Zn(II)/EDTA reaction, i.e.,

$$Zn^{2+} + Y^{4-} \rightleftharpoons ZnY^{2-} \qquad K_f = \frac{\left[ZnY^{2-}\right]}{\left[Zn^{2+}\right]\left[Y^{4-}\right]}$$

Now, let's write equations for the equilibrium Zn^{2+} and fully deprotonated EDTA in terms of concentration fractions and unchelated concentrations.

As noted earlier, equilibrium Zn^{2+} and Y^{4-} concentrations are products of the fractions of the unchelated concentrations in each form times the unchelated concentrations, i.e.,

$$\left[Zn^{2+}\right] = \gamma_0 C_{Zn,U} \qquad \left[Y^{4-}\right] = \alpha_{6,Y} C_{Y,U}$$

in which $C_{Zn,U}$ and $C_{Y,U}$ are unchelated Zn(II) and EDTA concentrations and γ_0 and $\alpha_{6,Y}$ are fractions of the unchelated concentrations present as Zn^{2+} and Y^{4-}, respectively.

Substituting these equations into the equation for the formation constant for the Zn(II)/EDTA chelate reaction, we obtain Equation 10.11a:

$$K_{f,Y} = \frac{\left[ZnY^{2-}\right]}{\left[Zn^{2+}\right]\left[Y^{4-}\right]} = \frac{\left[ZnY^{2-}\right]}{\gamma_0 C_{Zn,U} \alpha_{6,Y} C_{Y,U}} \qquad (10.11a)$$

in which $K_{f,Y}$ is the formation constant for the ZnY^{2-} chelate, $[ZnY^{2-}]$ is the chelate concentration, C_{Zn} and $C_{Y,U}$ are the diluted Zn(II) concentration and uncomplexed EDTA concentration, and γ_0 and $\alpha_{6,Y}$ are fractions of the unchelated concentrations present as Zn^{2+} and Y^{4-}, respectively.

By defining a new "constant," $K_{f,Y}''$ called a *conditional or effective constant*, Equation 10.11a can be rearranged into the following form (Equation 10.11b):

$$K_{f,Y}'' \equiv \gamma_0\,\alpha_{6,Y}K_f = \frac{\left[ZnY^{2-}\right]}{C_{Zn,U}C_{Y,U}} \tag{10.11b}$$

in which $K_{f,Y}''$ is a *conditional or effective constant* and other symbols are as defined above. In other words, the conditional constant is the intrinsic constant multiplied by the fractions, γ_0, of Zn(II) in the unchelated form, and α_{6Y}, of EDTA in the fully deprotonated form.

Conditional constants calculated in this way account for effects of the ammonia and hydrogen ion concentrations on behaviors of metal-ion/EDTA reactions. They also facilitate the use of unchelated metal-ion and EDTA concentrations in equilibrium calculations. We will talk later about concepts and equations associated with calculations of unchelated concentrations. For now, let's calculate a conditional constant for the Zn(II)/EDTA reaction (see Example 10.5).

Example 10.5 Calculate the conditional constant for the Zn(II)/EDTA reaction for a solution containing [L] = 0.20 M ammonia at pH = 10.0. For the Zn(II)/EDTA reaction: log K_f = 16.50; from Examples 10.3 and 10.4A: $\alpha_{6,Y}$ = 0.367 and γ_0 = 7.6 × 10^{-7}.

Therefore,

$$K_{f,Y}'' \equiv \gamma_0\alpha_6 K_f = \left(7.6\times10^{-7}\right)\left(0.367\right)10^{16.50} = \left(7.6\times10^{-7}\right)\left(0.367\right)\left(3.2\times10^{16}\right) = 8.9\times10^9$$

On the one hand, the inclusion of ammonia as an auxiliary ligand makes it possible to use the Zn(II)/EDTA reaction at higher pH's at which larger percentages of EDTA are in the fully dissociated form. On the other hand, the combined effects of ammonia and pH = 10.0 result in a conditional constant that is several orders of magnitude smaller than the intrinsic constant, i.e., $K_{f,Y}''$ = 8.8 × 10^9 vs. $K_{f,Y}$ = 3.2 × 10^{16}.

We will talk later about ways to quantify effects of conditional constants on the completeness of metal-ion/EDTA reactions. For now, let's talk about effects of magnitudes of formation constants on shapes of titration curves.

EFFECTS OF FORMATION CONSTANTS ON TITRATION CURVES

Figure 10.5 illustrates effects of formation constants from 1×10^8 to 1×10^{18} on shapes of titration curves for titrations of 25.0 mL of 1×10^{-3} metal ion concentrations with 1×10^{-3} M EDTA buffered near pH = 10.0 with 0.036 M ammonium chloride and 0.20 M ammonia. A template for the calculations used to obtain these results is in Cells A158–I176 on Sheet 5 of the complex ions program.

Rather than use constants for actual metal ions, I used constants for fictitious metal ions in order to get a better distribution. Not surprisingly, changes in pM values at equivalence points increase with increasing magnitudes of the constants.

COMPUTATIONAL FORMS OF EQUATIONS FOR UNCHELATED Zn(II) AND EDTA CONCENTRATIONS

As discussed earlier, unchelated Zn(II) and EDTA concentrations are the parts of the diluted concentrations of the reactants that remain when the Zn(II)/EDTA reaction has reached equilibrium. Unchelated Zn(II) and EDTA concentrations were represented in Equations 10.2a and 10.2b as differences between diluted concentrations and the chelate concentration. Whereas difference

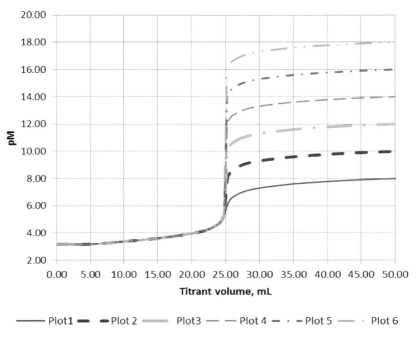

FIGURE 10.5 Effects of conditional formation constants on titration curves for 25.0-mL volumes of 1.0×10^{-3} M metal ions titrated with 1.0×10^{-3} M EDTA buffered near pH = 10 with 0.036 M ammonium chloride and 0.20 M ammonia.

Plots 1–6. Log K_{fy}'' = 8, 10, 12, 14, 16, and 18, respectively.

equations are used to calculate unchelated concentrations of excess reactants, more refined equations are needed to calculate unchelated concentrations of limiting reactants.

Using the titration of Zn(II) with EDTA as an example, computational forms of equations for unchelated Zn(II) and EDTA concentrations are described below for situations before, at, and after equivalence points.

Unchelated Concentrations before Equivalence Points

Zinc(II) is the excess reactant and EDTA is the limiting reactant before an equivalence point.

The unchelated Zn(II) concentration is the difference between the diluted Zn(II) concentration and the chelate concentration is expressed in Equation 10.12a:

$$C_{Zn,U} = C_{Zn} - \left[ZnY^{2-} \right] \tag{10.12a}$$

An equation for the uncomplexed EDTA concentration is obtained as a rearranged form, the equation for the conditional constant is expressed in Equation 10.12b:

$$K_{f,Y}'' = \frac{\left[ZnY^{2-} \right]}{C_{Zn,U} C_{Y,U}} \Rightarrow C_{Y,U} = \frac{\left[ZnY^{2-} \right]}{K_{f,Y}'' C_{ZnU}} \overset{Eq.\,10.12a}{=} \frac{\left[ZnY^{2-} \right]}{K_{f,Y}'' \left(C_{Zn} - \left[ZnY^{2-} \right] \right)} \tag{10.12b}$$

Unchelated Concentrations at Equivalence Points

Unchelated Zn(II) and EDTA concentrations are the same at any equivalence point, i.e., $C_{Zn,U} = C_{Y,U}$ at an equivalence point. An equation for the unchelated Zn(II) and EDTA concentrations at an equivalence point obtained by substituting $C_{Zn,U} = C_{Y,U}$ into the equation for the conditional constant and rearranging the resulting equation is

$$K''_{f,Y} = \frac{\left[ZnY^{2-}\right]}{C_{Zn,U}C_{Y,U}} \overset{C_{Zn,U}=C_{Y,U}}{\Rightarrow} C_{Zn,U} = C_{Y,U} = \sqrt{\frac{\left[ZnY^{2-}\right]}{K''_{f,Y}}}$$ (10.12c)

in which all symbols are as defined above.

Unchelated Concentrations after Equivalence Points

EDTA is the excess reactant and Zn(II) is the limiting reactant after equivalence points. Equations for unchelated EDTA and Zn(II) concentrations after equivalence points can be written by analogy with equations before the equivalence point. The equation for the unchelated EDTA concentration is

$$C_{Y,U} = C_Y - \left[ZnY^{2-}\right]$$ (10.12d)

The equation for unchelated Zn(II) concentration is obtained as follows:

$$K''_{f,Y} = \frac{\left[ZnY^{2-}\right]}{C_{Zn,U}C_{Y,U}} \Rightarrow C_{Zn,U} = \frac{\left[ZnY^{2-}\right]}{K''_f C_{Y,U}} \overset{Eq.\,10.12d}{=} \frac{\left[ZnY^{2-}\right]}{K''_{f,Y}\left(C_Y - \left[ZnY^{2-}\right]\right)}$$ (10.12e)

We will talk later about applications of these equations for equilibrium calculations.

COMPLETENESS OF METAL-ION/EDTA REACTIONS

When we talked earlier about conditional constants I threatened to talk about effects of reduced values of constants on the degrees of completion of metal-ion/EDTA reactions. Degrees of completion of metal-ion/EDTA reactions are important because procedures such as titrations that depend on conversions of limiting concentrations to the chelate can't be any more reliable than the degree to which the limiting reactant is converted to the chelate. In fact, the minimum error for a concentration based on an assumption that a chelate-forming reaction goes to 100% completion can be no less than the percent of the total concentration that remains unchelated at equilibrium.

Let's use the reaction of Zn(II) with EDTA to develop approximate equations for percentages of limiting reactants that haven't been converted to chelate.

By definition, the unchelated concentration of the limiting reactant is the part of the limiting reactant that hasn't been converted to the chelate. Therefore, the percent of the limiting reactant that hasn't been converted to the chelate is 100 times the unchelated concentration divided by the diluted concentration of the limiting reactant, i.e.,

$$\text{Unchelated }\% = 100\frac{C_{Lim,U}}{C_{Lim}}$$ (10.13a)

in which $C_{Lim,U}$ and C_{Lim} are the unchelated and diluted concentrations of the limiting reactant.

Now, let's adapt this equation to situations for which (a) Zn(II) is the excess reactant and EDTA is the limiting reactant, (b) both Zn(II) and EDTA are limiting reactants, and (c) EDTA is the excess reactant and Zn(II) is the limiting reactant.

Excess Zn(II), Limiting EDTA Concentration

Starting with Equation 10.12b and assuming for the moment that virtually all of the EDTA is converted to the chelate, i.e., that $[ZnY^{2-}] \cong C_Y$, Equation 10.13a can be rewritten as

$$C_{Y,U} \overset{Eq.\,10.12b}{\cong} \frac{\left[ZnY^{2-}\right]}{K''_{f,Y}\left(C_{Zn} - \left[ZnY^{2-}\right]\right)} \overset{\left[ZnY^{2-}\right] \cong C_Y}{\cong} \frac{C_Y}{K''_{f,Y}\left(C_{Zn} - C_Y\right)}$$

in which $C_{Y,U}$ is the unchelated EDTA concentration, $[ZnY^{2-}]$ is the chelate concentration, K''_{fY} is the conditional constant and C_{Zn} and C_Y are diluted Zn(II) and EDTA concentrations.

After replacing $C_{Lim,U}$ in Equation 10.13a with the last term in the foregoing equation, the resulting equation can be written and rearranged into the form as shown in Equation 10.13b:

$$\text{Unchelated } \% \cong 100 \frac{1}{K''_{f,Y}\left(C_{Zn} - C_Y\right)} \tag{10.13b}$$

in which K''_{fY} is the conditional constant and C_{Zn} and C_Y are diluted concentrations (see Example 10.6A).

> ### Example 10.6A Given a solution containing diluted concentrations, $C_{Zn} = 1.0 \times 10^{-3}$ M Zn(II) and $C_Y = 0.25 \times 10^{-3}$ M EDTA and $K''_{fY} = 8.9 \times 10^9$, the percent incomplete is approximated as
>
> $$\text{Incomplete } \% \cong 100 \frac{1}{K''_{f,Y}\left(C_{Zn} - C_Y\right)} \cong 100 \frac{1}{8.9 \times 10^9 \left(1.0 \times 10^{-3} - 0.25 \times 10^{-3}\right)} \cong 1.5 \times 10^{-5}$$

In other words, the percentage of EDTA that isn't chelated with Zn(II) is very, very small and other factors will be responsible for any significant errors (See Exercise 10.6).

EXERCISE 10.6

(A) Rearrange Equation 10.13b into a form that can be used to calculate the minimum conditional constant needed to ensure that a given percent of the limiting reactant will remain unchelated. (B) For concentrations in Example 10.6, show that the minimum conditional constant needed to ensure that no more than 0.1% of the EDTA remains unchelated is $K''_{fY} \cong 1.3 \times 10^6$.

Equivalent Zn(II) and EDTA Concentrations

Starting with Equation 10.12c and substituting $[ZnY^{2-}] \cong C_Y$, the unchelated EDTA concentration can be approximated as follows:

$$C_{Y,U} = \sqrt{\frac{\left[ZnY^{2-}\right]}{K''_{f,Y}}} \overset{\left[MY^{2-}\right] \cong C_M}{\cong} \sqrt{\frac{C_Y}{K''_{f,Y}}}$$

After replacing $C_{Lim,U}$ in Equation 10.13a with the last term in the foregoing equation, the result can be rewritten in the form as shown in Equation 10.13c:

$$\text{Unchelated } \% \cong 100 \frac{C_{Y,U}}{C_Y} \cong \frac{100}{C_Y}\sqrt{\frac{C_Y}{K''_{f,Y}}} \cong 100\sqrt{\frac{1}{K''_{f,Y}C_Y}} \tag{10.13c}$$

in which K''_{fy} is the conditional constant and C_Y is the diluted EDTA concentration.

Example 10.6B illustrates an application of Equation 10.13c.

Example 10.6B Given diluted Zn(II) and EDTA concentrations of $C_{Zn} = C_Y = 1.0 \times 10^{-3}$ M with $K''_{fY} = 8.9 \times 10^9$, calculate the percentage of the concentrations not converted to the chelate.

$$\text{Unchelated \%} \cong 100\sqrt{\frac{1}{K''_f C_Y}} \cong 100\sqrt{\frac{1}{\left(8.9 \times 10^9\right)\left(1.0 \times 10^{-3}\right)}} \cong 0.034\%$$

In other words, very small percentages of both Zn(II) and EDTA remain unchelated with each other.

EXERCISE 10.7

(A) Rearrange Equation 10.13c into a form that can be used to calculate the minimum conditional constant needed to ensure that any percent of the unchelated concentrations will remain unchelated with each other. (B) Given concentrations in Example 10.6B, show that the minimum conditional constant for no more than 0.1% of Zn(II) or EDTA to remain unchelated is $K''_{fY} \geq 1.0 \times 10^9$.

Using the procedure in Exercise 10.7, it can be shown that a conditional constant equal to or larger than 1×10^7 is required to ensure that no more than 1% of a limiting concentration of 10^{-3} M remains unchelated at the equivalence point.

Excess EDTA, Limiting Zn(II)

Equation 10.13b can be adapted to situations for which EDTA is the excess reactant and Zn(II) is the limiting reactant by interchanging C_{Zn} and C_Y (Equation 10.13d).

$$\text{Unchelated \%} \cong 100\frac{1}{K''_{f,Y}\left(C_Y - C_{Zn}\right)} \qquad (10.13d)$$

in which K''_{fY} is the conditional constant and C_{Zn} and C_Y are diluted concentrations.

How General and How Good Are the Foregoing Predictions?

Although described here in the context of the Zn(II)/EDTA reaction, analogous equations apply for other pairs of metal ions and chelating ligands. An exact equation for the chelate concentration to be discussed later was used to test the reliability of Equations 10.13b–10.13d. It was shown that errors for the equations are less than 10% for metal ion and/or EDTA concentrations larger than 1.0×10^{-4} and conditional constants larger than 1×10^5.

APPROXIMATE CALCULATIONS

On the one hand, the primary emphasis in this chapter is on iterative procedures that avoid the need for simplifying assumptions required for conventional approximations. On the other hand, for sake of completeness, I will solve a three-part example using conventional approximations. Anyone not interested in approximation procedures can skip to the section in which metal ions that don't form ammonia complexes are discussed.

Example 10.7 Given solutions buffered at pH = 10.0 with 0.036 M ammonium chloride and 0.20 M ammonia, calculate equilibrium ZnY^{2-}, Zn^{2+}, and Y^{4-} concentrations and pZn after adding (A) 15.0 mL, (B) 25.0 mL, and (C) 35.0 mL of 1.00×10^{-3} M EDTA to 25.0 mL of a solution containing 1.00×10^{-3} M Zn(II). From Examples 10.3, 10.4a, and 10.5: $\alpha_{6Y} = 0.367$, $\gamma_0 = 7.6 \times 10^{-7}$, and $K''_{fY} = 8.9 \times 10^9$.

Conditions in this example, particularly the ratio of the Zn(II) and ammonia concentrations, have been selected to give relatively small errors for both equilibrium concentrations and pZn values. I will guide you through the calculations for the three parts of the example and trust you to do the substitutions and calculations.

Example 10.7A illustrates approximate calculations for a situation before the equivalence point, i.e., for $V_Y C_Y^0 < V_{Zn} C_{Zn}^0$.

Example 10.7A Calculate ZnY^{2-}, Zn^{2+}, and Y^{4-} concentrations and pZn after adding 15.0 mL of 1.0×10^{-3} M EDTA to 25.0 mL of 1.0×10^{-3} M Zn(II) assuming that the solution is buffered at pH = 10.0. From Examples 10.3, 10.4a, and 10.5: $\alpha_{6Y} = 0.367$, $\gamma_0 = 7.6 \times 10^{-7}$, and $K_{fY}'' = 8.9 \times 10^9$.

Step 1A: Calculate diluted concentrations.

$$C_{Zn} = \frac{V_{Zn} C_{Zn}^0}{V_{Zn} + V_Y} = \frac{(25.0)(1.0 \times 10^{-3})}{(25.0 + 15.0)} \qquad C_Y = \frac{V_Y C_Y^0}{V_{Zn} + V_Y} = \frac{(15.0)(1.0 \times 10^{-3})}{(25.0 + 15.0)}$$

$$= 6.25 \times 10^{-4} \qquad\qquad\qquad = 3.75 \times 10^{-4}$$

Step 2A: Assume that the chelate concentration is approximately equal to the limiting concentration.

$$\left[ZnY^{2-} \right] \cong C_{Lim} \cong C_Y \cong 3.75 \times 10^{-4} M$$

Step 3A: Calculate unchelated Zn(II) and EDTA concentrations

$$C_{Zn,U} = C_{Zn} - \left[ZnY^{2-} \right] \cong 2.50 \times 10^{-4} M \qquad C_{Y,U} = \frac{\left[ZnY^{2-} \right]}{K_{f,Y}'' C_{ZnU}} \cong 1.71 \times 10^{-10}$$

Step 4A: Calculate Zn^{2+} and Y^{4-} concentrations

$$\left[Zn^{2+} \right] = \gamma_0 C_{Zn,U} \cong 1.93 \times 10^{-10} M \qquad\qquad \left[Y^{4-} \right] = \alpha_6 C_{Y,U} \cong 6.17 \times 10^{-11} M$$

Step 5A: Calculate pZn

$$pZn = -\log \left[Zn^{2+} \right] \cong 9.72$$

Example 10.7B illustrates approximate calculations for a situation at the equivalence point, i.e., for $V_Y C_Y^0 = V_{Ca} C_{Ca}^0$.

Example 10.7B Calculate ZnY^{2-}, Zn^{2+}, and Y^{4-} concentrations and pZn after adding 25.0 mL of 1.0×10^{-3} M EDTA to 25.0 mL of 1.0×10^{-3} M Zn(II) assuming that both solutions are buffered at pH = 10.0 with a total ammonia concentration of $C_L = 0.235$ M. From Examples 10.3, 10.4a, and 10.5: $\alpha_{6Y} = 0.367$, $\gamma_0 = 7.6 \times 10^{-7}$, and $K_{fY}'' = 8.9 \times 10^9$.

Steps 1B and 2B: I will trust you to confirm the quantities below.

Diluted concentrations   Chelate concentration

$$C_{Zn} = C_Y = 5.00 \times 10^{-4} \qquad\qquad \left[ZnY^{2-} \right] \cong 5.0 \times 10^{-4} M$$

Step 3B: Calculate unchelated Zn(II) and EDTA concentrations.

$$C_{Zn,U} = \sqrt{\frac{[ZnY^{2-}]}{K''_{f,Y}}} \cong 2.39 \times 10^{-7} \qquad C_{Y,U} \cong \sqrt{\frac{[ZnY^{2-}]}{K''_{f,Y}}} = 2.39 \times 10^{-7}$$

Step 4B: Calculate Zn^{2+} and Y^{4-} concentrations

$$[Zn^{2+}] = \gamma_0 C_{Zn,U} \cong 1.82 \times 10^{-13} M \qquad [Y^{4-}] = \alpha_6 C_{Y,U} \cong 8.62 \times 10^{-8} M$$

Step 5B: Calculate pZn

$$pZn = -\log[Zn^{2+}] \cong 12.74$$

Example 10.7C illustrates approximate calculations for a situation after the equivalence point, i.e., for $V_Y C_Y^0 > V_{Ca} C_{Ca}^0$.

Example 10.7C Calculate ZnY^{2-}, Zn^{2+}, and Y^{4-} concentrations and pZn after adding 35.0 mL of 1.0×10^{-3} M EDTA to 25.0 mL of 1.0×10^{-3} M Zn(II) assuming that both solutions are buffered at pH = 10.0. From Examples 10.3, 10.4a, and 10.5: $\alpha_{6Y} = 0.367$, $\gamma_0 = 7.6 \times 10^{-7}$, and $K''_{fY} = 8.9 \times 10^9$.

Steps 1C and 2C: I will trust you to confirm the quantities below.

Diluted concentrations Chelate concentration

$$C_{Zn} = 4.17 \times 10^{-4} \qquad C_Y = 5.83 \times 10^{-4} \qquad [ZnY^{2-}] \cong 4.17 \times 10^{-4} M$$

Step 3C: Calculate unchelated EDTA and Zn(II) concentrations

$$C_{Y,U} = C_Y - [ZnY^{2-}] \cong 1.67 \times 10^{-4} \qquad C_{Zn,U} \cong \frac{[ZnY^{2-}]}{K''_f C_{YU}} \cong 2.85 \times 10^{-10}$$

Step 4C: Calculate Zn^{2+} and Y^{4-} concentrations

$$[Zn^{2+}] = \gamma_0 C_{Zn,U} \cong 2.19 \times 10^{-16} M \qquad [Y^{4-}] = \alpha_6 C_{Y,U} \cong 6.02 \times 10^{-5} M$$

Step 5C: Calculate pZn

$$pZn = -\log[Zn^{2+}] \cong 15.67$$

As noted above, conditions for all parts of this example were selected to ensure approximation errors less than 5%. Let's talk about a situation for which approximation errors are much larger than those associated with Example 10.7.

APPROXIMATION ERRORS

Calculations in Example 10.7 are based on assumptions that:

a. Virtually all of the limiting metal-ion or EDTA concentration is converted to the EDTA chelate.

b. Changes in the auxiliary ligand concentration are negligibly small during the equilibration reactions.

c. Changes in the hydrogen ion concentration during equilibration reactions are negligibly small.

Example 10.8 is used here to illustrate types of errors that can be encountered if approximation procedures are applied indiscriminately.

> **Example 10.8 Given solutions buffered near pH = 10 with 0.036 M ammonium chloride and 0.20 M ammonia, calculate and plot approximation errors for (a) the ammonia concentration, (b) the hydrogen ion concentration, (c) the Cd^{2+} concentration, and (d) pCd calculated using conventional approximations for solutions prepared by adding 0–50 mL of 0.050 M EDTA in 2.5-mL increments to 25.0 mL of 0.050 M Cd(II).**

Programs on Sheet 4 of the complex ions program are designed to compare concentrations calculated using approximate and iterative procedures analogous to those discussed in the preceding chapter. We will talk later about the basis for that program. For now, let's use the program to solve Example 10.8.

A template in Cells A160–I176 of Sheet 4 of the complex ions program contains all the information needed to solve this example for $V_Y = 0.00$ mL of the EDTA volume in Cell F39. To use the template, copy it and paste it into Cells A28–I44 on Sheet 4 using the Paste (P) option. For 0.0 mL of EDTA, set C29 to 1 and wait for D46 to change from "Solving" to "Solved." If the iteration counter in G46 should stop changing before D46 changes to "Solved," press F9 one or more times until G46 stops changing and D46 changes to "Solved."

Calculations for other EDTA volumes from 2.5 to 50 mL are done by setting C29 to 0, setting F39 to the desired volume, resetting C29 to 1 and waiting until the iteration counter stops changing and D46 changes from "Solving" to "Solved." Percentage errors for each EDTA volume for (a) the hydrogen ion concentration, (b) the ammonia concentration, (c) the Cd^{2+} concentration, and (d) pCd are calculated in Cells K46, O46, K41, and R41, respectively.

Percentage errors for the ammonia, hydrogen-ion, and Cd^{2+} concentrations as well as for pCd are plotted in Figure 10.6 as Plots 1–4, respectively.

Some of the errors are quite large ranging from more than 200% for ammonia concentration and −100% for the hydrogen ion concentration before any EDTA is added, i.e., $V_Y = 0$. All errors decrease as the EDTA volume is increased up to the equivalence point at 25 mL. Some errors increase with increasing EDTA volumes after the equivalence point. Let's talk first about errors for EDTA volumes up to the equivalence point at 25.0 mL.

As we discuss the errors, remember that negative errors mean that the approximation option has underestimated the quantities and positive errors mean that the approximation option has overestimated the quantity.

Before the Equivalence Point

Plot 1 corresponds to errors in the ammonia concentration. As shown below, Cd(II) reacts with ammonia to produce complexes with one to four ammonias per Cd(II).

$$Cd^{2+} + L \underset{}{\overset{K_{f1}}{\rightleftarrows}} CdL^{2+} + L \underset{}{\overset{K_{f2}}{\rightleftarrows}} CdL_2^{2+} + L \underset{}{\overset{K_{f3}}{\rightleftarrows}} CdL_3^{2+} + L \underset{}{\overset{K_{f4}}{\rightleftarrows}} CdL_4^{2+}$$

This will decrease the ammonia concentration in the solution. The iterative option accounts for these changes in the ammonia concentration; the approximate option doesn't.

As EDTA is added to the solution, it reacts with Cd(II) to produce the CdY^{2-} chelate. This reduces the Cd(II) concentration available to react with ammonia. Therefore, errors in the ammonia concentration decrease with increasing EDTA volumes.

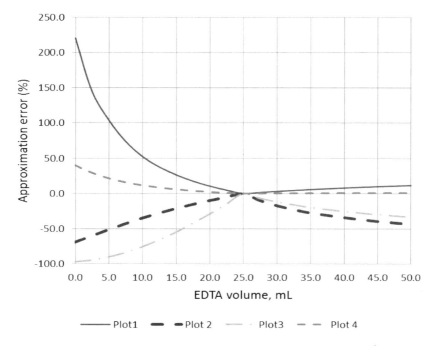

FIGURE 10.6 Approximation errors for equilibrium ammonia, hydrogen-ion and Cd^{2+} concentrations, and pCd calculated using conventional approximations for Example 10.8.

Plots 1–4: Approximation errors for ammonia, hydrogen ion, and Cd^{2+} concentrations and pCd, respectively.

Plot 2 corresponds to approximation errors for the hydrogen ion concentration. These errors result from the fact that the approximate option doesn't account for changes in the ammonia concentration. The deprotonation constant equation for the ammonium ion can be written and rearranged as follows:

$$K_{a,B} = \frac{[H^+][NH_3]}{[NH_4^+]} \Rightarrow [H^+] \cong K_{a,B} \frac{[NH_4^+]}{[NH_3]}$$

As noted above, reactions of Cd(II) with ammonia decrease the ammonia concentration causing the ratio $[NH_4^+]/[NH_3]$ to increase, which results in an increase in the hydrogen ion concentration. The iterative option accounts for these changes; the approximation option doesn't. The resulting errors decrease with increasing EDTA volumes as less and less Cd(II) is available to react with ammonia.

Plot 3 corresponds to errors in the Cd^{2+} concentration. Approximation errors for the Cd^{2+} concentrations result from changes in the ammonia concentration during the complex-formation reaction. It is assumed in conventional approximations that the ammonia concentration remains constant during the reaction. In reality, the ammonia concentration decreases during the reaction. The net result is that the approximation procedure underestimates the Cd^{2+} concentration. The iterative option accounts for effects of changing ammonia concentrations; the approximation option doesn't.

Errors for pCd values in Plot 4 result from errors in Cd^{2+} concentrations. Errors for pCd values are smaller than those for Cd^{2+} concentrations because the logarithmic function, pCd = −log $[Cd^{2+}]$, tends to suppress differences among concentrations. Signs of the errors are opposite those of concentration errors because of the negative sign of the pCd term. Errors decrease with increasing EDTA volumes because errors for Cd^{2+} concentrations decrease with increasing volumes.

After the Equivalence Point

All errors after the equivalence point result primarily from partial dissociation of excess H_2Y^{2-}.

$$H_2Y^{2-} \rightleftharpoons H^+ + HY^{3-}$$

Results of calculations using both the acid-base and complex ions programs show that HY^{3-} is the predominant form of EDTA after the equivalence point for conditions used for these calculations. This reaction produces an increase in the hydrogen ion concentration. The iterative option accounts for these changes; the approximate option doesn't. Without going into detail, increases in hydrogen ion concentration after the equivalence point cause larger errors in other quantities.

Some Perspectives

The Cd(II) and ammonia concentrations, 0.050 M and 0.20 M, respectively, used for these calculations are near the upper limit of the ratio of metal ion concentrations to ammonia concentrations for which the iterative program will function satisfactorily. Whereas concentration errors are quite large for this combination, the errors decrease with decreasing ratios of metal-on and ammonia concentrations. For solutions buffered at or near pH = 10 without EDTA, approximation errors are generally less than about 5% for metal-ion to ammonia concentration ratios less than or equal to about 0.005, i.e., $C_M/C_L \leq 0.005$. This ratio of metal-ion to ammonia concentrations for which approximation errors are 5% or less increases with increasing EDTA concentrations up to equivalence points.

As noted earlier, the pCd errors are significantly smaller than the other errors for all EDTA volumes. This point, which applies for other metal ions as well, has practical significance relevant to titrations of metal ions with EDTA. On the one hand, there may be small differences between titration curves based on approximate and iterative calculations at low and high EDTA volumes. On the other hand, the differences are very small for EDTA volumes just before, at, and just after an equivalence point. The net result is that titration curves based on either approximate or iterative options can be used equally well to judge the practicality of titrations of metal ions with EDTA.

METAL IONS THAT DON'T FORM AMMONIA COMPLEXES

To this point we have focused on metal ions such as Zn(II) and Cd(II) that form ammonia complexes. Let's talk next about how the concepts and equations developed for these situations can be adapted to ions such as Ca(II) and Mg(II) that don't form ammonia complexes.

We could treat ions that don't form ammonia complexes completely independently of ions that do form ammonia complexes as is usually done. However, as shown below, there is a more efficient way to treat these ions.

UNIFIED TREATMENT OF IONS THAT DO AND DON'T FORM AMMONIA COMPLEXES

From a computational point of view, the only difference between metal ions that do and don't form ammonia complexes involves the fractions, γ_j, of the ions complexed with different numbers, j, of ammonia per metal ion. Let's use Ni(II) and Ca(II) as examples. On the one hand, fractions, $\gamma_0-\gamma_6$, of Ni(II) in different forms are all less than 1. On the other hand, the fraction, γ_0, for Ca(II) present as Ca^{2+} is equal to 1 and fractions $\gamma_1-\gamma_6$ are all equal to zero. These differences make it very easy to adapt concepts and equations discussed above for metal ions that do form ammonia complexes to metal ions that don't form ammonia complexes.

For calculations using hand-held calculators, for metal ions that don't form ammonia complexes, you simply set $\gamma_0 = 1$ and $\gamma_1-\gamma_6 = 0$. For spreadsheet programs to be discussed later, the user sets a cell in the program to be discussed to "Yes" for ions that form ammonia complexes or to "No" for ions that don't form ammonia complexes and the program does the rest.

When the appropriate cell is set to "Yes," it tells the program to do calculations as described for ions that form ammonia complexes. When the cell is set to "No," it tells the program to set the coordination number to 0 which tells the program to set all cumulative formation constants for ammonia complexes to zero. As shown below, setting all cumulative formation constants to zero has the effect of setting γ_0 to 1 and other fractions to 0.

$$\gamma_0 = \frac{1}{1+\beta_1[L]+\beta_2[L]^2+\beta_3[L]^3+\beta_4[L]^4+\beta_5[L]^5+\beta_6[L]^6} \overset{\beta_1\cdots\beta_6=0}{\rightrightarrows} 1$$

Given that numerator terms for successive fractions, $\gamma_1-\gamma_6$, are successive denominator terms, setting all β's to zero also has the effect of setting fractions, $\gamma_1-\gamma_6$, to zero.

CORRELATION WITH CONVENTIONAL APPROACHES

The conventional approach to ions that do and don't form complexes with ammonia is to treat them independently. This is done by defining one conditional constant, $K'_{f,Y}$, for ions that don't form ammonia complexes and another constant, $K''_{f,Y}$, for ions that do form ammonia complexes. The two constants are expressed as follows (see Equation 10.14a and 10.14b):

$$K'_{f,Y} = \alpha_{6,Y}K_{f,Y} \qquad (10.14a) \qquad\qquad K''_{f,Y} = \gamma_0\alpha_{6,Y}K_{f,Y} \qquad (10.14b)$$

All symbols are as defined earlier.

Clearly, the two constants have the same value when γ_0 is equal to 1.

You are free to use both equations if you prefer. However, I will use Equation 10.14b for ions that do and don't form ammonia complexes with fractions $\gamma_0-\gamma_6$ set to their proper values for different situations. This means that regardless of the metal ion involved, all equations discussed above involving conditional constants are the same for both groups of metal ion. As examples, using Ca(II) and Ni(II) as examples, relationships between conditional constants and unchelated concentrations are as shown in Equations 10.15a and 10.15b:

$$K''_{f,Y} = \gamma_0\alpha_{6,Y}K_{f,Y} = \frac{\left[CaY^{2-}\right]}{C_{Ca,U}C_{Y,U}} \qquad (10.15a) \qquad K''_{f,Y} = \gamma_0\alpha_{6,Y}K_{f,Y} = \frac{\left[NiY^{2-}\right]}{C_{Ni,U}C_{Y,U}} \qquad (10.15b)$$

All symbols are as discussed earlier.

EXAMPLES

Example 10.9 illustrates the calculation of a conditional constant for the reaction of Ca(II) with EDTA.

Example 10.9 Calculate the conditional constant for the reaction of Ca(II) with EDTA in a solution buffered at pH = 10.0 with 0.036 M ammonium chloride and 0.20 M ammonia. For the Ca(II)/EDTA reaction: log $K_{f,Y}$ = 10.69. From Example 10.3: α_{6Y} = 0.367; From the preceding discussion γ_0 = 1.

$$K''_{f,Y} = \gamma_0\alpha_6 10^{\log K_f} = 1\times 0.367\times 10^{10.69} = 1.80\times 10^{10}$$

On the one hand, having calculated the conditional constant, all other calculations of equilibrium concentrations follow the same pattern as in Example 10.7. On the other hand, for sake of completeness, I will guide you through the process for an example involving Ca(II).

> **Example 10.10 Calculate CaY^{2-}, Ca^{2+}, and Y^{4-} concentrations and pCa in solutions prepared by adding (A) 15.0 mL, (B) 25.0 mL, and (C) 35.0 mL of 1.0×10^{-3} M EDTA to 25.0 mL of 1.0×10^{-3} M Ca(II) assuming that the solutions are buffered at pH = 10.0 with a buffer containing 0.036 M ammonium chloride and 0.20 M ammonia. From Example 10.3: $\alpha_{6Y} = 0.367$; from Example 10.9: $K''_{fY} = 1.8 \times 10^{10}$.**

I have used the same volumes and initial concentrations as in Example 10.7 so that you can compare similarities and differences between calculated concentrations and pM values between the two situations.

Example 10.10A illustrates calculations for a situation before the equivalence point, i.e., for $V_Y C_Y^0 < V_{Ca} C_{Ca}^0$.

> **Example 10.10A Calculate CaY^{2-}, Ca^{2+}, and Y^{4-} concentrations and pCa after adding 15.0 mL of 1.0×10^{-3} M EDTA to 25.0 mL of 1.0×10^{-3} M Ca(II) assuming that the solutions buffered at pH = 10.0 with a buffer containing 0.036 M ammonium chloride and 0.20 M ammonia.**

Step 1A: Calculate diluted concentrations.

$$C_{Ca} = \frac{V_{Ca} C_{Ca}^0}{V_{Ca} + V_Y} = \frac{(25.0)(1.0 \times 10^{-3})}{(25.0 + 15.0)} \qquad C_Y = C_{Lim} = \frac{V_Y C_Y^0}{V_{Ca} + V_Y} = \frac{(15.0)(1.0 \times 10^{-3})}{(25.0 + 15.0)}$$

$$= 6.25 \times 10^{-4} \qquad\qquad\qquad = 3.75 \times 10^{-4}$$

Step 2A: Assume that the chelate concentration is approximately equal to the limiting concentration.

$$\left[CaY^{2-} \right] \cong C_{Lim} \cong 3.75 \times 10^{-4} M$$

Step 3A: Calculate unchelated Ca(II) and EDTA concentrations

$$C_{Ca,U} = C_{Ca} - \left[CaY^{2-} \right] \cong 2.50 \times 10^{-4} M \qquad C_{Y,U} = \frac{\left[CaY^{2-} \right]}{K''_{f,Y} C_{Ca,U}} \cong 8.29 \times 10^{-11}$$

Step 4A: Calculate Ca^{2+} and Y^{4-} concentrations

$$\left[Ca^{2+} \right] = \gamma_0 C_{Ca,U} \overset{\gamma_0 = 1}{\cong} 2.50 \times 10^{-4} M \qquad \left[Y^{4-} \right] = \alpha_{6,Y} C_{Y,U} \cong 2.99 \times 10^{-11} M$$

Step 5A: Calculate pCa

$$pCa = -\log \left[Ca^{2+} \right] \cong 3.60$$

Given that $\gamma_0 = 1$, the Ca^{2+} concentration is the same as the uncomplexed Ca(II) concentration in this and other parts of this example, i.e., $[Ca^{2+}] = \gamma_0 C_{Ca,U} = C_{Ca,U}$.

Example 10.10B illustrates calculations for a situation at the equivalence point, i.e., for $V_Y C_Y^0 = V_{Ca} C_{Ca}^0$.

Example 10.10B Calculate CaY^{2-}, Ca^{2+}, and Y^{4-} concentrations and pCa after adding 25.0 mL of 1.0×10^{-3} M EDTA to 25.0 mL of 1.0×10^{-3} M Ca(II) assuming that the solutions are buffered at pH = 10.0 with a buffer containing 0.0176 M ammonium chloride and 0.10 M ammonia.

Steps 1B–3B: I will trust you to confirm the quantities below using procedures from Example 10.10A.

Diluted Ca(II) concentration   Diluted EDTA concentration   Chelate concentration

$$C_{Ca} = 5.00 \times 10^{-4} \qquad\qquad C_Y = 5.00 \times 10^{-4} \qquad\qquad \left[CaY^{2-}\right] \cong 5.0 \times 10^{-4} M$$

Step 4B: Calculate unchelated Ca(II) and EDTA concentrations

$$C_{Ca,U} = \sqrt{\frac{\left[CaY^{2-}\right]}{K''_{f,Y}}} \cong 1.67 \times 10^{-7} \qquad\qquad C_{Y,U} = \sqrt{\frac{\left[CaY^{2-}\right]}{K''_{f,Y}}} \cong 1.67 \times 10^{-7}$$

Step 5B: Calculate equilibrium Ca^{2+} and Y^{4-} concentrations

$$\left[Ca^{2+}\right] = \gamma_0 C_{Ca,U} \overset{\gamma_0 = 1}{\cong} 1.67 \times 10^{-7} M \qquad\qquad \left[Y^{4-}\right] = \alpha_{6,Y} C_{Y,U} \cong 6.00 \times 10^{-8} M$$

Step 6B: Calculate pCa

$$pCa = -\log\left[Ca^{2+}\right] \cong 6.78$$

Example 10.10C illustrates calculations for a situation after the equivalence point, i.e., for $V_Y C_Y^0 > V_{Ca} C_{Ca}^0$.

Example 10.10C Calculate CaY^{2-}, Ca^{2+}, and Y^{4-} concentrations and pCa after adding 35.0 mL of 1.0×10^{-3} M EDTA to 25.0 mL of 1.0×10^{-3} M Ca(II) assuming that the solution is buffered at pH = 10.0.

Steps 1C–3C: I will trust you to confirm the quantities below using procedures from Example 10.10A.

Diluted Ca(II) concentration   Diluted EDTA concentration   Chelate concentration

$$M, C_{Ca} = 4.17 \times 10^{-4} \ M \qquad\qquad C_Y = 5.83 \times 10^{-4} \ M \qquad\qquad \left[CaY^{2-}\right] \cong 4.17 \times 10^{-4} \ M$$

Step 4C: Calculate unchelated Ca(II) and EDTA concentrations

$$C_{Ca,U} = \frac{\left[CaY^{2-}\right]}{K''_{f,Y} C_{Y,U}} \cong 1.39 \times 10^{-10} \qquad\qquad C_{Y,U} = C_Y - \left[CaY^{2-}\right] \cong 1.67 \times 10^{-4}$$

Step 5C: Calculate equilibrium Ca^{2+} and Y^{4-} concentrations.

$$\left[Ca^{2+}\right] = \gamma_0 C_{Ca,U} \overset{\gamma_0 = 1}{\cong} 1.39 \times 10^{-10} M \qquad\qquad \left[Y^{4-}\right] = \alpha_{6,Y} C_{Y,U} \cong 6.02 \times 10^{-5} \ M$$

Step 6C: Calculate pCa

$$pCa = -\log\left[Ca^{2+}\right] \cong 9.86$$

Analogous procedures apply for reactions of EDTA with other metal ions that don't form complexes with ammonia.

Let's talk next about an iterative program to solve problems involving reactions of metal ions that do and don't form complexes with ammonia.

AN ITERATIVE PROGRAM

The program described here is on Sheet 4 of the complex ions program. Instructions for using the program are in Cells A1–I24. The input section of the program is in Cells A28–I44. Some quantities used in the iterative calculations are calculated in Cells A45–I66. Iterative calculations are done in Cells V27–BR56. A modified form of this program on Sheet 5 is used to plot titration curves as described earlier.

The program on Sheet 4 is designed to do iterative calculations for situations involving reactions of EDTA with metal ions such as Ca(II) and Mg(II) that don't form complexes with ammonia as well as ions such as Zn(II) and Ni(II) that do form complexes with ammonia. Whereas ammonium ion and ammonia are the buffer components for both groups of ions, ammonia also serves as the auxiliary ligand for ions that form ammonia complexes.

Example 10.11 is used as the basis for this discussion.

> **Example 10.11 Calculate equilibrium concentrations of reactants and products in solutions prepared by adding (A) 15.0 mL, (B) 25.0 mL, and (C) 35.0 mL of a solution containing 1.0×10^{-3} M disodium, dihydrogen EDTA, $Na_2H_2Y \cdot 2H_2O$, to 25 mL of a solution containing 1.0×10^{-3} M nickel chloride, $NiCl_2$, assuming that both solutions are buffered near pH = 10 with 0.036 M ammonium chloride and 0.200 M ammonia.**

Let's begin with an exact equation for the chelate concentration after which we will talk about the rationale for the iterative procedure.

EXACT EQUATION FOR THE CHELATE CONCENTRATION, $[NiY^{2-}]$

Given that many formation constants for monoligand chelates are quite large, it is common practice to approximate the chelate concentration by equating it to the limiting metal ion or ligand concentration, i.e., the smaller of the metal ion and ligand concentrations. However, for more exacting calculations, it is relatively easy to use difference equations for unchelated metal-ion and EDTA concentrations with the equation for the conditional formation constant to develop an exact equation for the chelate concentration. The relevant equations are repeated below for your convenience.

$$C_{Ni,U} = C_{Ni} - \left[NiY^{2-}\right]; \quad C_{Y,U} = C_Y - \left[NiY^{2-}\right]; \quad K''_{f,Y} = \frac{\left[NiY^{2-}\right]}{C_{Ni,U}C_{Y,U}}$$

After substituting equations for unchelated Ni(II) and EDTA concentrations into the equation for the conditional constant, the resulting equation can be rearranged into Equation 10.16a:

$$\left[NiY^{2-}\right]^2 - \left(C_{Ni} + C_Y + \frac{1}{K''_{f,Y}}\right)\left[NiY^{2-}\right] + C_{Ni}C_Y = 0 \tag{10.16a}$$

I will trust you to do the substitutions and rearrangements to confirm this equation.

For reasons I won't try to explain, it is the negative root of the quadratic equation that gives chelate concentrations. For example, the NiY^{2-} chelate concentration is expressed as

$$\left[NiY^{2-}\right] = \frac{1}{2}\left[\left(C_{Ni} + C_Y + \frac{1}{K''_{f,Y}}\right) - \sqrt{\left(C_{Ni} + C_Y + \frac{1}{K''_{f,Y}}\right)^2 - 4C_{Ni}C_Y}\right] \tag{10.16b}$$

in which $[NiY^{2-}]$ is the chelate concentration, C_{Ni} and C_Y are diluted Ni(II) and EDTA concentrations, and K''_{fY} is the conditional constant that accounts for fractions of Ni(II) in the Ni^{2+} form and EDTA in the fully deprotonated form (See Exercise 10.8).

EXERCISE 10.8

Given a solution containing concentrations, $[L] = 0.20$ M ammonia, $C_{Ni} = C_Y = 1.0 \times 10^{-3}$ M each of Ni^{2+} and EDTA at pH = 10.0, and a conditional constant of $K''_{fY} = 1.8 \times 10^{10}$, show that the equilibrium chelate concentration is $[NiY^{2-}] = 1.0 \times 10^{-3}$ M.

For any who may wish to confirm the conditional constant, logarithms of formation constants for the Ni(II)/ammonia complexes are: log $K_{f1} = 2.72$, log $K_{f2} = 17$, log $K_{f3} = 1.66$, log $K_{f4} = 1.12$, log $K_{f5} = 0.67$, log $K_{f6} = -0.03$.

For situations for which the conditional constant is much larger than the Ni(II) and EDTA concentrations, Equation 10.16b simplifies to Equation 10.16c:

$$\left[NiY^{2-} \right] \cong \frac{1}{2}\left[\left(C_{Ni} + C_Y \right) - \sqrt{\left(C_{Ni} + C_Y \right)^2 - 4C_{Ni}C_Y} \right] \qquad \left(\text{If } K''_{fY} \gg C_{NI}, C_Y \right) \quad (10.16c)$$

It follows that chelate concentrations for situations involving effective formation constants much larger than the metal-ion and EDTA concentrations are independent of hydrogen ion concentrations. This observation has important implications for a charge-balance equation used in the iterative program.

RATIONALE FOR AN ITERATIVE PROGRAM TO SOLVE EXAMPLE 10.11 AND OTHER ANALOGOUS PROBLEMS

For solutions prepared using alkali-metal salts of anions and chloride or nitrate salts of cations, the iterative program is based on the fact that concentrations of all reactants and products except the alkali metal ions and chloride or nitrate ions depend on the hydrogen ion concentration. For example, as discussed earlier, the equilibrium concentration of fully deprotonated EDTA is the fraction, α_{6Y}, of EDTA in the fully deprotonated form times the unchelated EDTA concentration, $C_{Y,U}$, i.e., $[Y^{4-}] = \alpha_{6Y}C_{Y,U}$. Obviously, α_{6Y}, depends on the hydrogen ion concentration. Similar arguments apply for other reactants and products.

IONIC REACTIONS ASSOCIATED WITH EXAMPLE 10.11

Given that we will use a charge-balance equation to solve the problem, let's begin with reactions used to identify ions that will contribute to the charge-balance equation. Reactions associated with Example 10.11 are summarized in Table 10.3. One of each ion is in bold type to help us confirm that we account for all ions in the solution.

These reactions are used below to write an ionic form of the charge-balance equation.

IONIC FORM OF THE CHARGE-BALANCE EQUATION, CBE

The ionic form of the charge-balance equation is written as the algebraic sum of products of signed charges on the ions times equilibrium concentrations. Given the complexity of the equation, it is written below in four parts. Part 1 is the algebraic sum of selected ions; Part 2 includes Ni^{2+} and the ammonia complexes of Ni(II); Part 3 contains the different acid-base forms of EDTA and Part 4 is the concentration of the Ni(II)/EDTA chelate.

$$\text{Part 1: } \left[H^+ \right] - \left[OH^- \right] + \left[Na^+ \right] - \left[Cl^- \right] + \left[NH_4{}^+ \right] \qquad (10.17a)$$

TABLE 10.3

Reactions Associated with Example 10.11

$$H_2O \overset{K_w}{\rightleftharpoons} H^+ + OH^-$$

$$Na_2H_2Y \rightarrow 2Na^+ + H_2Y^{2-}$$

$$NiCl_2 \rightarrow Ni^{2+} + 2Cl^-$$

$$NH_4Cl \rightarrow NH_4^+ + Cl^-$$

$$NH_4^+ \rightleftharpoons H^+ + NH_3$$

$$H_6Y^{2+} \overset{\kappa_1}{\rightleftharpoons} H^+ + H_5Y^+ \overset{\kappa_2}{\rightleftharpoons} H^+ + H_4Y^0 \overset{\kappa_3}{\rightleftharpoons} H^+ + H_3Y^{-1} \overset{\kappa_4}{\rightleftharpoons} H^+ + H_2Y^{-2} \overset{\kappa_5}{\rightleftharpoons} H^+ + HY^{-3} \overset{\kappa_6}{\rightleftharpoons} H^+ + Y^{-4}$$

$$Ni^{2+} + L \overset{\beta_1}{\rightleftharpoons} NiL^{2+} + L \overset{\beta_2}{\rightleftharpoons} NiL_2^{2+} + L \overset{\beta_3}{\rightleftharpoons} NiL_3^{2+} + L \overset{\beta_4}{\rightleftharpoons} NiL_4^{2+} + L \overset{\beta_5}{\rightleftharpoons} NiL_5^{2+} + L \overset{\beta_6}{\rightleftharpoons} NiL_6^{2+}$$

$$Ni^{2+} + Y^{4-} \overset{K_f}{\rightleftharpoons} NiY^{2-}$$

Notes: (1) L is NH_3, (2) Y^{4-} is fully dissociated EDTA, (3) κ_1–κ_6 are cumulative acid dissociation constants for EDTA, and (4) β_1–β_6 are cumulative complex formation constants for the Ni(II)/ammonia complexes.

$$\text{Part 2:} + 2\left(\left[Ni^{2+}\right] + \left[NiL^{2+}\right] + \left[NiL_2^{2+}\right] + \left[NiL_3^{2+}\right] + \left[NiL_4^{2+}\right] + \left[NiL_5^{2+}\right] + \left[NiL_6^{2+}\right]\right) \quad (10.17b)$$

$$\text{Part 3:} + 2\left[H_6Y^{2+}\right] + \left[H_5Y^+\right] + 0\left[H_4Y\right] - \left[H_3Y^-\right] - 2\left[H_2Y^{2-}\right] - 3\left[HY^{3-}\right] - 4\left[Y^{4-}\right] \quad (10.17c)$$

$$\text{Part 4:} - 2\left[NiY^{2-}\right] \quad (10.17d)$$

The complete charge-balance equation is the algebraic sum of the four parts.

I will trust you to confirm that (a) the concentration of each charged ion is included in the equation and (b) that each term is the product of the charge on the ion times the concentration of the ion. The reason the Ni(II)/EDTA chelate concentration is given special attention in Part 4 of the equation is discussed later.

Concentration symbols in Equations 10.17a–10.17d are identified in Table 10.4.

TABLE 10.4

Symbols Used in the Ionic Form of the Charge-Balance Equation

$[H^+]$	Equilibrium hydrogen ion concentration.
$[OH^-]$	Equilibrium hydroxide ion concentration.
$[Na^+]$	Equilibrium sodium ion concentration.
$[Cl^-]$	Equilibrium chloride ion concentration.
$[NH_4^+]$	Equilibrium ammonium ion concentration, i.e., protonated form of ammonia, the ligand, L.
$[H_2Y^{2-}] - [Y^{4-}]$	Equilibrium concentrations of the seven acid-base forms of EDTA.
$[Ni^{2+}] - [NiL_6^{2+}]$	Equilibrium concentrations of Ni^{2+} and the six ammonia complexes.
$[NiY^{2-}]$	Equilibrium nickel(II)/EDTA chelate concentration.

Intermediate Forms of the Charge-Balance Equation

Intermediate forms of the four parts of the charge-balance equation are written below in terms of diluted concentrations, concentration fractions, and signed coefficients on the ions in balanced reactions. Equations are annotated to identify ions represented by the different terms.

$$\text{Part 1: } \left[H^+\right] - \overbrace{\frac{K_w}{\left[H^+\right]}}^{[OH^-]} - \overbrace{\left(C_{NH_4Cl} + z_{Ni(II)}C_{NiCl_2}\right)}^{\left[Cl^-\right]} + \overbrace{z_{H_nY}C_{Na_2H_2Y}}^{\left[Na^+\right]} + \overbrace{\alpha_{0,L}C_{L,U}}^{\left[NH_4^+\right]} \tag{10.18a}$$

$$\text{Part 2: } \overbrace{z_{Ni}\left(\gamma_0 + \gamma_1 + \gamma_2 + \gamma_3 + \gamma_4 + \gamma_5 + \gamma_6\right)C_{Ni,U}}^{\left(\left[Ni^{2+}\right],\ \left[NiL^{2+}\right]\cdots\left[NiL_5^{2+}\right],\ \left[NiL_6^{2+}\right]\right)} \tag{10.18b}$$

$$\text{Part 3: } \overbrace{\left(2\alpha_{0,Y} + \alpha_{1,Y} + 0\alpha_{2,Y} - \alpha_{3,Y} - 2\alpha_{4,Y} - 3\alpha_{5,Y} - 4\alpha_{6,Y}\right)C_{Y,U}}^{\left(\left[H_6Y^{2+}\right],\ \left[H_5Y^+\right]\cdots\left[H_3Y^{3-}\right],\ \left[Y^{4-}\right]\right)} \tag{10.18c}$$

$$\text{Part 4: } -2\overbrace{\left[NiY^{2-}\right]}^{Eq.\ 10.16b} \tag{10.18d}$$

We will talk about cell locations in which the different equations are implemented after we talk about the algorithm for the hydrogen ion concentration.

Algorithm for the Hydrogen Ion Concentration

Whereas the algorithm used to vary the hydrogen ion concentration in Column W is similar to that used in the acid-base programs, there are some differences.

First, given that solutions to be considered are buffered, we will have initial estimates of hydrogen ion concentrations reasonably close to equilibrium concentrations. Therefore, it isn't necessary to do a decade search first.

Second, given that some of the ammonia is used to form complexes, the ammonia concentration will be equal to or less than that at the start of the reaction. Therefore, the hydrogen ion concentration will be equal to or larger than that of the buffer. Accordingly, the starting hydrogen ion concentration is set at 95% of that of the buffer by setting I28 to 0.95.

Finally, different degrees of resolution are used in different steps in the process. Numbers in Cells X25, Y25, and Z25 control the resolution in Steps 1–19, 20–25, and 26–30, respectively. Numbers 1, 2, or 3 in these cells correspond to resolution of the hydrogen ion concentration to 10%, 1%, and 0.1%, respectively, of the hydrogen ion concentration of the buffer. Most results reported herein are done with X25, Y25, and Z25 set to 1, 2, and 2, respectively. However, don't be afraid to change these numbers between 1 and 3 to obtain desired compromises between convergence time and resolution for different situations. The larger each number, the larger the number of significant figures to which the hydrogen ion concentration is resolved and the longer it will take for the program to converge on a final value of the hydrogen ion concentration.

Computational Form of the Charge-Balance Equation

Unless stated otherwise, this discussion applies for parts of the program on Sheet 4 of the complex ions program. Whereas column references are the same for the program on Sheet 5, row numbers are not.

The four parts of the charge-balance equation as they are implemented in Cell X27 on Sheet 4 of the program are summarized in Equations 10.19a–10.19d.

$$\text{Part 1: } W27 - \frac{\$H\$42}{W27} - \left(\$B\$65 + \$F\$42*\$B\$66\right) + \$G\$42*\$D\$66 + BE27 \tag{10.19a}$$

$$\text{Part 2: } + \$F\$42*SUM\left(AV27: BB27\right) \tag{10.19b}$$

$$\text{Part 3: } + SUM\left(\left(C\$53: I\$53\right)*\left(BG27: BM27\right)\right) \tag{10.19c}$$

$$\text{Part 4: } + \$G\$29*\left(\$D\$42 + \$F\$42\right)*BN27 \tag{10.19d}$$

I will leave it to you to correlate cell/column locations with quantities in Equations 10.18a–10.18d. Analogous forms are used in other cells in Column X of the programs on Sheets 4 and 5. Now comes the most unsettling part of this text.

Cell G29 in the complex ions program is used to determine if the chelate concentration is or isn't included in the charge-balance equation. Results described in this chapter were obtained with G29 set to 0 to exclude the chelate concentration from the charge-balance equation. Reasons are that results obtained by excluding chelate concentrations are consistent with results calculated using independent methods known to give accurate results; results obtained by setting G29 to 1 to include chelate concentrations are not consistent with results obtained using independent methods. Let me explain.

One set of conditions for which conventional approximations give accurate results for virtually any set of metal ion and EDTA concentrations is at the equivalence point. For large values of conditional constants, virtually all of the metal ion and EDTA will be chelated with each other. This means that there will be very little free EDTA to influence the hydrogen ion concentration and very little free metal ion to react with ammonia. Consequently, the ratio of the ammonium ion and ammonia concentrations, and therefore, the hydrogen ion concentration, should be the same as in the original buffer.

Calculations with the chelate concentration excluded from the charge-balance equation satisfy this condition; calculations with chelate concentration included don't. These are reasons why I have done calculations with G29 set to 0 to exclude the chelate term from the charge-balance equation.

SOME LIMITATIONS ON RATIOS OF AMMONIA AND METAL-ION CONCENTRATIONS

When an auxiliary ligand is used there will always be some lower limit of the ratio of the initial ligand and metal-ion concentrations for which the convergence time of the iterative program will be acceptable. As noted earlier, my experience using ammonia as the auxiliary ligand for a variety of metal ions at pH = 10 is that *the initial ammonia concentration should always be at least four times larger than the initial metal-ion concentration*, i.e., $C_L^0/C_M^0 \geq 4$, for acceptable convergence times. The concentration, C_L^0, in this inequality is the ammonia concentration and not the sum of the ammonia and ammonium ion concentrations. The lower limit of the ratio stays about the same with increasing pH but decreases with decreasing pH.

DETECTING END POINTS FOR EDTA TITRATIONS USING COLOR INDICATORS

Options for detecting end points in metal-ion/EDTA titrations include colorimetry, potentiometry, and visual detection using color indicators[1]. We will focus here on the use of two-color indicators, i.e., indicators that have one color before the equivalence point and another color after the equivalence point. Indicators to be discussed here change from red to blue at end points.

Erichrome Black T, abbreviated EBT herein, is a commonly used indicator for selected metal-ion/EDTA titrations. EBT is an uncharged triprotic acid that dissociates in three steps. Representing the indicator by the generic symbol, In, the three deprotonation steps are as follows:

$$H_3In \rightarrow H^+ + H_2In^- \overset{K_{a2,In}}{\rightleftharpoons} H^+ + HIn^{2-} \overset{K_{a3,In}}{\rightleftharpoons} H^+ + In^{3-}$$

The first deprotonation step behaves as a strong acid and the second and third steps behave as weak acids.

Zinc ion reacts with the fully deprotonated form of the indicator to form a red complex ion, i.e.,

$$Zn^{2+} + In^{3-} \overset{K_{f,In}}{\rightleftharpoons} \underset{Red}{ZnIn^-}$$

The net reaction at the equivalence point at or near pH = 10 is as follows:

$$\underset{Red}{ZnIn^-} + HY^{3-} \rightarrow ZnY^{2-} + \underset{Blue}{HIn^{2-}}$$

We detect the end-point in a titration as a red to blue color change. The range of pZn values from the start to the end of the color change is called the *transition range* for the indicator. The range of EDTA volumes, $\Delta V_{Y,Ep}$, required to produce the red to blue color change determines the *sharpness* of the end point. The difference between the end-point volume and the equivalence-point volume determines the error in the Zn(II) concentration.

Our task here is to learn to do the calculations needed to predict the range of volumes corresponding to the transition range of the indicator, the closeness of the end-point volume to the equivalence-point volume and the percentage error in the Zn(II) concentration.

An overall problem (Example 10.12) can be stated as follows.

Example 10.12 For the titration of a 50.0-mL sample of 0.010 M Zn(II) with 0.010 M EDTA using EBT as a color indicator in solutions buffered near pH = 10 with 0.018 M ammonium chloride and 0.10 M ammonia, calculate (A) the range of EDTA volumes required to produce an observable red to blue color change, (B) the end-point volume, and (C) the percentage error in the Zn(II) concentration. Using a program on Sheet 9 of the complex ions program it can be shown that the fraction of Zn(II) present as Zn^{2+} is $\gamma_0 = 1.17 \times 10^{-5}$ and the conditional constant for the Zn(II)/EDTA reaction is $K''_{fY} = 1.34 \times 10^{11}$.

Whereas some calculations associated with this example are best done using Zn^{2+} concentrations other calculations are best done using pZn values. We will discuss procedures used to calculate both quantities as we work through the solution to this example.

We will divide the problem into a series of smaller steps and then put everything together in a stepwise solution to the example at the end. Having discussed the concepts, equations, and procedures required to solve this example we will talk about a spreadsheet program to do the calculations for this and other situations.

Let's begin with the EDTA volume at the equivalence point.

EDTA Volume at the Equivalence Point

The EDTA volume at the equivalence point is the volume of Zn(II) times the initial Zn(II) concentration divided by the initial EDTA concentration, i.e.,

$$V_{Y,Eq} = \frac{V_{Zn}C_{Zn}^0}{C_Y^0} = \frac{(50.0 \text{ mL})(0.010 \text{ M})}{0.010 \text{ M}} = 50.0 \text{ mL}$$

in which $V_{Y,Eq}$ is the equivalence point volume, V_{Zn} is the volume of Zn(II), C_{Zn}^0 is the initial Zn(II) concentration and C_Y^0 is the EDTA concentration.

Diluted Zn(II) Concentration at the Equivalence Point

The diluted Zn(II) concentration at the equivalence point is the volume of Zn(II) times the initial Zn(II) concentration divided by the sum of the EDTA and Zn(II) volumes, i.e.,

$$C_{Zn,Eq} = \frac{V_{Zn}C_{Zn}^0}{(V_{Zn} + V_Y)} = \frac{(50.0 \text{ mL})(0.010 \text{ M})}{(50.0 + 50.0)} = 0.0050 \text{ M}$$

Symbols are as defined above.

Equilibrium Zn²⁺ Concentration and pZn at the Equivalence Point

Based on Equation 10.12c, the unchelated Zn(II) concentration is expressed as follows:

$$C_{Zn,U} = \sqrt{\frac{\left[ZnY^{2-} \right]}{K_{f,Y}''}}$$

in which $C_{Zn,U}$ is the unchelated Zn(II) concentration, $[ZnY^{2-}]$ is the chelate concentration, and $K_{f,Y}''$ is the conditional constant for the Zn(II)/EDTA reaction. Writing the Zn^{2+} concentration as $\gamma_0 C_{Zn,U}$ and assuming that virtually all of the Zn(II) concentration is converted to the chelate, the Zn^{2+} concentration and pZn are approximated as shown in Equation 10.20b and Exercise 10.9:

$$\left[Zn^{2+} \right]_{Eq} \cong \gamma_0 \sqrt{\frac{C_{Zn}}{K_f''}} \quad (10.20a) \qquad\qquad pZn_{Eq} = -\log\left(\gamma_0 \sqrt{\frac{C_{Zn}}{K_f''}} \right) \quad (10.20b)$$

EXERCISE 10.9

Given $C_{Zn} = 0.0050$ M at the equivalence point, $\gamma_0 = 1.17 \times 10^{-5}$ and $K_{f,Y}'' = 1.34 \times 10^{11}$, show that the Zn^{2+} concentration and pZn at the equivalence point are $[Zn^{2+}]_{Eq} \cong 2.26 \times 10^{-12}$ M and $pZn_{Eq} \cong 11.64$.

We will use the equivalence-point pZn later to determine which of two equations should be used to calculate EDTA volumes at the start and end of the transition range of the indicator.

RANGE OF Zn²⁺ CONCENTRATIONS AND pZn VALUES REQUIRED TO PRODUCE A DETECTABLE COLOR CHANGE

As noted above, the colored forms of the indicator are the red complex ion, $ZnIn^-$, and the blue conjugate base of EBT, HIn^-. It is generally assumed that the ratio of concentrations of the colored forms of the indicator, i.e., $[ZnIn^-]/[HIn^-]$, must change from 10/1 at the start of a color change to 1/10 at the end of a color change for most observers to detect a color change effectively. After developing some mathematical relationships, this requirement is used as a basis for calculating the range of Zn^{2+} concentrations and pZn values required to produce an observable color change.

Let's begin with some properties of the indicator, EBT.

Reactions and Equilibrium Constants for Several Indicators, Including EBT

Constants for selected metallochromic indicators are included in Table 10.5.

We will focus here on the use of Erichrome black T, EBT, as an indicator for titrations of Zn(II), Mg(II), and Ca(II) with EDTA.

Fully protonated EBT is an uncharged triprotic weak acid often represented as H_3In. Whereas the first deprotonation step behaves like a strong acid, the second and third steps behave as weak acids with deprotonation constants of $K_{a2,In} = 5.0 \times 10^{-7}$ and $K_{a3,In} = 2.5 \times 10^{-12}$. We are primarily interested here in the third deprotonation step, i.e.,

$$\underset{Blue}{HIn^{2-}} \overset{K_{a3,In}}{\rightleftharpoons} H^+ + In^{3-} \qquad K_{a3,In} = \frac{\left[H^+\right]\left[In^{3-}\right]}{\left[HIn^{2-}\right]} = 2.5 \times 10^{-12}$$

Zinc ion reacts with the deprotonated form of the indicator to produce a complex ion with a formation constant of 7.95×10^{12}, i.e.,

$$Zn^{2+} + In^{3-} \rightleftharpoons \underset{Red}{ZnIn^-} \qquad K_{f,In} = \frac{\left[ZnIn^-\right]}{\left[Zn^{2+}\right]\left[In^{3-}\right]} = 7.95 \times 10^{12}$$

As shown in the reactions, the singly protonated form of EBT, HIn^{2-}, is blue and the Zn(II) complex, $ZnIn^-$, is red.

TABLE 10.5
Formation Constants for Selected Metal-Ion/Indicator Complexes and Deprotonation Constants for Second and Third Deprotonation Steps of the Indicators[a]

Metal ion	Indicator	Log $K_{f,Y}$	$K_{M,In}$	$K_{a2,In}$	$K_{a3,In}$
Zn(II)	EBT	16.5	7.95×10^{12}	5.0×10^{-7}	2.5×10^{-12}
Mg(II)	EBT	8.8	1.0×10^{7}	Same	Same
Ca(II)	EBT	10.7	2.5×10^{5}	Same	Same
Ca(II)	Calcon[b]	10.7	4.0×10^{5}	4.0×10^{-8}	3.2×10^{-14}
Mg(II)	Calmagite	8.8	1.0×10^{8}	7.2×10^{-9}	4.47×10^{-13}

[a] All indicators change from red to blue at end points of titrations with EDTA.
[b] It is recommended that the indicator be used at pH = 12.3 corresponding to $[H^+] = 5.0 \times 10^{-13}$ M.

Combining the Deprotonation and Complex-Formation Reactions

Our goal here is to obtain a reaction and equilibrium constant that represents the behavior of the indicator near the equivalence point. As a starting point I will remind you of two features of reactions and equilibrium constants. First, if two reactions are added together, the resulting equilibrium constant is the product of the constants for the individual reactions. Second, if the direction of a reaction is reversed, the equilibrium constant for the reversed reaction is the reciprocal of the constant for the forward reaction.

Writing the sum of the deprotonation and complex-formation reactions, canceling the In^{3-} terms on both sides of the resulting reaction and writing the product of the two constants, we obtain Equation 10.21a:

$$\underset{Blue}{\underline{HIn^{2-}}} + Zn^{2+} \overset{K_{a3,In}}{\rightleftharpoons} H^+ + \underset{Red}{\underline{ZnIn^-}} \qquad K_{a3,In}K_{f,In} = \frac{\left[H^+\right]\left[ZnIn^-\right]}{\left[HIn^{2-}\right]\left[Zn^{2+}\right]} \tag{10.21a}$$

The direction of this reaction is the reverse of the order in which changes occur near the equivalence point. Therefore, let's reverse the reaction and write the reciprocal of the combined equilibrium constants to obtain the following.

$$H^+ + \overset{Before\ Eq.\ Pt.}{\underset{Red}{\underline{ZnIn^-}}} \rightleftharpoons Zn^{2+} + \overset{After\ Eq.\ Pt.}{\underset{Blue}{\underline{HIn^{2-}}}} \qquad \frac{1}{K_{f,In}K_{a3,In}} = \frac{\left[Zn^{2+}\right]\left[HIn^{2-}\right]}{\left[H^+\right]\left[ZnIn^-\right]} \tag{10.21b}$$

As long as there is excess Zn(II) in the solution the solution will contain the $ZnIn^-$ complex ion and will be red. When sufficient EDTA has been added to react with all the Zn(II), any additional EDTA, having a larger attraction for Zn(II) than EBT, will steal Zn(II) from the indicator and the indicator will be protonated to form HIn^- producing a blue solution.

Equation 10.21b is easily rearranged into the following form for the Zn^{2+} concentration (see also Exercise 10.10).

$$\left[Zn^{2+}\right] = \frac{\left[H^+\right]}{K_{f,In}K_{a3,In}}\frac{\left[ZnIn^-\right]}{\left[HIn^{2-}\right]} \tag{10.21c}$$

EXERCISE 10.10

Assuming a solution buffered at pH = 10.0, use Equation 10.21c to show that the Zn^{2+} concentrations at the start and end of the EBT color change are $[Zn^{2+}]_{Strt} = 5.0 \times 10^{-11}$ and $[Zn^{2+}]_{End} = 5.0 \times 10^{-13}$ and that $pZn_{Strt} = 10.3$ and $pZn_{End} = 12.3$. (*Reminders*: The ratio, $[ZnIn^-]/[HIn^{2-}]$, is assumed to change from 10/1 at the start of the color change to 1/10 at the end of the color change; $K_{f,In} = 7.95 \times 10^{12}$; $K_{a3,In} = 2.5 \times 10^{-12}$.)

In other words, the red to blue color change will start when $[Zn^{2+}]_{Strt} = 5.0 \times 10^{-11}$ M and end at $[Zn^{2+}]_{End} = 5.0 \times 10^{-13}$ M corresponding to $pZn_{Strt} = 10.3$ and $pZn_{End} = 12.3$. The assumptions regarding ratios of the two forms of the indicator result in a 100-fold change in concentrations between the start and end of the color change and a two pZn unit change between the two points.

RANGE OF EDTA VOLUMES CORRESPONDING TO THE TRANSITION RANGE OF THE INDICATOR

EDTA volumes at the start and end of a color change are used in two ways herein. The difference between EDTA volumes at the start and end of a color change, ΔV_Y, tells us the range of volumes required to produce an observable color change. The smaller the volume change required to produce the color change, the more reliably the end point can be detected.

We will talk shortly about equations used to calculate EDTA volumes at the start and end of a color change. First however, let's use some graphical results to illustrate how EDTA volumes at the start and end of a color change depend on the pZn values at those points.

Plots of EDTA Volumes vs. pZn

I used the program on Sheet 4 of the complex ions program to calculate pZn values for several EDTA volumes just before and just after the equivalence point for conditions in Example 10.12. EDTA volumes are plotted vs. pZn values in Figure 10.7 along with dashed lines corresponding to pZn values at the start and end of the transition range for the indicator i.e., $pZn_{Strt} = 10.3$ and $pZn_{End} = 12.3$.

I estimate EDTA volumes at the first and second crossing points as $V_{Y,Strt} \cong 49.96$ and $V_{Y,End} = 50.02$ mL. This corresponds to a volume difference of $\Delta V_Y \cong 50.02 - 49.96 \cong 0.055$ mL or about one drop of the EDTA solution. In other words we can expect a very sharp end point. The end-point volume, $V_{ep} \cong 50.02$ mL, differs by less than 0.1% from the equivalence-point volume, $V_{Eq} = 50.00$ mL.

On the one hand, plots such as this can help us visualize effects of the transition range for the indicator on the titrant volume required to produce an observable color change. On the other hand, it is time consuming to collect and plot results for such plots. A simpler approach is to develop and use closed-form equations to calculate volumes corresponding to Zn^{2+} concentrations and pM values at the start and end of the transition range for each indicator.

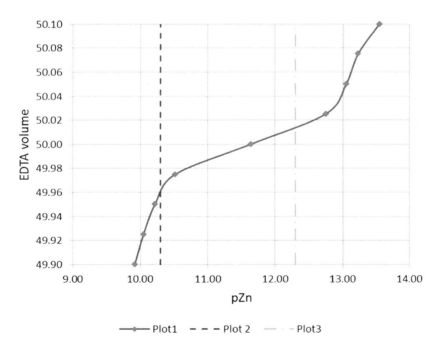

FIGURE 10.7 EDTA volume vs. pZn near the equivalence point for a titration of 50 mL of 0.010 Zn(II) with 0.010 M EDTA in solution buffered near pH 10 with 0.018 M ammonium chloride and 0.10 M ammonia using EBT as an indicator.

Plot 1: Volume vs. pZn. Plots 2 and 3: pZn values corresponding to the start and end of the transition range of EBT.

EDTA Volumes before and after Equivalence Points

Example 10.13 illustrates a process used to develop equations for EDTA volumes corresponding to pZn values before and after an equivalence point. The equations account for the fact that pZn values at the start and end of the transition range can occur before or after the equivalence point.

Example 10.13 Develop equations for EDTA volumes corresponding to pZn's before and after the equivalence point.

Before the equivalence point, $pZn < pZn_{Eq}$ After the equivalence pointp, $pZn > pZn_{Eq}$

Step 1: Write equations for unchelated Zn(II) concentrations.

$$\overset{Eq.10.12a}{C_{Zn,U} \,\hat{=}\, C_{Zn} - \left[ZnY^{2-} \right]} \qquad\qquad \overset{Eq.10.12e}{C_{Zn,U} \,\hat{\cong}\, \frac{\left[ZnY^{2-} \right]}{K_{f,Y}'' \left(C_Y - \left[ZnY^{2-} \right] \right)}}$$

Step 2: Assume that $[ZnY^{2-}] \cong C_Y$ before the equivalence point and $[ZnY^{2-}] \cong C_{Zn}$ after the equivalence point.

$$\overset{Eq.\ 10.12a}{C_{Zn,U} \,\hat{\cong}\, C_{Zn} - C_Y} \qquad\qquad \overset{Eq.\ 10.12e}{C_{Zn,U} \,\hat{\cong}\, \frac{C_{Zn}}{K_{f,Y}'' \left(C_Y - C_{Zn} \right)}}$$

Step 3: Substitute the following equations into equations in the preceding row.

$$\left[Zn^{2+} \right] = \gamma_0 C_{Zn,U} \Rightarrow C_{Zn,U} = \frac{\left[Zn^{2+} \right]}{\gamma_0}, \qquad C_{Zn} = \frac{V_{Zn} C_{Zn}^0}{V_{Zn} + V_Y}, \quad C_Y = \frac{V_Y C_Y^0}{V_{Zn} + V_Y}$$

Step 4: Rearrange the resulting equations into the forms in Equations 10.22a and 10.22b for points before and after the equivalence point.

$$V_{Y,Bfor} \cong V_{Zn} \frac{\gamma_0 C_{Zn}^0 - \left[Zn^{2+} \right]}{\left[Zn^{2+} \right] + \gamma_0 C_Y^0} \qquad (10.22a) \qquad\qquad V_{Y,Aft} \cong V_{Zn} \frac{C_{Zn}^0}{C_Y^0} \left(\frac{\gamma_0}{K_{f,Y}'' \left[Zn^{2+} \right]} + 1 \right) \qquad (10.22b)$$

As indicated in the example, Equations 10.22a and 10.22b correspond to pM values before and after an equivalence point, respectively.

Strategy for Choosing between Equations 10.22a and 10.22b

The strategy for choosing between Equations 10.22a and 10.22b is to compare pM values at the start and end of the transition range of the indicator to the pZn value at the equivalence point. Equation 10.22a is used for situations for which the start and/or end of the color change occur before the equivalence point, i.e., pM_{Strt} and/or $pM_{End} \le pM_{Eq}$. Equation 10.22b is used for situations for which the start and/or end of the color change occur after the equivalence point, pM_{Strt} and/or $pM_{End} > pM_{Eq}$.

As shown below for Example 10.12, pZn values for the start and end of the transition range are less than and larger than pZn_{Eq}.

$$pZn_{Strt} = 10.3 < pZn_{Eq} \qquad\qquad pZn_{Eq} = 11.64 \qquad\qquad pZn_{End} = 12.3 > pZn_{Eq}$$

For this example, this means that Equation 10.22a should be used to calculate the EDTA volume, $V_{Y,Strt}$, at the start of the color change and that Equation 10.22b should be used to calculate the EDTA volume, $V_{Y,End}$, at the end of the color change (see Exercise 10.11).

EXERCISE 10.11

For conditions in Example 10.12, show that EDTA volumes at the start and end of the EBT color change are $V_{Y,Strt} = 49.96$ mL and $V_{Y,End} = 50.01$ mL. Information needed to solve this exercise is summarized below.

Before and after the equivalence point: $V_{Zn} = 50.0$ mL, $C_Y^0 = C_{Zn}^0 = 0.010$ M, $\gamma_0 = 1.17 \times 10^{-5}$, $K''_{fY} = 1.34 \times 10^{11}$.

At the equivalence point: $pZn_{Eq} = 11.64$

Before the equivalence point: $[Zn^{2+}] = 5.0 \times 10^{-11}$, $pZn = 10.3$

After the equivalence point: $[Zn^{2+}]_{End} = 5.0 \times 10^{-13}$, $pZn_{End} = 12.3$

Zn(II) Concentration Corresponding to the End Point Volume

The Zn(II) concentration calculated using the EDTA volume at the end-point is the end-point volume times the EDTA concentration divided by the sample volume, i.e.,

$$C_{Zn,Calc}^0 = \frac{V_{Y,Ep}C_Y^0}{V_{Zn}} = \frac{(50.01\ \text{mL})(0.0100\ \text{M})}{50.00} = 0.0100\ \text{M}$$

This is virtually identical to the Zn(II) concentration in the original sample, i.e., $C_{Zn}^0 = 0.0100$ M. In other words, the end-point volume is an accurate representation of the diluted Zn(II) concentration.

We have now calculated all the most relevant quantities associated with Example 10.12. Let's organize it into a compact stepwise solution to the example.

Stepwise Solution to Example 10.12

A stepwise approach to solving Example 10.12 is summarized below.

> **Example 10.12 (From above)** For the titration of a 50.0-mL sample of 0.010 M Zn(II) with 0.010 M EDTA using EBT as a color indicator in solutions buffered near pH = 10 with 0.018 M ammonium chloride and 0.10 M ammonia, calculate (A) the range of EDTA volumes required to produce an observable red to blue color change, (B) the end-point volume and, (C) the percentage error in the Zn(II) concentration.

Using a program on Sheet 9 of the complex ions program it can be shown that the fraction of Zn(II) present as Zn^{2+} is $\gamma_0 = 1.17 \times 10^{-5}$ and the conditional constant for the Zn(II)/EDTA reaction is $K''_{fY} = 1.34 \times 10^{11}$.

Step 1: Calculate the diluted Zn(II) concentration.

$$C_{Zn} = V_{zn}C_{Zn}^0/(V_{Zn} + V_Y) = 5.0 \times 10^{-3}\ \text{M}$$

Step 2: Use the diluted Zn(II) concentration with γ_0 and K''_{fY} to calculate pZn at the equivalence point.

$$pZn_{Eq} \overset{\text{Eq. 10.20b}}{\cong} -\log\left(\gamma_0\sqrt{\frac{C_{Zn}}{K''_{f,Y}}}\right) \cong 11.64$$

Step 3: Calculate Zn^{2+} concentrations at the start and end of the transition range.

$$\left[Zn^{2+}\right]_{Strt} \overset{\text{Eq.10.21c}}{\cong} \frac{1.0\times10^{-10}}{\left(7.95\times10^{12}\right)\left(2.5\times10^{-12}\right)}\left(\frac{10}{1}\right) \qquad \left[Zn^{2+}\right]_{End} \overset{\text{Eq.10.21c}}{\cong} \frac{1.0\times10^{-10}}{\left(7.95\times10^{12}\right)\left(2.5\times10^{-12}\right)}\left(\frac{1}{10}\right)$$

$$= 5.0\times10^{-11} \qquad\qquad\qquad\qquad\qquad = 5.0\times10^{-13}$$

Step 4: Calculate pZn values at the start and end of the transition range.

$$pZn = -\log\left(5.0\times10^{-11}\right) = 10.3 \qquad\qquad pZn = -\log\left(5.0\times10^{-13}\right) = 12.3$$

Step 5: Given that pZn values at the start and end of the transition range are less than and greater than that at the equivalence-point, respectively, use Equations 10.22a and 10.22b to calculate EDTA volumes.

Start[a]:

$$V_{Y,Strt} \overset{\text{Eq.10.22a}}{\cong} V_{Zn}\frac{\gamma_0 C_{Zn}^0 - \left[Zn^{2+}\right]}{\left[Zn^{2+}\right]-\gamma_0 C_Y^0} \cong 50\frac{\left(1.17\times10^{-5}\right)0.010 - 5.0\times10^{-11}}{5.0\times10^{-11}-\left(1.17\times10^{-5}\right)0.010} \cong 49.96 \text{ mL}^{a}$$

End[a]:

$$V_{Y,End} \overset{\text{Eq.10.22b}}{\cong} V_{Zn}\frac{C_{Zn}^0}{C_Y^0}\left(\frac{\gamma_0}{K_{f,Y}''}\frac{1}{\left[Zn^{2+}\right]}+1\right) \cong 5.0\frac{0.010}{0.010}\left(\frac{1.17\times10^{-5}}{\left(1.34\times10^{-11}\right)\left(5.0\times10^{-13}\right)}+1\right) \cong 50.01 \text{ mL}^{a}$$

Step 6: Calculate the difference between volumes at the start and end of the transition range for EBT.

$$\Delta V_{Eq} = V_{Y,End} - V_{Y,Strt} = 50.01 \text{ mL} - 49.96 \text{ mL} = 0.05 \text{ mL}$$

Step 7: Calculate the diluted Zn(II) concentration corresponding to the end-point volume.

$$C_{Zn,Ep} = \frac{V_{Y,Ep}C_Y^0}{V_{Zn}} = \frac{\left(50.01 \text{ mL}\right)\left(0.010 \text{ M}\right)}{50.00} = 0.010 \text{ M}$$

[a] These calculations were done using a spreadsheet to obtain the desired number of significant figures.

The basis for each step of the calculation was discussed earlier and is not discussed further here. The two most important conclusions from these calculations are that (a) the calculated value of the end-point concentration is very close to the starting concentration and (b) the range of volumes for the transition range of the indicator is very small, about one drop of the EDTA solution, corresponding a very sharp end point. Having done a preliminary titration to locate the approximate end-point volume, it would be possible to locate the end point volume with a high degree of confidence. As is shown later, this isn't always the case.

One additional point merits comment. If both pZn values in Step 3 had been less than the pZn at the equivalence point, then Equation 10.22a would have been used to calculate both EDTA volumes in Step 5. If both pZn values in Step 3 had been larger than the pZn at the equivalence point, then Equation 10.22b would have been used to calculate both EDTA volumes.

A SPREADSHEET PROGRAM

A program to do these calculations and instructions for using the program are included on Sheet 9 of the complex ions program. Quantities including constants for the indicator are entered into Cells A30–I44. Except for the formation constant for the metal-ion indicator complex, $K_{f,In}$ in B30, the third deprotonation constant for the indicator, $K_{a3,In}$, in D30, and the EDTA volume in F39, entries in the part of the program in Cells A30–I44 are the same as for applications of the iterative programs on Sheets 4 and 5. As with those programs, I43 should be set to "Yes" for metal ions such as Zn(II) and Ni(II) that form ammonia complexes and to "No" for ions such as Ca(II) and Mg(II) that don't form ammonia complexes. Calculations of quantities related more directly to end-point volumes etc. are done in Cells L7–Q14. Symbols are as identified in text.

A template for Example 10.12 is in A74–I88 on Sheet 9 of the complex ions program. Templates for several other problems involving metal ions that do and don't form ammonia complexes are in A93–I183. Instructions for using the templates are included in Cells A71–I71 on Sheet 9.

RESULTS FOR SELECTED SITUATIONS

Two factors that influence the end-point error in a titration are the closeness of the end-point volume to the equivalence-point volume and the range of volumes required to produce the color change, ΔV_{End}. Some results used to illustrate these two features of an indicator are summarized in Table 10.6 for titrations of Zn(II), Mg(II), and Ca(II) with EDTA using EBT as the indicator. In all cases the equivalence-point volume is 50.0 mL.

As noted earlier, templates for these three situations are included in Cells A72–I126 on Sheet 9 of the complex ions program.

Notice first that all three end-point volumes are very close to the equivalence-point volumes corresponding to *apparent errors* from less than 0.1% for Zn(II) to about 1.1% for Mg(II). Notice next that volume changes, ΔV_{End}, from the start to the end of the color change for Zn(II) and Mg(II) are 0.050 mL and 0.94 mL, respectively. These volume changes represent about one drop of EDTA for the Zn(II) end point and about 20 drops for the Mg(II) end point.

Having done a preliminary titration to locate an approximate end-point volume, it should be possible to determine the end-point volume for Zn(II) with a high degree of reliability and the end-point volume for Mg(II) with a modest degree of reliability.

It is a very different story for Ca(II). On the one hand, the end-point volume differs by about 0.3% from the equivalence-point volume. On the other hand, the transition range for the indicator

---

TABLE 10.6

Calculated Results for 50.0 mL of 0.010 M Zn(II), Mg(II), and Ca(II) Titrated with 0.010 M EDTA in Solutions Buffered Near pH = 10.0 with 0.018 M Ammonium Chloride and 0.10 M Ammonia. For All Ions: $K_{a3,In} = 2.5 \times 10^{-12}$

	Input Information		Calculated Results for Start and End of Color Change			
Ion	log $K_{f,Y}$	$K_{M,In}$	V_{Strt}	V_{End}	ΔV_{End}	Error[a]
Zn(II)	16.6	7.95×10^{12}	49.96	50.01	0.052	<0.1%
Mg(II)	8.8	1.0×10^{7}	49.95	50.53	0.94	1.1%
Ca(II)	10.7	2.5×10^{5}	35.90	49.84	13.94	−0.3%

[a] Apparent errors if end points can be detected reliably.

occurs over a range of about 14 mL. Even if one could detect the start and end of the transition range reliably, can you imagine adding titrant dropwise for a range of 14 mL? I can't.

What are the reasons for differences between behaviors of Mg(II) and Ca(II)? On the one hand, given that the formation constant for the Ca(II)/EDTA reaction, $K_{f,Y} = 5.0 \times 10^{10}$, is larger than that for the Mg(II)/EDTA reaction, $K_{f,Y} = 6.3 \times 10^8$, the Ca(II)/EDTA reaction approaches 100% completion more closely than the Mg(II)/EDTA reaction. On the other hand, given that the formation constant for the Ca(II)/EBT reaction, $K_{f,In} = 2.5 \times 10^5$, is about 100-fold less than that for the Mg(II)/EBT reaction, $K_{f,In} = 1.0 \times 10^7$, the Ca(II)/EBT reaction approaches 100% completion less closely than the Mg(II)/EBT reaction. The result is that the red-to-blue color change for the Ca(II)/EBT reaction extends over a much wider range of EDTA volumes than the Mg(II)/EBT reaction.

In any event, chemists have devised a way to titrate Ca(II) reliably using EBT as the indicator. The procedure is to add a small amount of Mg(II) to the Ca(II) solution. Because Mg(II) has a smaller formation constant with EDTA than Ca(II), i.e., 6.3×10^8 vs. 5×10^{10}, the EDTA will react with virtually all of the Ca(II) before it reacts with any Mg(II). Because Mg(II) has a larger formation constant for the EBT complex than Ca(II), 1.0×10^7 vs 2.5×10^5, it gives a sharper end point than Ca(II) would without the Mg(II). Because the formation constant for Ca(II) is larger than that for Mg(II), the Ca(II) will react with EDTA before the Mg(II) reacts. Thus, the color change occurs at the point at which all the Ca(II) has reacted with EDTA.

A FINAL POINT

One other point merits comment. As shown below using Equations 10.20a and 10.21c for Ca^{2+} as an example, there is no direct relationship between the equation used to calculate the equivalence-point metal ion concentration and the metal ion concentration at which an indicator changes color.

$$\overbrace{\left[Ca^{2+}\right]_{Eq} \cong \gamma_0 \sqrt{\frac{C_{Ca}}{K_f''}}}^{\text{Equivalence-point equation}} \qquad (10.20a)$$

$$\overbrace{\left[Ca^{2+}\right] = \frac{\left[H^+\right]\left[CaIn^-\right]}{K_{f,In}K_{a3,In}\left[HIn^{2-}\right]}}^{\text{Indicator equation}} \qquad (10.21c)$$

This independence of the two sets of calculations can lead to some disconcerting results. For example, consider the titration of 50.0 mL of 2.0×10^{-3} Ca(II) with 2.0×10^{-3} M EDTA at pH = 10.0 using EBT as the indicator.

The equivalence-point volume of EDTA will be 50.0 mL and the diluted Ca(II) concentration at the equivalence point will be (50.0 mL)(2.0×10^{-3} M)/(50.0 mL + 50.0 mL) = 1.0×10^{-3} M. The Ca^{2+} concentration calculated by substituting $[H^+] = 1.0 \times 10^{-10}$, $K_{f,In} = 2.5 \times 10^5$, $K_{a3,In} = 2.5 \times 10^{-12}$, and $[CaIn^-]/[HIn^{2-}] = 10/1$ into Equation 10.21c is $[Ca^{2+}]_{Strt} = 1.6 \times 10^{-3}$ M.

In other words, the Ca^{2+} concentration, 1.6×10^{-3} M, calculated using the indicator equation is larger than the diluted concentration, $C_{Ca} = 1.0 \times 10^{-3}$ M, at the equivalence point. This of course isn't physically possible. Whereas this can be disconcerting, the interpretation is quite simple; EBT is not a satisfactory indicator for this titration without some modification of solution conditions. As discussed earlier, the problem can be solved by adding a small amount of Mg(II) to the solution.

SUMMARY

This chapter emphasizes the basis for and implementation of iterative approaches to calculations for reactions of metal ions with EDTA that eliminate the need for simplifying assumptions associated with conventional approximation procedures. Calculations are based on a charge-balance equation that in turn is based on the fact that concentrations of all reactants and products other than alkali metal ions and halide ions depend on the hydrogen ion concentration. Having resolved the hydrogen ion concentration that satisfies equilibrium constants for all reactions in a solution, the program then uses the hydrogen ion concentration to calculate concentration fractions and concentrations of other reactants and products.

We begin with a discussion of compromises in solution conditions imposed by the facts that whereas (a) very low hydrogen ion concentrations favor the fully deprotonated form of EDTA that forms chelates with metal ions (b) higher hydrogen ion concentrations are needed to prevent precipitation of metal hydroxide precipitates that would interfere with the chelate-forming reactions.

We continue with a description of a part of the spreadsheet program designed to let the user plot up to six titration curves on a single figure in order to visualize effects of different variables on shapes of titration curves. The program is used to illustrate effects of metal ion and EDTA concentrations, hydrogen ion concentrations, ammonia concentrations, and formation constants on shapes of titration curves.

The chapter continues with a discussion of background information related to reactions of metal ions with EDTA and with ammonia used as an auxiliary ligand to decrease metal ion concentrations at which hydroxide precipitates begin to form. Topics discussed include but are not limited to chelate-forming reactions and formation constants, relationships among unchelated and equilibrium metal-ion and EDTA concentrations, descriptions of excess and limiting reactants, effects of hydrogen ion concentrations on the fraction of EDTA in the fully deprotonated form, a description of how ammonia prevents precipitation of metal hydroxides, and conditional or effective formation constants.

We discuss simple equations used to calculate upper limits of metal ion concentrations for which precipitation of metal hydroxide can be avoided, and percentages of metal-ion and EDTA concentrations that remain unreacted at different points during titrations or other applications of metal-ion/EDTA reactions. We also discuss conventional approximations and errors in selected reactant and product concentrations and pM values resulting from simplifying assumptions on which conventional approximations are based.

We discuss a way that calculations for metal ions that don't form ammonia complexes can be treated as special cases of the calculations for metal ions that do form ammonia complexes. By exploiting differences between fractions of metal ions in different forms, it is possible to use the same equations and iterative programs for metal ions that do and don't form ammonia complexes by changing a cell in the program from "Yes" or "No" the metal ion does or doesn't form ammonia complexes.

We also discuss the basis for and implementation of the spreadsheet program used for iterative calculations in the chapter. Finally, we discuss concepts and equations associated with the use of a two-color indicator, Erichrome Black T, to detect end points in titrations of selected metal ions with EDTA.

REFERENCE

1. Reilley, C. N. and Schmid, R. W.; *Anal. Chem.*, **1959,** *31,* 887–897.

Index